PRACTICE MAKES PERFECT™

Precalculus

William D. Clark, Ph.D., and Sandra Luna McCune, Ph.D.

New York Chicago San Francisco Lisbon London Madrid Mexico City
Milan New Delhi San Juan Seoul Singapore Sydney Toronto

The McGraw-Hill Companies

1 2 3 4 5 6 7 8 9 10 11 12 13 14 15 16 17 QDB/QDB 1 9 8 7 6 5 4 3 2 1

ISBN 978-0-07-176178-9
MHID 0-07-176178-0

Library of Congress Control Number 2011928641

Interior design by Village Bookworks
Interior illustrations by Newgen Imaging Systems

This book is printed on acid-free paper.

PRACTICE MAKES PERFECT™

Precalculus

Contents

Preface

Practice Makes Perfect: Precalculus is designed to help you be successful in precalculus. It is not intended to introduce concepts; rather it is meant to reinforce ideas and concepts that you have previously encountered. The topics presented are those that a competent user of precalculus needs to know. You will find this practice workbook to be a useful supplementary text for your precalculus course. It can also serve as a refresher text if you are using it to review previously learned precalculus concepts and techniques.

As with most topics worth knowing, learning precalculus requires diligence and hard work. The foremost purpose of *Practice Makes Perfect: Precalculus* is to serve as a source of solved precalculus problems. We believe that the best way to develop understanding, while at the same time acquiring accuracy and speed in precalculus skills is to work numerous practice exercises. This book has more than 750 practice exercises from beginning to end. A variety of exercises and levels of difficulty are presented to provide reinforcement of precalculus knowledge, understanding, and skills. In each chapter, a concept discussion followed by example problems precedes each set of exercises, to serve as a concise review for readers already familiar with the topics covered. Concepts are broken into basic components to provide ample practice of fundamental skills.

To use *Practice Makes Perfect: Precalculus* in the most effective way, it is important that you work through every exercise. After working an exercise set, use the worked-out solutions to check your understanding of the concepts. We sincerely hope this book will help you acquire greater competence and confidence in using precalculus in your future mathematical endeavors.

REVIEW OF BASIC CONCEPTS

In Part I you will review basic concepts that you are expected to know when you begin the study of precalculus. Topics include the real and complex number systems, operations with the real and complex numbers, absolute value, and the Cartesian coordinate plane. It is not uncommon for students to begin precalculus with a weak understanding of these topics. Even though you might feel confident that you have sufficient skills in these areas, we recommend that you take the time to brush up on this prerequisite material. By doing so, you will enhance your performance in precalculus.

·1·

Real numbers

Real number system

A thorough foundation in the real number system will serve you well in precalculus—and in subsequent mathematical endeavors. Following are some important subsets in the real number system.

The set of **natural numbers**, N, consists of the counting numbers; that is,

$$N = \{1, 2, 3, \ldots\}$$

When you include the number 0 with the set of natural numbers, you have the set of **whole numbers**, W. Thus,

$$W = \{0\} \cup \{1, 2, 3, \ldots\} = \{0, 1, 2, 3, \ldots\}$$

If you expand the set of whole numbers to include the negatives of the natural numbers, you obtain the set of **integers**, $\mathbf{Z}$. Therefore,

$$\mathbf{Z} = \{\ldots, -3, -2, -1\} \cup \{0, 1, 2, 3, \ldots\} = \{\ldots, -3, -2, -1, 0, 1, 2, 3, \ldots\}$$

The set of **rational numbers**, $\mathbf{Q}$, is made up of numbers that can be expressed as a quotient of an integer divided by an integer other than zero. That is,

$$\mathbf{Q} = \left\{\frac{p}{q}, \text{ where } p \text{ and } q \text{ are integers, } q \neq 0\right\}$$

Note: Don't forget that division by zero is undefined, so $\frac{p}{0}$ has no meaning no matter what number you put in the place of p.

Fractions, decimals, and percents are rational numbers. All of the natural numbers, whole numbers, and integers are rational numbers as well because each number n contained in one of these sets can be written in the form $\frac{n}{1}$.

The **irrational numbers** are numbers that cannot be expressed as $\frac{p}{q}$, where p and q are integers and $q \neq 0$; or, equivalently, numbers whose decimal representations neither terminate nor repeat. Examples of irrational numbers are $\sqrt{7}$, $\sqrt[3]{24}$, e, and π. When using these numbers in computations, you approximate their values to a desired number of decimal places. For instance, to six decimal place accuracy, you have $\sqrt{7} \approx 2.645751$, $\sqrt[3]{24} \approx 2.884499$, $e \approx 2.718282$, and $\pi \approx 3.141593$. *Note*: When you are working with real numbers, even roots (that is, square roots, fourth roots, and so on) of negative numbers are not real numbers.

The set of **real numbers**, ***R***, is the union of the set of rational numbers and the set of irrational numbers. That is,

$$\boldsymbol{R} = \{\text{rational numbers}\} \cup \{\text{irrational numbers}\}$$

PROBLEM List all the sets in the real number system to which the given number belongs.

a. 0

b. 0.75

c. -25

d. $\sqrt{36}$

e. $\sqrt[3]{5}$

f. $\frac{2}{3}$

SOLUTION a. whole numbers, integers, rationals, reals

b. rationals, reals

c. integers, rationals, reals

d. natural numbers, whole numbers, integers, rationals, reals

e. irrationals, reals

f. rationals, reals

PROBLEM State whether the given number is rational or irrational.

a. $\sqrt[4]{100}$

b. $\sqrt{0.25}$

c. $\sqrt[3]{-8}$

d. $-e$

SOLUTION a. irrational

b. rational

c. rational

d. irrational

EXERCISE 1·1

For 1–10, list all the sets in the real number system to which the given number belongs.

1. 10
2. $\sqrt{0.64}$
3. $\sqrt[3]{\frac{8}{125}}$
4. $-\pi$
5. $-1{,}000$
6. $\sqrt{2}$
7. $-\sqrt{\frac{3}{4}}$
8. $-\sqrt{\frac{9}{4}}$
9. 1
10. $\sqrt[3]{0.001}$

For 11–15, state whether the given number is rational or irrational.

11. $\sqrt{25}$
12. $\sqrt[3]{15}$
13. $\sqrt[3]{-\frac{64}{125}}$
14. $\sqrt{41}$
15. $-\frac{\pi}{2}$

Intervals on the real number line

You can represent the real numbers on the real number line (see Figure 1.1). Every point on the number line corresponds to a real number; and, conversely, every real number corresponds to a point on the number line.

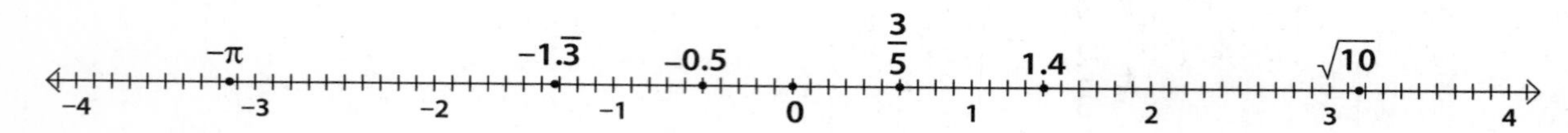

Figure 1.1 Real number line

You use intervals to show sets of numbers on the real number line. **Open intervals** do not include the endpoints. **Closed intervals** include both endpoints. **Half-open** (or **half-closed**) **intervals** include only one endpoint. Finite intervals are **bounded intervals**. Intervals that extend indefinitely to the right or left are **unbounded intervals**.

Interval notation is a concise way to describe the intervals. In interval notation, you place parentheses or square brackets around the specified endpoints of the interval. A square bracket means the endpoint is included, and a parenthesis means the endpoint is not included. For unbounded intervals, you use the symbols ∞ and $-\infty$ to represent the notion of extending indefinitely to the right or left, respectively.

You graph the interval on the number line by shading the number line to show the numbers included in the interval. You use a solid circle to indicate an endpoint is included, and an open circle to indicate an endpoint is not included. Table 1.1 summarizes intervals and interval notation.

PROBLEM Represent the nonnegative numbers, using interval notation. State the interval type.

SOLUTION $[0, \infty)$ unbounded, half-open.

Use interval notation to represent the indicated interval. State the interval type.

1. real numbers
2. negative numbers
3. all numbers less than 3.5
4. $-10 \leq x \leq 30$
5. (number line with solid point at 2.75)

Table 1.1 Intervals and interval notation

Interval	Notation and type	Graph
$x<b$	$(-\infty,b)$ unbounded, open	b
$x>a$	(a,∞) unbounded, open	a
$x\le b$	$(-\infty,b]$ unbounded, half-open	b
$x\ge a$	$[a,\infty)$ unbounded, half-open	a
$a<x<b$	(a,b) bounded, open	a b
$a\le x<b$	$[a,b)$ bounded, half-open	a b
$a<x\le b$	$(a,b]$ bounded, half-open	a b
$a\le x\le b$	$[a,b]$ bounded, closed	a b

Properties of real numbers

For much of precalculus, you will be working with the set of real numbers along with the operations of **addition**, denoted by the plus symbol (+), and **multiplication**, denoted by the raised center dot (·). Of course, for any two real numbers *a* and *b*, you can also denote $a \cdot b$ by *ab*, *a*(*b*), (*a*)*b*, or (*a*)(*b*).

The set of real numbers has the properties shown in Table 1.2, on the next page, for all real numbers *a*, *b*, and *c* under the operations of addition and multiplication.

PROBLEM State the field property that is illustrated in each of the following:

a. $0+1.25=1.25$

b. $(\pi+\sqrt{2})\in R$ *Note*: The symbol $\in$ is read "is an element of."

c. $\frac{3}{4}\cdot\frac{5}{6}=\frac{5}{6}\cdot\frac{3}{4}$

SOLUTION a. additive identity property

b. closure property for addition

c. commutative property for multiplication

Table 1.2 Properties of the real numbers

Property	Explanation
Closure Property: $(a+b)$ and $(a \cdot b)$ are real numbers.	When you add or multiply two real numbers, you get a real number as the answer.
Commutative Property: $a+b=b+a$ and $a \cdot b = b \cdot a$.	You can reverse the order of the numbers when you add or multiply without changing the answer.
Associative Property: $(a+b)+c=a+(b+c)$ and $(ab)c=a(bc)$.	When you have three numbers and one operation (+ or ·), the final answer will be the same regardless of the way you group the numbers to perform the operation.
Additive Identity Property: There exists a real number 0, called the *additive identity*, such that $a+0=a$ and $0+a=a$.	The additive identity zero is a real number, and its sum with any real number is the number.
Multiplicative Identity Property: There exists a real number 1, called the *multiplicative identity*, such that $a \cdot 1 = a$ and $1 \cdot a = a$.	The multiplicative identity 1 is a real number, and its product with any real number is the number.
Additive Inverse Property: For every real number a, there is a real number called its *additive inverse*, denoted $-a$, such that $a+-a=0$ and $-a+a=0$.	Every real number has an additive inverse that is a real number, which when added to the given number, gives 0.
Multiplicative Inverse Property: For every nonzero real number a, there is a real number called its *multiplicative inverse*, denoted a^{-1} or $\frac{1}{a}$, such that $a \cdot a^{-1} = a \cdot \frac{1}{a} = 1$ and $a^{-1} \cdot a = \frac{1}{a} \cdot a = 1$.	Every real number, *except zero*, has a multiplicative inverse that is a real number which when multiplied by the given number, yields 1.
Distributive Property: $a(b+c)=a \cdot b + a \cdot c$ and $(b+c)a = b \cdot a + c \cdot a$.	When you have a number times a sum, you can either first add and then multiply, or first multiply and then add. Either way, the final answer is the same.

Besides the field properties, you should keep in mind that the number zero has the following unique characteristic:

Zero Factor Property: If a real number is multiplied by zero, the product is zero; that is, $a \cdot 0 = 0 \cdot a = 0$.

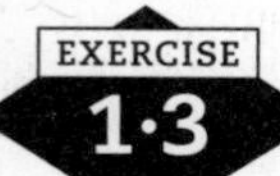

State the property of the real numbers that is illustrated.

1. $0.25(1500) \in R$
2. $\frac{2}{5}+\frac{3}{4}=\frac{3}{4}+\frac{2}{5}$
3. $43 \cdot \frac{1}{43}=1$
4. $\left(1.3+\frac{1}{3}\right) \in R$
5. $43+(7+25)=(43+7)+25$
6. $60(10+3)=600+180$

7. $-\sqrt{41}+\sqrt{41}=0$

8. $-999\cdot 0=0$

9. $3x+\sqrt{5}x=\left(3+\sqrt{5}\right)x$

10. $\left(0.6\cdot\frac{3}{4}\right)\frac{4}{3}=0.6\left(\frac{3}{4}\cdot\frac{4}{3}\right)$

11. $\left(y^4-9\right)(x+1)(0)\left(z^2+2z+1\right)=0$

12. $\left(-v^5+9u\right)+-10=-10+\left(-v^5+9u\right)$

13. $\left(8x^3\right)(-7\cdot 3)=\left(8x^3\cdot -7\right)(3)$

14. $(z+5)\cdot\frac{1}{(z+5)}=1;\ (z+5)\neq 0$

15. $5(2a+b)=10a+5b$

Absolute value

The absolute value of a real number is its distance from zero on the number line. For example, as shown in Figure 1.2, the absolute value of –8 is 8 because –8 is 8 units from zero.

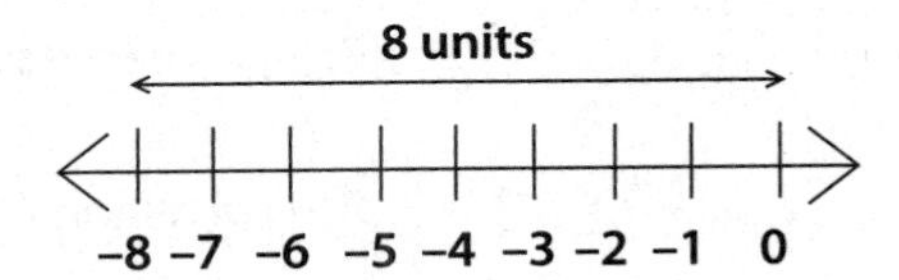

Figure 1.2 The absolute value of –8

By definition the absolute value of a real number x, denoted $|x|$, is given by

$$|x|=\begin{cases} x \text{ if } x\geq 0 \\ -x \text{ if } x<0 \end{cases}$$

Absolute value always has a *nonnegative* (*positive* or *zero*) value. Thus, for every real number x, its absolute value is either x or $-x$, whichever is a *nonnegative* number, that is, whichever one is farther to the right on the number line.

PROBLEM Simplify each of the following:

a. $|-30|$

b. $|0.4|$

c. $\left|-2\frac{1}{3}\right|$

d. $|x|$, when x is negative

e. $-|-30|$

SOLUTION

a. 30

b. 0.4

c. $2\frac{1}{3}$

d. $-x$

e. -30

PROBLEM Insert <, >, or = in the blank to make a true statement.

a. $-|-9|$ ______ -7

b. $\left|-\frac{1}{2}\right|$ ______ $-\left|-\frac{1}{2}\right|$

c. $|x|$ ______ $|-x|$

SOLUTION a. <

b. >

c. =

EXERCISE 1·4

For 1–10, simplify.

1. $|-45|$
2. $|5.8|$
3. $\left|-5\frac{2}{3}\right|$
4. $-|-60|$
5. $|0|$
6. $\left|-\frac{\pi}{2}\right|$
7. $|a|$, when a is negative
8. $|-a|$, when a is negative
9. $-|a|$, when a is negative
10. $-|-a|$, when a is negative

For 11–15, insert <, >, or = in the blank to make a true statement.

11. $-|-6|$ ______ $-|2|$
12. $-\left|-\frac{1}{4}\right|$ ______ $-\left|\frac{1}{2}\right|$
13. $|-1.25|$ ______ $|1.25|$
14. $-|-a|$ ______ $|a|$
15. $|-5|$ ______ $|-20|$

Operations with real numbers

Computations using real numbers are performed by using the common arithmetic operations familiar to you (addition, subtraction, multiplication, and division) on the absolute values (which are always positive or zero) of the numbers. For algebraic addition of real numbers, use the following two rules:

Rules for algebraic addition of real numbers

Same sign: To add two real numbers that have the same sign, add their absolute values and affix the common sign to the sum.

Different signs: To add two real numbers that have opposite signs, subtract the lesser absolute value from the greater absolute value and give the result the same sign as the number with the greater absolute value; if the two numbers have the same absolute value, their sum is 0.

EXAMPLES

- $5.75+2.18=$

SOLUTION $5.75+2.18=7.93$

- $-\frac{5}{7}+\frac{3}{7}=$

SOLUTION $-\frac{5}{7}+\frac{3}{7}=-\frac{2}{7}$

For algebraic multiplication of real numbers, use the following two rules:

Rules for algebraic multiplication of real numbers

Same sign: To multiply two real numbers that have the same sign, multiply their absolute values and leave the product positive.

Different signs: To multiply two real numbers that have different signs, multiply their absolute values and affix a minus sign to the product.

EXAMPLES

- $\frac{1}{3}\cdot 27=$

SOLUTION $\frac{1}{\cancel{3}_1}\cdot \cancel{27}^9=9$

- $-0.06(5000)=$

SOLUTION $-0.06(5000)=-300$

You do not have separate rules for algebraic subtraction or algebraic division of real numbers. Instead, you define these two operations in terms of algebraic addition and multiplication, respectively, as follows:

Algebraic subtraction: For two real numbers a and b, $a-b=a+-b$; that is, to subtract a number, add its additive inverse.

Algebraic division: For two real numbers a and b $(b\neq 0)$, $a\div b=\frac{a}{b}=a\cdot b^{-1}$; that is, to divide by a nonzero number, multiply by its multiplicative inverse.

EXAMPLES

◆ $-100-20=$

SOLUTION $-100-20=-100+-20=-120$

◆ $\frac{-500}{50}=$

SOLUTION $\frac{-500}{50}=\ ^{-10}\cancel{500}\cdot\frac{1}{\cancel{50}_1}=-10$

EXERCISE **1·5**

Compute as indicated.

1. $-80+-40=$
2. $0.7+-1.4=$
3. $\left(-\frac{5}{6}\right)\left(\frac{2}{5}\right)=$
4. $\frac{18}{-3}=$
5. $(-100)(-8)=$
6. $\left(4\frac{1}{2}\right)\left(3\frac{3}{5}\right)=$
7. $\frac{-1\frac{1}{3}}{-\frac{1}{3}}=$
8. $450.95-65.83=$
9. $\frac{3}{4}-\left(-\frac{5}{8}\right)=$
10. $0.8\div-0.01=$

Exponentiation

In mathematical expressions, you show **exponentiation** by a small raised number, called the **exponent**, written to the upper right of a quantity, which is called the **base** for the exponential

Table 1.3 Types of exponents

Type of exponent	Meaning
Positive integer	If x is any real number and n is a positive integer, then $x^n=\underbrace{x\cdot x\cdot x\cdot\ldots\cdot x}_{n\text{ factors of }x}$.
Zero	For any real number x (except zero), $x^0=1$.
Unit fraction	If x is a real number and n is a natural number, then $x^{1/n}=\sqrt[n]{x}$; that is, x is the nth root such that $\left(x^{1/n}\right)^n=x$, provided that when n is even, $x\geq 0$ and $x^{1/n}$ is the nonnegative root.
Negative number	If x is any real number (except zero) and $-n$ is a negative number, then $x^{-n}=\frac{1}{x^n}$ and $\frac{1}{x^{-n}}=x^n$.
Rational number	If x is any real number and p and q are integers with $q\neq 0$, then $x^{p/q}=\left(x^{1/q}\right)^p=\left(\sqrt[q]{x}\right)^p$ or $x^{p/q}=\left(x^p\right)^{1/q}=\sqrt[q]{x^p}$ provided, in all cases, that $x\geq 0$ when q is even and division by zero or 0^0 does not occur.

expression. For example, in the exponential expression 2^5, 5 is the exponent and 2 is the base. Table 1.3 summarizes types of exponents you should know when you begin precalculus.

You have the following rules for exponents.

Rules for exponents

For real numbers x and y and integers m, n, and p:

$$x^m x^n = x^{m+n}, \frac{x^m}{x^n} = x^{m-n}, (x^n)^p = x^{np}, (xy)^p = x^p y^p, \left(\frac{x}{y}\right)^p = \frac{x^p}{y^p},$$

$$\left(\frac{x}{y}\right)^{-n} = \left(\frac{y}{x}\right)^n$$

provided, in all cases, that *neither even roots of negative quantities nor division by zero occurs.*

For real numbers x and y and natural number n:

$$(x+y)^n = \underbrace{(x+y)(x+y)\ldots(x+y)}_{n \text{ times}}$$

according to the rules for multiplying binomials.

EXAMPLES

◆ Evaluate $(-3)^4$.

SOLUTION $(-3)^4 = -3 \cdot -3 \cdot -3 \cdot -3 = 81$

◆ Evaluate -3^4.

SOLUTION $-3^4 = -(3 \cdot 3 \cdot 3 \cdot 3) = -81$

Note: An exponent applies only to the quantity to which it is immediately attached.

◆ Evaluate $16^{\frac{5}{2}}$.

SOLUTION $16^{\frac{5}{2}} = \left(16^{\frac{1}{2}}\right)^5 = 4^5 = 1{,}024$

◆ Simplify $x^3 x^5 y^4 y$.

SOLUTION $x^3 x^5 y^4 y = x^{3+5} y^{4+1} = x^8 y^5$

◆ Simplify $\frac{a^5 b^7}{a^2 b^3}$.

SOLUTION $\frac{a^5 b^7}{a^2 b^3} = a^{5-2} b^{7-3} = a^3 b^4$

◆ Simplify $(x+3)^2$.

SOLUTION $(x+3)^2 = (x+3)(x+3) = x^2 + 6x + 9$

EXERCISE

1·6

For 1–10, evaluate the given expression.

1. $-2^5 =$
2. $36^{\frac{1}{2}} =$
3. $\left(\frac{5}{3}\right)^{-3} =$
4. $\frac{6^{-2}}{3^{-4}} =$
5. $\frac{2^{-5}}{3^{-2}} =$
6. $2^3 - 4^3 =$
7. $-64^{\frac{2}{3}} =$
8. $\pi^0 =$
9. $\left(\frac{2}{3}\right)^{-1} =$
10. $\frac{1}{2^{-1}+3^{-1}}$

For 11–15, simplify the given expression. Assume that all variables are positive.

11. $a^{\frac{1}{2}}a^{\frac{3}{4}} =$
12. $\frac{b^6}{b^2}$
13. $\left(y^3\right)^4 =$
14. $\left(\frac{x}{y}\right)^{-3}$
15. $(x+3)^3 =$
16. $\left(x^3y^6\right)^{\frac{1}{3}} =$
17. $\frac{\left(a^2\right)^3 b^8}{a^5\left(b^3\right)^2} =$
18. $\left(x^{-2}\right)^4 y^7 y^{-9} =$
19. $\frac{\left(a^2b^{-5}\right)^{-4}}{\left(a^5b^{-2}\right)^{-3}} =$
20. $\left(\left(x^{-2}y^3z\right)^3\right)^{-2} =$

Order of operations

You must follow the order of operations to simplify mathematical expression. Use the mnemonic "Please Excuse My Dear Aunt Sally"—abbreviated as PE(MD)(AS) to help you remember the following:

First do computations inside Parentheses (or other grouping symbols); then Evaluate exponential expressions; next perform Multiplication and Division, in the order in which these operations occur from left to right; and, finally, perform Addition and Subtraction, in the order in which these operations occur from left to right.

EXAMPLE

◆ Simplify $100 - 8\cdot 3^2 + 63 \div (2+5)$.

SOLUTION $= 100 - 8\cdot 3^2 + 63 \div (7)$ First compute inside parentheses.

$= 100 - 8\cdot 9 + 63 \div 7$ Next evaluate exponents.

$= 100 - 72 + 9$ Then multiply and divide.

$= 37$ Finally, add and subtract.

When simplified, the numerical expression $100 - 8\cdot 3^2 + 63 \div (2+5) = 37$.

homework 1-5

EXERCISE 1·7

Simplify.

1. $(5+7)6-10$
2. $(-7)^2(6-8)$
3. $(2-3)(-20)$
4. $3(-2)-\dfrac{10}{-5}$
5. $9-\dfrac{20+22}{6}-2^3$
6. $-2^2\cdot-3-(15-4)^2$
7. $5(11-3-6\cdot2)^2$
8. $-10-\dfrac{-8-(3\cdot-3+15)}{2}$
9. $\dfrac{7^2-8\cdot5+3^4}{3\cdot2-36\div12}$
10. $\dfrac{13-2(3^2\cdot2-3\cdot2^3)}{10^2}$

·2· Complex numbers

Complex number system

The set of **complex numbers** comprise the numbers that can be written in the **standard form** $x + yi$, where x and y are real numbers and $i^2 = -1$. If $z = x + yi$ is a complex number, the coefficients x and y are the **real part** and **imaginary part**, respectively, of z. Two complex numbers $x + yi$ and $u + vi$ are equal if and only if $x = u$ and $y = v$.

Complex numbers of the form $0 + yi$ ($y \neq 0$) are **pure imaginary** numbers. For example, $0 - 3i$, $0 + \frac{\sqrt{41}}{2}i$, and $0 + \frac{3\pi}{2}i$ are pure imaginary numbers.

Complex numbers of the form $x + 0i$ are **real** numbers. Thus, the real numbers are a subset of the complex numbers. In other words, all the rational and irrational numbers are complex numbers.

If either the real part or the imaginary part of a complex number is zero, you can omit that part when writing the number—except that $0 + 0i = 0$. For example, $0 + 5i = 5i$ and $7 + 0i = 7$.

You can represent the complex numbers in the **complex plane**, where the horizontal axis is the **real axis** and the vertical axis is the **imaginary axis**. Figure 2.1 shows complex numbers in the complex plane.

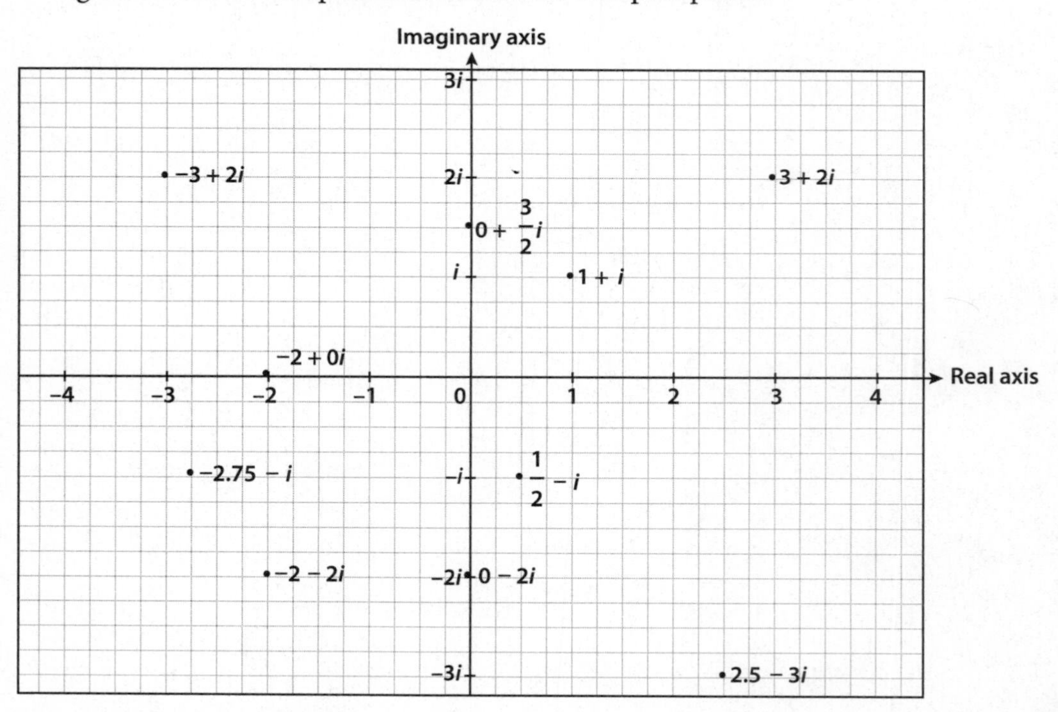

Figure 2.1 Complex numbers in the complex plane

Given that $i^2 = -1$ it follows that the **imaginary unit *i*** has the property that $i = \sqrt{-1}$. Capitalizing on this property, you have that for any positive real number a, $\sqrt{-a} = \sqrt{a}i$. For example, $\sqrt{-36} = 6i$, $\sqrt{-\frac{9}{16}} = \frac{3}{4}i$, and $\sqrt{-41} = \sqrt{41}i$. *Note*: To avoid confusion, you write numbers like $\sqrt{41}i$ as $i\sqrt{41}$.

PROBLEM List all the sets in the complex number system to which the given number belongs.

a. 0

b. $\sqrt{-64}$

c. $\frac{5\pi}{2}$

d. $-4i$

e. $\frac{3}{4}+\frac{1}{2}i$

SOLUTION a. whole numbers, integers, rationals, reals, complex numbers

b. pure imaginary numbers, complex numbers

c. irrationals, reals, complex numbers

d. pure imaginary numbers, complex numbers

e. complex numbers

PROBLEM State whether the given number is rational, irrational, or pure imaginary.

a. $\sqrt{3}$

b. $\sqrt[3]{-8}$

c. $\sqrt{-3}$

d. $i\sqrt{5}$

e. $-\frac{\sqrt{3}}{2}i$

SOLUTION a. irrational

b. rational

c. pure imaginary

d. pure imaginary

e. pure imaginary

For 1–5, fill in the blank(s) to make a true statement.

1. In the complex number $x+yi$, __________ is the real part and __________ is the imaginary part.
2. In the complex number $x+yi$, both x and y are __________ numbers and i^2 equals __________.

3. Two complex numbers $x+yi$ and $u+vi$ are equal if and only if __________ and __________.

4. Complex numbers of the form __________ are pure imaginary numbers.

5. Complex numbers of the form $x+0i$ are __________ numbers.

For 6–15, list all the sets in the complex number system to which the given number belongs.

6. -300

7. $\frac{1}{2}-\frac{\sqrt{3}}{2}i$

8. $-\pi+2i$

9. $\sqrt{2}$

10. $-\sqrt{\frac{3}{4}}$

11. $\sqrt{-2}$

12. $4.5+0.7i$

13. $\sqrt{-\frac{25}{36}}$

14. $-\frac{4}{3}+\frac{\pi}{3}i$

15. i

For 16–20, state whether the given number is rational, irrational, or pure imaginary.

16. $\sqrt{81}$

17. $\sqrt{-81}$

18. $\sqrt[3]{-\frac{64}{125}}$

19. $0.75i$

20. $\frac{\pi}{2}i$

Addition and subtraction of complex numbers

Addition of two complex numbers $x+yi$ and $u+vi$ is defined as follows:

$$(x+yi)+(u+vi)=(x+u)+(y+v)i$$

As you can see, the real part of the sum is the sum of the real parts of the two complex numbers, and the imaginary part of the sum is the sum of the imaginary parts. Here are examples.

- $(1+6i)+(5+2i)=(1+5)+(6+2)i=6+8i$
- $(-3+10i)+(5+4i)=(-3+5)+(10+4)i=2+14i$
- $(-5-i)+(-3-4i)=(-5-3)+(-1-4)i=-8-5i$
- $\left(\frac{1}{2}+\frac{3}{4}i\right)+\left(-\frac{3}{2}-\frac{1}{4}i\right)=\left(\frac{1}{2}-\frac{3}{2}\right)+\left(\frac{3}{4}-\frac{1}{4}\right)i=-\frac{2}{2}+\frac{2}{4}i=-1+\frac{1}{2}i$

Under the operation of addition, the set of complex numbers has the following field properties:

Closure Property: $(x+yi)+(u+vi)=(x+u)+(y+v)i$ is a complex number.

Commutative Property: $(x+yi)+(u+vi)=(u+vi)+(x+yi)$.

Associative Property: $[(x+yi)+(u+vi)]+(a+bi)=(x+yi)+[(u+vi)+(a+bi)]$.

Additive Identity Property: There exists a complex number 0 such that $(x+yi)+(0)=x+yi$ and $(0)+(x+yi)=x+yi$.

Additive Inverse Property: For every complex number $x+yi$, there exists a complex number $-x+-yi$ such that $(x+yi)+(-x+-yi)=0+0i$ and $(-x+-yi)+(x+yi)=0+0i$.

Note: Proof of these properties relies on the corresponding field property for the real numbers.

Subtraction of two complex numbers is defined as follows. For two complex numbers $(x+yi)$ and $(u+vi)$:

$$(x+yi) - (u+vi) = (x+yi) + (-u+-vi) = (x+-u) + (y+-v)i = (x-u) + (y-v)i$$

In other words, to subtract a complex number, add its additive inverse. Here are examples:

- $(1+5i) - (3+2i) = (1+5i) + (-3+-2i) = (1-3) + (5-2)i = -2+3i$
- $(2-i) - (-3-7i) = (2-i) + (3+7i) = (2+3) + (-1+7)i = 5+6i$

You don't have to memorize these definitions because you can perform the computations as you would with binomials, being sure to express your final answer in the standard form $(x+yi)$. Here are examples:

- $(-3+10i)+(5+4i) = -3+10i+5+4i = -3+5+10i+4i = 2+14i$
- $(-5-i)+(-3-4i) = -5-i-3-4i = -5-3-i-4i = -8-5i$
- $(1+5i)-(3+2i) = 1+5i-3-2i = 1-3+5i-2i = -2+3i$
- $(2-i)-(-3-7i) = 2-i+3+7i = 2+3-i+7i = 5+6i$

EXERCISE 2·2

homework 6-10

Perform the indicated computation.

1. $(-3-5i) + (4+5i)$
2. $(-1-i) - (-3-4i)$
3. $(1-i\sqrt{3})+(3+2i\sqrt{3})$
4. $(2.8-1.5i) - (-3.5+7i)$
5. $\left(-\frac{1}{2}+\frac{3}{4}i\right)+\left(-\frac{3}{2}-\frac{1}{4}i\right)$
6. $(9-5i) + (0+0i)$
7. $(5-i) - (-4i)$
8. $(4+2i) - (-4-2i)$
9. $(4+2i) + (-4-2i)$
10. $\left(6+\frac{1}{6}i\right)+\left(-\frac{1}{2}-\frac{5}{6}i\right)$

Multiplication of complex numbers and complex conjugates

Multiplication of two complex numbers $(x+yi)$ and $(u+vi)$ is defined as follows:

$$(x+yi)(u+vi) = (xu-yv) + (xv+yu)i$$

Here is an example:

- $(-3+10i)(5+4i) = (-3\cdot5-10\cdot4)+(-3\cdot4+10\cdot5)i = (-15-40)+(-12+50)i = -55+38i$

As with addition and subtraction, you don't have to memorize this definition because you can perform the computation as you would with binomials, being sure to replace i^2 with -1

wherever it occurs and to express your final answer in standard form $x+yi$. Here are examples:

- $(-3+10i)(5+4i) = -3\cdot 5 + -3\cdot 4i + 10i\cdot 5 + 10i\cdot 4i = -15 + -12i + 50i + 40i^2 =$ $-15 - 12i + 50i + 40(-1) = -15 - 12i + 50i - 40 = -55 + 38i$
- $(3-2i)(-5-4i) = 3\cdot -5 + 3\cdot -4i + -2i\cdot -5 + -2i\cdot -4i = -15 + -12i + 10i + 8i^2 =$ $-15 - 12i + 10i + 8(-1) = -15 - 12i + 10i - 8 = -23 - 2i$
- $2i(10+3i) = 2i\cdot 10 + 2i\cdot 3i = 20i + 6i^2 = 20i + 6(-1) = 20i + -6 = -6 + 20i$
- $(3+2i)(3-2i) = 3\cdot 3 + 3\cdot -2i + 2i\cdot 3 + 2i\cdot -2i = 9 + -6i + 6i + -4\cdot i^2 = 9 - 4\cdot(-1) =$ $9 + 4 = 13$

The last example illustrates a special situation. The complex numbers $3+2i$ and $3-2i$ are complex conjugates of each other. When you multiply them, you obtain a real number as the product.

In general, the complex numbers $x+yi$ and $x-yi$ are **complex conjugates** of each other, and their product is the real number x^2+y^2; that is:

$$(x+yi)(x-yi) = x^2 + y^2$$

Here are examples:

- $(10+5i)(10-5i) = 10^2 + 5^2 = 100 + 25 = 125$
- $(-3-4i)(-3+4i) = (-3)^2 + 4^2 = 9 + 16 = 25$

Under the operation of multiplication, the set of complex numbers has the following field properties:

Closure Property: $(x+yi)(u+vi) = (xu - yv) + (xv + yu)i$ is a complex number.

Commutative Property: $(x+yi)(u+vi) = (u+vi)(x+yi)$.

Associative Property: $[(x+yi)(u+vi)](a+bi) = (x+yi)[(u+vi)(a+bi)]$.

Multiplicative Identity Property: There exists a complex number 1 such that $(x+yi)(1) = x+yi$ and $(1)(x+yi) = x+yi$.

Multiplicative Inverse Property: For every nonzero complex number $x+yi$, there exists a complex number $\frac{x}{x^2+y^2} + \frac{-y}{x^2+y^2}i$ such that $(x+yi)\left(\frac{x}{x^2+y^2} + \frac{-y}{x^2+y^2}i\right) = 1$ and $\left(\frac{x}{x^2+y^2} + \frac{-y}{x^2+y^2}i\right)(x+yi) = 1$.

Additionally, for the operations of addition and multiplication, you have the following property for complex numbers:

Distributive Property: $(x+yi)[(a+bi)+(c+di)] = (x+yi)(a+bi) + (x+yi)(c+di)$.

Note: Proof of each of these properties relies on the corresponding field property for the real numbers.

You can use the definition of multiplication and the fact that $i^2 = -1$ to compute whole-number powers of the imaginary unit i. Here are examples:

$$i^1 = i;\ i^2 = -1;\ i^3 = i \cdot i^2 = i \cdot -1 = -i;\ i^4 = i^2 \cdot i^2 = -1 \cdot -1 = 1;\ i^5 = i \cdot i^4 = i \cdot 1 = i;$$
$$i^6 = i^2 \cdot i^4 = -1 \cdot 1 = -1;\ \ i^7 = i^3 \cdot i^4 = -i \cdot 1 = -i;\ i^8 = i^4 \cdot i^4 = 1 \cdot 1 = 1$$

As you look at this list, you see in the answers a pattern of $i, -1, -i, 1, i, -1, -i, 1, \ldots$ This pattern will continue through higher powers of i. In general, $(i^4)^n = 1$, for any integer n, so, for example, $i^{103} = (i^4)^{25} i^3 = 1 \cdot -i = -i$.

EXERCISE 2·3

homework 2-10 skip 6

Perform the indicated computation.

1. $(4+5i)(10-i)$
2. $(-2-3i)(4-8i)$
3. $(5-i\sqrt{3})(2+i)$
4. $(8+2i)(8-2i)$
5. $(\sqrt{2}-i\sqrt{5})(\sqrt{2}+i\sqrt{5})$
6. $(-2+3i)(4-8i)$
7. $(5-i\sqrt{3})(1+i\sqrt{3})$
8. $(x+yi)\left(\dfrac{x}{x^2+y^2}+\dfrac{-y}{x^2+y^2}i\right)$
9. $(2+i)[(3-4i)+(5+7i)]$
10. i^{402}

Division of complex numbers

Division of two complex numbers $x+yi$ and $u+vi$ is performed as follows:

$$\frac{x+yi}{u+vi} = \frac{(x+yi)}{(u+vi)} \cdot \frac{(u-vi)}{(u-vi)} = \frac{xu+yv}{u^2+v^2} + \frac{yu-xv}{u^2+v^2} i$$

Here is an example:

- $$\frac{1+2i}{3+4i} = \frac{1 \cdot 3 + 2 \cdot 4}{3^2+4^2} + \frac{2 \cdot 3 - 1 \cdot 4}{3^2+4^2} i = \frac{3+8}{9+16} + \frac{6-4}{9+16} i = \frac{11}{25} + \frac{2}{25} i$$

Again, it is not necessary to memorize the definition. You can accomplish the division by multiplying the numerator and denominator by the complex conjugate of the denominator as shown here:

- $$\frac{1+2i}{3+4i} = \frac{(1+2i)(3-4i)}{(3+4i)(3-4i)} = \frac{1 \cdot 3 + 1 \cdot -4i + 2i \cdot 3 + 2i \cdot -4i}{3^2+4^2} = \frac{3-4i+6i-8i^2}{25} = \frac{3-4i+6i+8}{25}$$
$$= \frac{11+2i}{25} = \frac{11}{25} + \frac{2}{25} i$$

Note: Remember to write your answer in $x+yi$ form.

homework 1-5 skip 2

EXERCISE
2·4

Perform the indicated computation.

1. $\frac{1-i}{2+4i}$

2. $\frac{4-2i}{2+3i}$

3. $\frac{1}{2+3i}$

4. $\frac{2i}{5+3i}$

5. $\frac{3-2i}{3+2i}$

The Cartesian coordinate system

·3·

Terminology for the Cartesian coordinate system

The **Cartesian coordinate system** is a rectangular coordinate system defined by two real number lines, one horizontal and one vertical, intersecting at right angles at their zero points. The two real number lines are the **coordinate axes**. The **horizontal axis**, commonly the ***x*-axis**, has positive direction to the right; and the **vertical axis**, commonly the ***y*-axis**, has positive direction upward. The two axes determine the plane of the rectangular coordinate system. Their point of intersection is called the **origin**.

With this coordinate system each point P in the plane is identified by an **ordered pair** (x, y) of real numbers x and y, called its **coordinates**. Specifically, there is a one-to-one correspondence between points in the Cartesian coordinate plane and ordered pairs of real numbers.

The order in an ordered pair is important; that is, if $x \neq y$, then $(x,y) \neq (y,x)$. The first coordinate, x, corresponds to the *directed* (right or left) distance of the point from the vertical axis. The second coordinate, y, corresponds to the *directed* (up or down) distance of the point from the horizontal axis, as shown in Figure 3.1.

Two ordered pairs are equal if and only if their corresponding coordinates are equal; that is, $(a, b) = (c, d)$ if and only if $a = c$ and $b = d$.

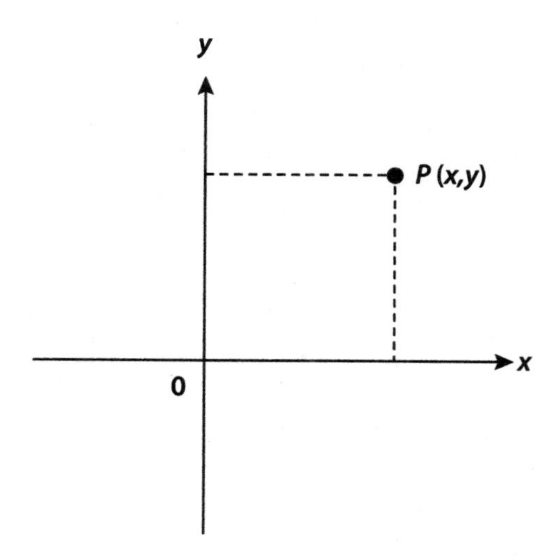

Figure 3.1 Point in a Cartesian coordinate plane

The axes divide the Cartesian coordinate plane into four **quadrants**. They are numbered with Roman numerals **I, II, III,** and **IV**, beginning in the upper right and proceeding counterclockwise, as shown in Figure 3.2.

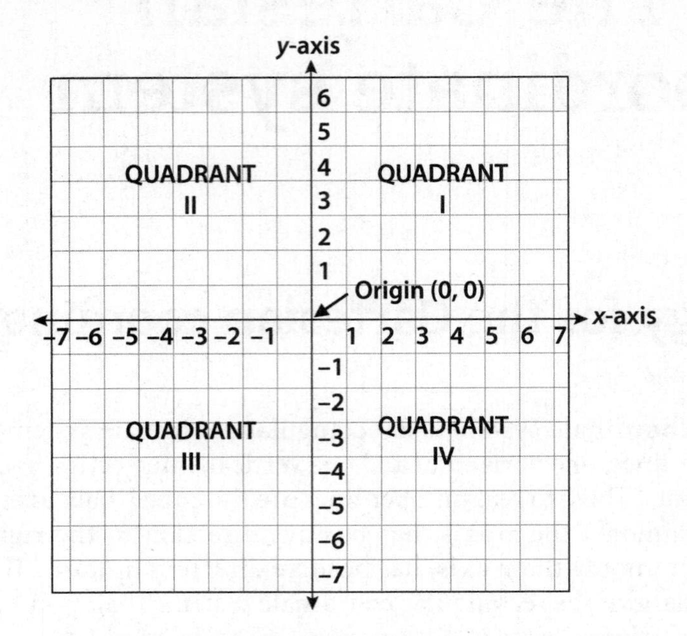

Figure 3.2 Quadrants in the complex plane

In quadrant I, both the x-coordinate and the y-coordinate are positive; in quadrant II, the x-coordinate is negative, and the y-coordinate is positive; in quadrant III, both the x-coordinate and the y-coordinate are negative; and in quadrant IV, the x-coordinate is positive, and the y-coordinate is negative. Points that have zero as one or both of the coordinates lie on the axes. If the x-coordinate is zero, the point lies on the y-axis. If the y-coordinate is zero, the point lies on the x-axis. If both coordinates of a point are zero, the point is at the origin.

PROBLEM What ordered pair of integers corresponds to the point A in the coordinate plane shown?

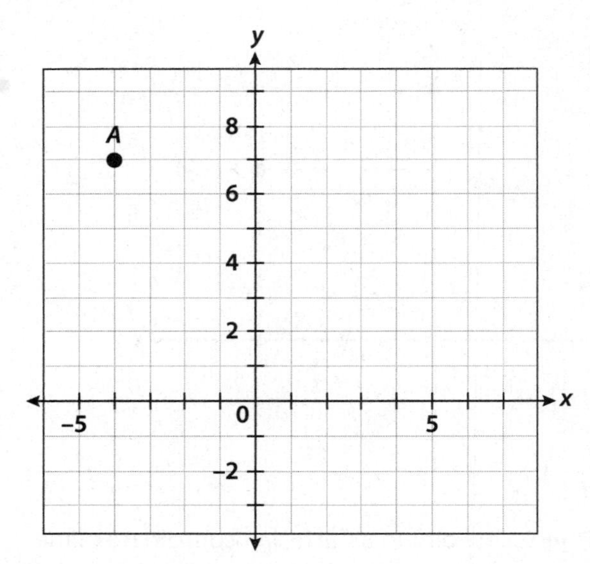

SOLUTION The point A is 4 units to the left and 7 units up from the origin. The ordered pair $(-4, 7)$ corresponds to the point A.

EXERCISE

3·1

For 1–8, indicate whether the statement is true or false.

1. The intersection of the coordinate axes is the origin.
2. (2, 3) = (3, 2)
3. $\left(\frac{2}{3}, \frac{1}{2}\right) = \left(\frac{4}{6}, \frac{5}{10}\right)$
4. The coordinates of a point in the plane can be rational or irrational.
5. The point $\left(\frac{3}{4}, -5\right)$ is in quadrant II.
6. The point $\left(-\frac{\sqrt{2}}{2}, -\frac{\sqrt{2}}{2}\right)$ is in quadrant III.
7. The point (5, 0) is in quadrant I.
8. Ordered pairs of the form (0, *y*) correspond to points that lie on the *y*-axis.

For 9–10, state the ordered pair of integers corresponding to each point in the coordinate plane shown.

9. *A*: ___________ *B*: ___________ *C*: ___________.

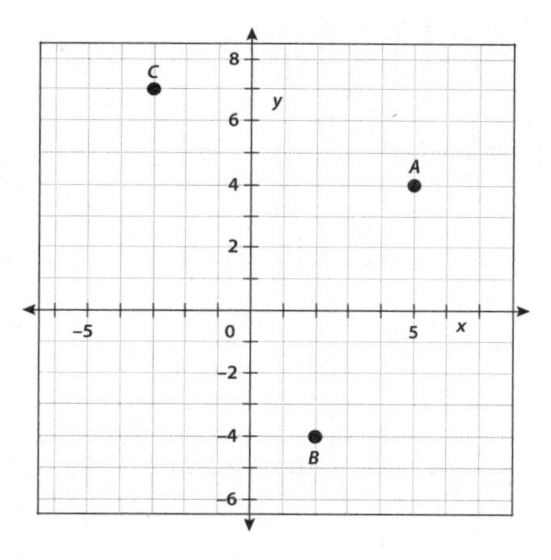

10. *A*: ___________ *B*: ___________ *C*: ___________.

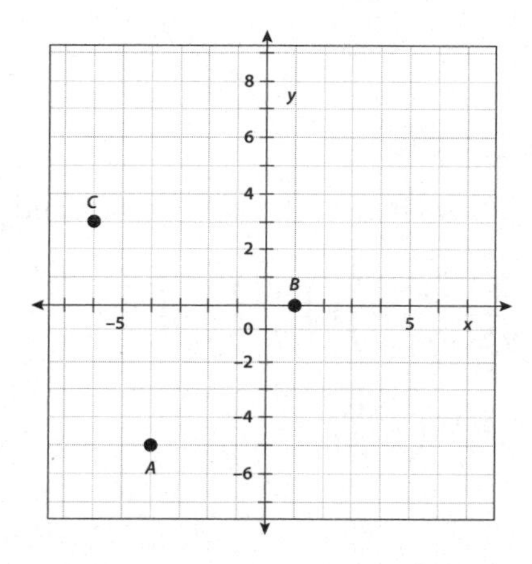

Distance between two points

The distance d between two points (x_1, y_1) and (x_2, y_2) in a coordinate plane is given by the formula:

$$\text{Distance} = d = \sqrt{(x_2 - x_1)^2 + (y_2 - y_1)^2}$$

Here is an example:

PROBLEM Find the distance between the points (−1, 4) and (5, −3).

SOLUTION Let $(x_1, y_1) = (-1, 4)$ and $(x_2, y_2) = (5, -3)$. Then $x_1 = -1$, $y_1 = 4$, $x_2 = 5$, and $y_2 = -3$.

$$d = \sqrt{(x_2 - x_1)^2 + (y_2 - y_1)^2} = \sqrt{((5)-(-1))^2 + ((-3)-(4))^2}$$

$$= \sqrt{(5+1)^2 + (-3-4)^2} = \sqrt{(6)^2 + (-7)^2} = \sqrt{36+49} = \sqrt{85}$$

The distance between (−1, 4) and (5, −3) is $\sqrt{85}$ units.

EXERCISE 3·2

Find the distance between each pair of points.

1. (1, 4), (5, 7)
2. (−1, −3), (1, 0)
3. (1.5, 4.5), (−3.5, −2.5)
4. $\left(\frac{4\sqrt{3}}{2}, -\frac{1}{4}\right), \left(\frac{3\sqrt{3}}{2}, \frac{1}{4}\right)$
5. $(\pi, 1), \left(\frac{\pi}{2}, 0\right)$

Midpoint between two points

The midpoint between two points (x_1, y_1) and (x_2, y_2) in a coordinate plane is the point with coordinates:

$$\left(\frac{x_1 + x_2}{2}, \frac{y_1 + y_2}{2}\right)$$

Here is an example:

PROBLEM Find the midpoint between (−1, 4) and (5, −3).

SOLUTION Let $(x_1, y_1) = (-1, 4)$ and $(x_2, y_2) = (5, -3)$. Then $x_1 = -1$, $y_1 = 4$, $x_2 = 5$, and $y_2 = -3$.

$$\text{Midpoint} = \left(\frac{x_1 + x_2}{2}, \frac{y_1 + y_2}{2}\right) = \left(\frac{-1+5}{2}, \frac{4-3}{2}\right) = \left(\frac{4}{2}, \frac{1}{2}\right) = \left(2, \frac{1}{2}\right)$$

The midpoint between (−1, 4) and (5, −3) is $\left(2, \frac{1}{2}\right)$.

EXERCISE
3·3

Find the midpoint between each pair of points.

1. $(1, 4), (5, 7)$
2. $(-1, -3), (1, 0)$
3. $(1.5, 4.5), (-3.5, -2.5)$
4. $\left(\frac{4\sqrt{3}}{2}, -\frac{1}{4}\right), \left(\frac{3\sqrt{3}}{2}, \frac{1}{4}\right)$
5. $(\pi, 1), \left(\frac{\pi}{2}, 0\right)$

Slope of a line

To calculate the slope m of a line, find any two distinct points (x_1, y_1) and (x_2, y_2), on the line and use the formula:

$$\text{Slope} = m = \frac{y_2 - y_1}{x_2 - x_1}$$

Note: Be sure to subtract the coordinates in the same order in both the numerator and the denominator.

From the formula, you can see that the slope is the ratio of the change in y-coordinates (the *rise*) and the change in x-coordinates (the *run*). Thus, slope $= \frac{\text{rise}}{\text{run}}$. Figure 3.3 illustrates the rise and run for the slope of the line that contains points $P_1(x_1, y_1)$ and $P_2(x_2, y_2)$.

In general, lines that slant upward to the right have positive slopes, and lines that slant downward to the right have negative slopes. Following are examples.

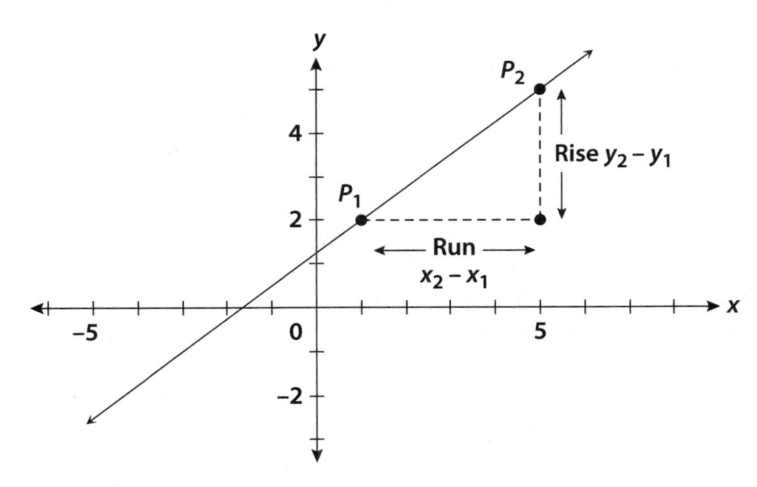

Figure 3.3 Rise and run

PROBLEM Find the slope of the line that contains the points shown.

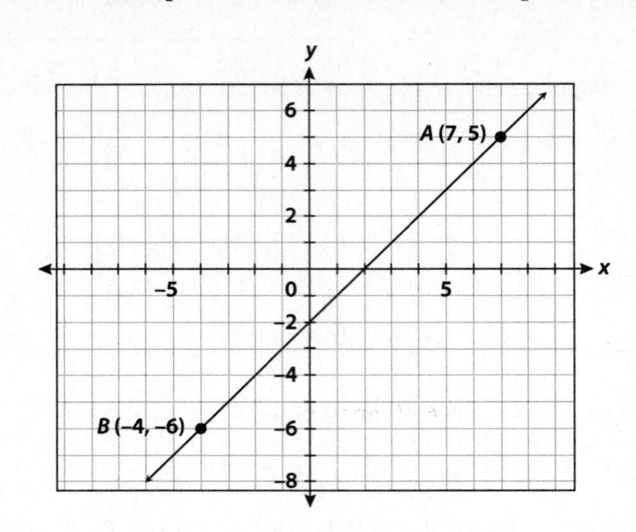

SOLUTION Let $(x_1, y_1) = (7, 5)$ and $(x_2, y_2) = (-4, -6)$. Then $x_1 = 7, y_1 = 5, x_2 = -4$, and $y_2 = -6$.

$$m = \frac{y_2 - y_1}{x_2 - x_1} = \frac{(-6)-(5)}{(-4)-(7)} = \frac{-6-5}{-4-7} = \frac{-11}{-11} = 1$$

The slope of the line that contains (7, 5) and (−4, −6) is 1. *Note*: The line slants upward to the right, so its slope is positive.

PROBLEM Find the slope of the line that contains the points shown.

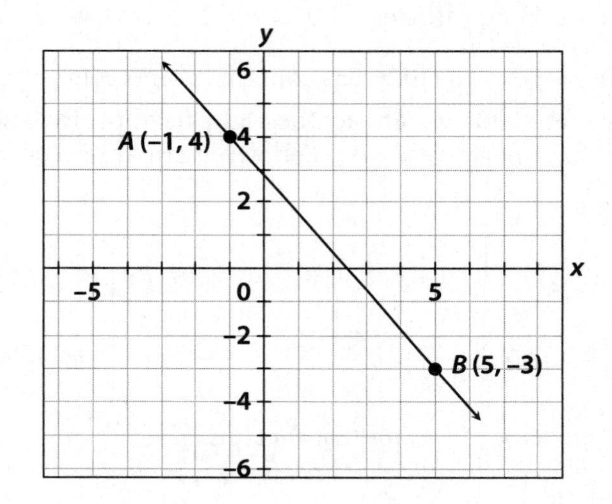

SOLUTION Let $(x_1, y_1) = (-1, 4)$ and $(x_2, y_2) = (5, -3)$. Then $x_1 = -1, y_1 = 4, x_2 = 5$, and $y_2 = -3$.

$$m = \frac{y_2 - y_1}{x_2 - x_1} = \frac{(-3)-(4)}{(5)-(-1)} = \frac{-3-4}{5+1} = \frac{-7}{6} = -\frac{7}{6}$$

The slope of the line that contains (−1, 4) and (5, −3) is $-\frac{7}{6}$. *Note*: The line slants downward to the right, so its slope is negative.

For horizontal lines, the slope is zero. For vertical lines, the slope is undefined. Here are examples:

PROBLEM Find the slope of the line that contains the points shown.

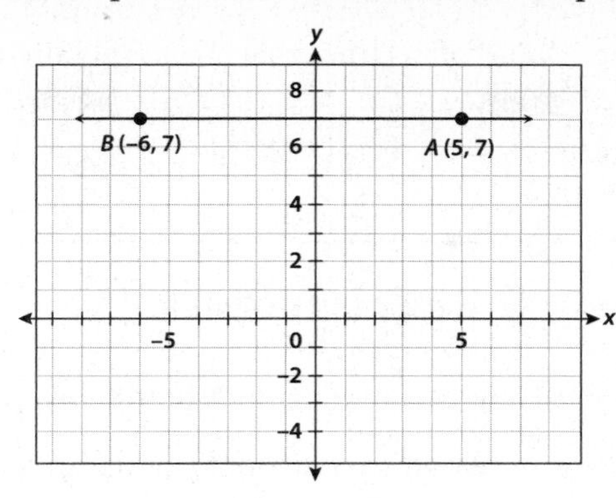

SOLUTION Let $(x_1, y_1) = (-6, 7)$ and $(x_2, y_2) = (5, 7)$. Then $x_1 = -6$, $y_1 = 7$, $x_2 = 5$, and $y_2 = 7$.

$$m = \frac{y_2 - y_1}{x_2 - x_1} = \frac{(7)-(7)}{(5)-(-6)} = \frac{7-7}{5+6} = \frac{0}{11} = 0$$

The slope of the line that contains (−6, 7) and (5, 7) is 0. *Note:* The line is horizontal, so it has zero slope.

PROBLEM Find the slope of the line that contains the points shown.

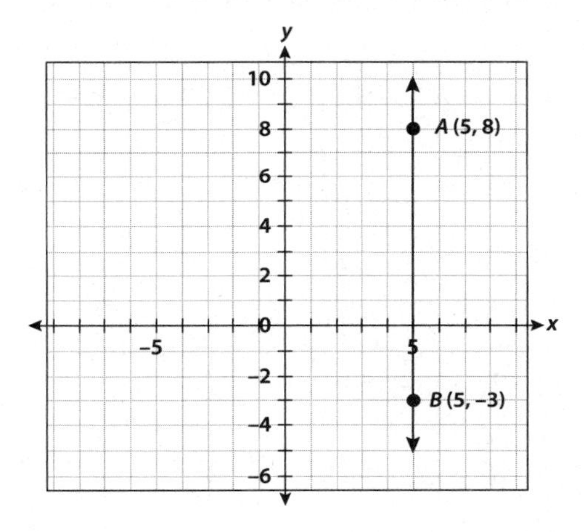

SOLUTION Let $(x_1, y_1) = (5, 8)$ and $(x_2, y_2) = (5, -3)$. Then $x_1 = 5$, $y_1 = 8$, $x_2 = 5$, and $y_2 = -3$.

$$m = \frac{y_2 - y_1}{x_2 - x_1} = \frac{(-3)-(8)}{(5)-(5)} = \frac{-3-8}{5-5} = \frac{-11}{0} = \text{undefined}$$

The slope of the line that contains (5, 8) and (5, −3) is undefined. *Note:* The line is vertical, so its slope is undefined.

It can be shown that if two lines are parallel, their slopes are equal; and if two lines are perpendicular, their slopes are negative reciprocals of each other. Here are examples:

PROBLEM (a) Find the slope of a line that is parallel to the line passing through the points (−3, 4) and (−1, −2); (b) find the slope of a line that is perpendicular to the line passing through the points (−3, 4) and (−1, −2).

SOLUTION The line that passes through (−3, 4) and (−1, −2) has slope:

$$m = \frac{y_2 - y_1}{x_2 - x_1} = \frac{(-2)-(4)}{(-1)-(-3)} = \frac{-2-4}{-1+3} = \frac{-6}{2} = -3$$

a. The slope of a line that is parallel to the given line has the same slope, –3.

b. The slope of a line that is perpendicular to the given line has slope $= -\frac{1}{m} = -\frac{1}{-3} = \frac{1}{3}$.

For 1–7, fill in the blank to make a true statement.

1. The change in *y*-coordinates between two distinct points on a line is the __________.
2. The change in *x*-coordinates between two distinct points on a line is the __________.
3. Lines that slant downward to the right have __________ slopes.
4. Lines that slant upward to the right have __________ slopes.
5. Horizontal lines have __________ slope.
6. The slope of a line that is parallel to a line with a slope of $\frac{2}{3}$ is __________.
7. The slope of a line that is perpendicular to a line with a slope of $\frac{3}{4}$ is __________.

For 8–15, find the slope of the line that contains the two given points.

8. (–2, 5), (4, –1)
9. $\left(-2, -\frac{2}{3}\right), \left(\frac{1}{2}, 2\right)$
10. (–8, 0), (0, 5)
11. (–1, 4), (–1, 0)
12. (3, 4), (0, 0)
13. (–9, 5), (4, 5)
14. (–2, 4), (4, 0)
15. Points *A* and *B* in the figure shown.

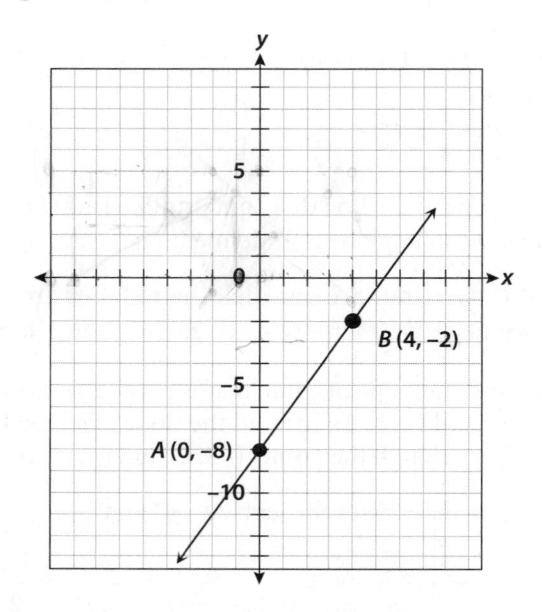

PRECALCULUS ALGEBRA

Part II presents algebra concepts that are essential for success in precalculus. The topics are concentrated around the concept of function, a concept that is the cornerstone of modern mathematics. You begin with basic function concepts and end with functions and problem solving. Mastery of Part II will enrich your understanding not only in precalculus but also in calculus and other areas of higher-level mathematics.

Basic function concepts

Cartesian product, relations, and functions

The set of all possible ordered pairs of real numbers is a **Cartesian product**, denoted $\boldsymbol{R} \times \boldsymbol{R}$, or simply $\boldsymbol{R}^2$. You cannot list the elements of $\boldsymbol{R} \times \boldsymbol{R}$ in a roster; however, you can use set notation to describe it as $\boldsymbol{R} \times \boldsymbol{R} = \{(x, y) | x \in R \text{ and } y \in R\}$. Graphically, $\boldsymbol{R}^2$ is represented by the Cartesian coordinate plane.

A **relation** $\mathcal{R}$ in $\boldsymbol{R}^2$ is any subset of $\boldsymbol{R}^2$. You can define relations in various ways. For instance, for a relation that consists of a *finite* number of ordered pairs, you can define the relation by listing or showing its ordered pairs in a set, in a table, as a diagram, or as a graph in a coordinate plane. You also might define the relation by giving a verbal description or an equation. When the number of ordered pairs is *infinite*, you usually define the relation by an equation, a graph, or set notation.

The **domain** of a relation $\mathcal{R}$ is the set of all first components from the ordered pairs in $\mathcal{R}$, and the **range** of $\mathcal{R}$ is the set of all second components.

PROBLEM Identify the domain, D_S, and range, R_S, for the relation S:

$$S = \{(0,0), (1,1), (1,-1), (4,2), (4,-2), (9,3), (9,-3)\}$$

SOLUTION $D_S = \{0,1,4,9\}$, $R_S = \{-3,-2,-1,0,1,2,3\}$

In the problem shown, each element in the domain of the relation S is related to two elements in its range. In precalculus, you are particularly concerned with relations in which each element in the domain is related to *one and only one* element in the range. Such relations are functions.

A **function** is a set of ordered pairs for which each first component is paired with *one and only one* second component. Thus, a function is a special relation in which no two ordered pairs have the same first component but different second components. Symbolically, this definition means that if (x,a) and (x,b) are ordered pairs in a function, then $a = b$. *Note*: In this book, only functions that are subsets of $\boldsymbol{R}^2$ are considered. This means that both the domains and the ranges of the functions consist of real numbers. Furthermore, equations herein that define functions will use only real numbers as coefficients or constants.

PROBLEM Both of the following sets S and T are relations, but only one is a function. Which one is a function? Explain.

$$S = \{(-2,1), (1,3), (-2,-1), (4,2)\}$$
$$T = \{(-5,3), (1,3), (-2,-1), (4,2)\}$$

SOLUTION The relation T is a function because each first component is paired with one and only one second component. The relation S is not a function because the first component -2 is paired with two different second components, namely, 1 and -1.

PROBLEM Let $J = \{(x,y)|y = x^2\}$ and $K = \{(x,y)|y^2 = x\}$. Both J and K are relations, but only one is a function. Which one is a function? Explain.

SOLUTION The relation J is a function because, for every ordered pair in J, each first component, x, in the domain of J is paired with a unique second component, namely, x^2. The relation K is not a function because if x is a first component in the domain of K, then K relates x to two different second components, namely, $-\sqrt{x}$ and $\sqrt{x}$. For example, both $(4,-2)$ and $(4,2)$ are in K.

Commonly, single letters such as *f*, *g*, and *h* designate functions. For the function *f*, the ordered pairs are $(x, f(x))$ or (x, y), where $y = f(x)$. In the function defined by $y = f(x)$, x is the **independent variable**, and y is the **dependent variable**. The variable y is "dependent" on x in the sense that you substitute a value of x into $y = f(x)$ to find y, the **value** of f at x. Since the value of y is determined by the value of x, variable y is a function of x. The value that is substituted for x is the **argument** for the function *f*, and $y = f(x)$ is the **image** of x under *f*.

PROBLEM Let $f = \{(x,y)|y = 2x-1\}$, and find (a) $f\left(-\frac{1}{2}\right)$, (b) $f(0)$, and (c) $f(4)$.

SOLUTION
a. $f\left(-\frac{1}{2}\right) = 2\left(-\frac{1}{2}\right) - 1 = -1 - 1 = -2$

b. $f(0) = 2(0) - 1 = 0 - 1 = -1$

c. $f(4) = 2(4) - 1 = 8 - 1 = 7$

Note: You can use informal language to refer to functions as long as no confusion will arise. For instance, you might use the phrase "the function $y = 2x - 1$" or "the function $f(x) = 2x - 1$" to refer to "the function *f* defined by $f(x) = 2x - 1$."

EXERCISE 4·1

For 1–6, indicate whether the statement is true or false.

1. All relations are functions.
2. All functions are relations.
3. $\{(-1,4),(-3,4),(0,4),(5,4)\}$ is a function.
4. $\{(x,y)|y^2 = 3x\}$ is a function.
5. If $(4,a)$ and $(4,b)$ are elements of a function, then $a = b$.
6. The domain of a function is a subset of that function.

For 7–10, fill in the blank(s) to make a true statement.

7. In the function $f = \{(x,y)|y = 2x-1\}$, x is the _______________ (dependent, independent) variable, and y is the _______________ (dependent, independent) variable.
8. In the function $f = \{(-5,3),(1,3),(-2,-1),(4,2)\}$, __________ is the value of f at 1.

9. In the expression $f\left(-\frac{1}{2}\right)=-2$, the argument of f is __________.

10. Given the function $f=\{(x,y)|y=2x-1\}$, the image of 4 in f is __________.

For 11–15, evaluate and simplify $y=f(x)$ as indicated.

11. $y=f\left(\frac{3}{4}\right)$ when $f(x)=8x-10$

12. $y=f(3)$ when $f(x)=x^2+1$

13. $y=f(-1)$ when $f(x)=4x^5+2x^4-3x^3-5x^2+x+5$

14. $y=f\left(\frac{\pi}{2}\right)$ when $f(x)=|x|$

15. $y=f(2)$ when $f(x)=\frac{4x-5}{x^2+1}$

For 16–20, $f(x)=\frac{2x-3}{x+1}$. Evaluate and simplify $f(x)$ for the argument given.

16. $f(0)$

17. $f\left(\frac{\pi}{2}\right)$

18. $f(5a)$

19. $f(b-1)$

20. $f(x^2-1)$

Domain and range of functions

For some functions, you can list the ordered pairs. For these functions, you can easily identify the domain and range. For example, for the function $f=\{(-5,3),(1,3),(-2,-1),(4,2)\}$, the domain is $D_f=\{-5,-2,1,4\}$ and the range is $R_f=\{-1,2,3\}$.

When a function f is defined by an equation $y=f(x)$, with no domain specified, the **domain** is the set of all real numbers x for which the equation is defined and for which each x value yields a corresponding y value that is a *real* number. To determine the domain, start with the real numbers and exclude all values for x, if any, that would make the equation undefined over the real numbers. Here are examples of finding the domain of a function.

PROBLEM Find the domain of the function f defined by $f(x)=(x+1)^2$.

SOLUTION No domain is specified, but $f(x)=(x+1)^2$ is defined for all real numbers. Thus, $D_f=\{x|x\in R\}$.

PROBLEM Find the domain of the function f defined by $f(x)=\frac{3}{x+4}$.

SOLUTION No domain is specified, but $f(x)=\frac{3}{x+4}$ is undefined when $x=-4$ because division by zero would occur. Thus, $D_f=\{x|x\neq-4\}$.

PROBLEM Find the domain of the function f defined by $f(x)=\sqrt{x-5}$.

SOLUTION No domain is specified, but $f(x)=\sqrt{x-5}$ yields a real number only when $x \geq 5$. Thus, $D_f=\{x|x \geq 5\}$.

Note: Routinely, division by zero and even roots of negative numbers create domain problems; however, be aware that other problematic situations can arise. For instance, the domain for the logarithm function, which will be discussed in Chapter 11, "Exponential and logarithmic functions," cannot include 0 or negative values for x.

The **range** of a function f defined by an equation $y=f(x)$ is the set of all real numbers y for which y is the image of at least one x value in the domain of f. You determine the range in a manner similar to that used for determining the domain. In some cases, you might need to try first solving the equation $y=f(x)$ explicitly for x. Here are examples of finding the range of a function.

PROBLEM Find the range of the function f defined by $f(x)=(x+1)^2$.

SOLUTION The domain of f is $D_f=(-\infty, \infty)$. For all real numbers x, $y=f(x)=(x+1)^2$ is nonnegative. Thus, $R_f=[0,\infty)$.

PROBLEM Find the range of the function f defined by $f(x)=\dfrac{3}{x+4}$.

SOLUTION The domain of f is $D_f=\{x|x \neq -4\}$. To find the range of f, solve $y=f(x)=\dfrac{3}{x+4}$ explicitly for x to obtain $x=\dfrac{3-4y}{y}$. Because $x=\dfrac{3-4y}{y}$ is undefined when $y=0$, $R_f=\{y|y \neq 0\}$.

PROBLEM Find the range of the function f defined by $f(x)=\sqrt{x-5}$.

SOLUTION The domain of f is $D_f=\{x|x \geq 5\}$. For all real numbers x, $y=f(x)=\sqrt{x-5}$ is nonnegative. Thus, $R_f=\{y|y \geq 0\}$.

Note: You might want to use a graphing calculator to look at the graph of a function to help you determine its domain and range. When using a graphing calculator to explore a function, use trial and error and the Zoom feature to find a good viewing window; otherwise, you might be misled by the graph displayed.

EXERCISE 4·2

Find the domain and range for each of the following functions.

1. $f=\left\{(3,12),(4,\sqrt{2}),\left(10,-\frac{3}{4}\right),(5.2,-1)\right\}$
2. f defined by $f(x)=\sqrt{2x+3}$
3. g defined by $g(x)=\dfrac{5}{x-1}$
4. h defined by $h(x)=|3x+5|$
5. v defined by $v(x)=\sqrt{x^2+1}$

Equality of functions

The functions f and g are **equal**, written $f = g$, if and only if their domains are equal and they contain exactly the same set of ordered pairs; that is, $D_f = D_g$ and $f(x) = g(x)$ for all x in their common domain.

PROBLEM Given $f = \left\{(3, 12), (4, \sqrt{2}), (5.2, -1), \left(10, -\frac{3}{4}\right)\right\}$ and $g = \left\{(3, 12), (4, \sqrt{2}), \left(10, -\frac{3}{4}\right), (5.2, -1)\right\}$. Does $f = g$? Explain.

SOLUTION Yes, $f = g$. $D_f = D_g = \{3, 4, 5.2, 10\}$ and f and g contain exactly the same set of ordered pairs.

PROBLEM Suppose $f = \{(x, y) | y = x^3\}$ with domain $D_f = \{x | x \geq 0\}$ and $g = \{(x, y) | y = x^3\}$ with domain $D_g = \{x | x < 0\}$. Does $f = g$? Explain.

SOLUTION No, because these two functions have different domains $f \neq g$.

EXERCISE 4·3

State whether the given functions are equal. Explain your reasoning.

1. $f = \{(-3, 1), (0, 5), (1, -3), (4, -5)\}$, $g = \{(-3, -6), (0, 7), (1, -2), (4, -5)\}$
2. $g = \{(2, 3), (4, 5), (6, 7), (8, 9)\}$, $h = \{(3, 2), (5, 4), (7, 6), (9, 8)\}$
3. $f = \{(x, y) | y = x^2\}$ with domain $D_f = \{x | x \text{ is an integer}\}$, $g = \{(x, y) | y = x^2\}$ with domain $D_g = \{x | x \in R\}$
4. $f = \left\{(x, f(x)) \middle| f(x) = \frac{3}{x+2}\right\}$, $g = \left\{(x, g(x)) \middle| g(x) = \frac{3(x-2)}{(x-2)(x+2)}\right\}$
5. $f = \{(x, y) | y = 5x + 3\}$, $g = \{(t, g(t)) | g(t) = 5t + 3\}$

Arithmetic of functions

You can add, subtract, multiply, and divide functions by performing the algebraic operations on the equations that define the functions. The definitions for this "arithmetic" of functions are as follows.

Given that both $f(x)$ and $g(x)$ exist, for all real numbers x such that $x \in D_f \cap D_g$:

The **sum** of f and g is the function $f+g$, defined by $(f+g)(x)=f(x)+g(x)$.

The **difference** of f and g is the function $f-g$, defined by $(f-g)(x)=f(x)-g(x)$.

The **product** of f and g is the function fg, defined by $(fg)(x)=f(x)\cdot g(x)$.

The **quotient** of f and g is the function $\frac{f}{g}$, defined by $\left(\frac{f}{g}\right)(x)=\frac{f(x)}{g(x)}$, where $g(x)\neq 0$.

PROBLEM Let $f(x)=x^2+1$ and $g(x)=\sqrt{x}-3$.

a. Find D_f and D_g. Then find the domains of $f+g, f-g, fg$, and $\frac{f}{g}$.

b. Write simplified expressions for $(f+g)(x), (f-g)(x), (fg)(x)$, and $\left(\frac{f}{g}\right)(x)$.

c. If possible, evaluate $(f+g)(3), (f-g)(-3), (fg)(0)$, and $\left(\frac{f}{g}\right)(1)$.

SOLUTION a. $f(x)=x^2+1$ is defined for all real numbers, so $D_f=\{x|x\in R\}$; $g(x)=\sqrt{x}-3$ yields a real number when $x\geq 0$, so $D_g=\{x|x\geq 0\}$. The domains of $f+g$, $f-g$ and fg include all real numbers x that are in both D_f and D_g, so their domains are $\{x|x\geq 0\}$.

The domain of $\frac{f}{g}$ must also exclude x-values for which $g(x)=\sqrt{x}-3=0$. This occurs when $x=9$ so the domain of $\frac{f}{g}$ is $\{x|x\geq 0, x\neq 9\}$.

b. $(f+g)(x)=f(x)+g(x)=(x^2+1)+(\sqrt{x}-3)=x^2+\sqrt{x}-2$

$(f-g)(x)=f(x)-g(x)=(x^2+1)-(\sqrt{x}-3)=x^2+1-\sqrt{x}+3=x^2-\sqrt{x}+4$

$(fg)(x)=f(x)\cdot g(x)=(x^2+1)(\sqrt{x}-3)=x^2\sqrt{x}-3x^2+\sqrt{x}-3$

$\left(\frac{f}{g}\right)(x)=\frac{f(x)}{g(x)}=\frac{x^2+1}{\sqrt{x}-3}$

c. $(f+g)(3)=f(3)+g(3)=(3^2+1)+(\sqrt{3}-3)=10+\sqrt{3}-3=7+\sqrt{3}$

$(f-g)(-3)$ is undefined because -3 is not in the domain of the function $f-g$.

$(fg)(0)=f(0)\cdot g(0)=(0^2+1)(\sqrt{0}-3)=(1)(-3)=-3$

$$\left(\frac{f}{g}\right)(1)=\frac{f(1)}{g(1)}=\frac{1^2+1}{\sqrt{1}-3}=\frac{2}{\sqrt{1}-3}=\frac{2}{-2}=-1$$

When you simplify expressions that contain functional notation, evaluate the functions *before* performing other operations. Here is an example:

- Given $f(x)=3x$ and $g(x)=\sqrt{x}-3$:

$$(fg)(4)=f(4)\cdot g(4)=\underbrace{(3\cdot 4)(\sqrt{4}-3)}_{\text{Evaluate } f(x) \text{ and } g(x) \text{ before multiplying}}=(12)(2-3)=-12$$

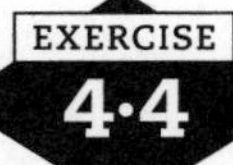

EXERCISE 4·4

For 1–5, use $f(x)=\frac{1}{x}$ *and* $g(x)=x^3$.

1. Find D_f and D_g.
2. Find the domains of $f+g, f-g, fg,$ and $\frac{f}{g}$.
3. Write simplified expressions for $(f+g)(x), (f-g)(x), (fg)(x),$ and $\left(\frac{f}{g}\right)(x)$.
4. If possible, evaluate $(f+g)(-5), (f-g)(2), (fg)(0),$ and $\left(\frac{f}{g}\right)(-3)$.
5. If possible, evaluate $(f+g)\left(\frac{1}{2}\right), (f-g)(\pi), (fg)(\sqrt[3]{5}),$ and $\left(\frac{f}{g}\right)(-1)$.

For 6–10, use $f(x)=\sqrt{x}+2$ *and* $g(x)=\sqrt{x}-2$.

6. Find D_f and D_g.
7. Find the domains of $f+g, f-g, fg,$ and $\frac{f}{g}$.
8. Write simplified expressions for $(f+g)(x), (f-g)(x), (fg)(x),$ and $\left(\frac{f}{g}\right)(x)$.
9. If possible, evaluate $(f+g)(4), (f-g)(-4), (fg)\left(\frac{1}{4}\right),$ and $\left(\frac{f}{g}\right)(4)$.
10. If possible, evaluate $(f+g)(\pi^2), (f-g)(0.25), (fg)(16),$ and $\left(\frac{f}{g}\right)(1)$.

Composition of functions

The **composition** of two functions *f* and *g* is the function $f \circ g$ defined by:

$$(f \circ g)(x) = f(g(x)), \text{ provided that } g(x) \in D_f$$

Note: Read $f(g(x))$ as "*f* of *g* of *x*."

The domain of $f \circ g$ is all *x* in the domain of *g* such that $g(x)$ is defined and $g(x)$ is in the domain of *f*.

To find $(f \circ g)(x)$, the function *g* is applied to *x* and produces from it the number $g(x)$, and then *f* is applied to $g(x)$. Notice that it is necessary that $g(x)$ be in the domain of *f* because *f* uses it to produce $f(g(x))$.

PROBLEM Suppose $f(x) = x^2$ and $g(x) = \sqrt{x+5}$.

a. Write a simplified expression for $(f \circ g)(x)$.
b. Determine the domain of $f \circ g$.
c. If possible, evaluate $(f \circ g)(4), (f \circ g)(0)$, and $(f \circ g)(-7)$.

SOLUTION a. $(f \circ g)(x) = f(g(x)) = f(\sqrt{x+5}) = (\sqrt{x+5})^2 = x+5.$

b. The domain of $f \circ g$ is all *x* in the domain of *g* such that $g(x)$ is defined and $g(x)$ is in the domain of *f*. The function *g* defined by $g(x) = \sqrt{x+5}$ has the domain $D_g = \{x | x \geq -5\}$. In the composition, $g(x)$ must be in the domain of *f*. The domain of *f* is all real numbers, so $g(x)$ is definitely in the domain of *f*. Therefore, the domain of $f \circ g$ is the same as the domain of *g*; that is, the domain of $f \circ g = \{x | x \geq -5\}$.

c. $(f \circ g)(4) = f(g(4)) = f(\sqrt{4+5}) = f(\sqrt{9}) = f(3) = 3^2 = 9$
$(f \circ g)(0) = f(g(0)) = f(\sqrt{0+5}) = f(\sqrt{5}) = (\sqrt{5})^2 = 5$
$(f \circ g)(-7)$ is undefined because –7 is not in the domain of *g*.

Note: For two functions *f* and *g*, the product function *fg* is fundamentally different from the composition function $f \circ g$. For instance, if $f(x) = 2x$ and $g(x) = x^2$:

$$fg(x) = f(x) \cdot g(x) = (2x) \cdot (x^2) = 2x^3$$

but

$$(f \circ g)(x) = f(g(x)) = f(x^2) = 2(x^2) = 2x^2$$

Composition of functions is not commutative; that is, in general:

$$(f \circ g)(x) \neq (g \circ f)(x)$$

PROBLEM Suppose $f(x)=\sqrt{x}$ and $g(x)=2x+1$.

a. Write simplified expressions for $(f\circ g)(x)$ and $(g\circ f)(x)$.

b. Determine the domains of $f\circ g$ and $g\circ f$.

c. If possible, evaluate $(f\circ g)(4), (g\circ f)(4), (f\circ g)\left(-\frac{1}{4}\right)$ and $(g\circ f)\left(-\frac{1}{4}\right)$.

SOLUTION a. $(f\circ g)(x)=f(g(x))=f(2x+1)=\sqrt{2x+1}$

$(g\circ f)(x)=g(f(x))=g(\sqrt{x})=2(\sqrt{x})+1=2\sqrt{x}+1$

b. The domain of $f\circ g$ is all x in the domain of g such that $g(x)$ is defined and $g(x)$ is in the domain of f. The domain for the function g defined by $g(x)=2x+1$ is all real numbers. In the composition, $g(x)$ must be in the domain of f. The domain for the function f defined by $f(x)=\sqrt{x}$ is $\{x|x\geq 0\}$, so for $g(x)$ to be in the domain of f, $2x+1\geq 0$, or $x\geq -\frac{1}{2}$. Therefore, the domain of $f\circ g=\left\{x\middle|x\geq -\frac{1}{2}\right\}$.

The domain of $g\circ f$ is all x in the domain of f such that $f(x)$ is defined and $f(x)$ is in the domain of g. The domain for the function f defined by $f(x)=\sqrt{x}$ is $\{x|x\geq 0\}$. In the composition, $f(x)$ must be in the domain of g. The domain of g is all real numbers, so $f(x)$ is definitely in the domain of g. Therefore, the domain of $g\circ f$ is the same as the domain of f; that is, the domain of $g\circ f=\{x|x\geq 0\}$.

c. $(f\circ g)(4)=f(g(4))=f(2(4)+1)=f(9)=\sqrt{9}=3$

$(g\circ f)(4)=g(f(4))=g(\sqrt{4})=g(2)=2(2)+1=5$

$(f\circ g)\left(-\frac{1}{4}\right)=f\left(g\left(-\frac{1}{4}\right)\right)=f\left(2\left(-\frac{1}{4}\right)+1\right)=f\left(\frac{1}{2}\right)=\sqrt{\frac{1}{2}}=\frac{\sqrt{2}}{2}$

$(g\circ f)\left(-\frac{1}{4}\right)$ is undefined because $-\frac{1}{4}$ is not in the domain of $g\circ f$.

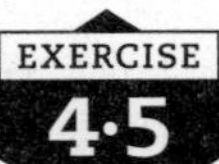

For 1–5, use the given $f(x)$ and $g(x)$ to (a) write a simplified expression for $(f\circ g)(x)$ and (b) determine the domain of $f\circ g$.

1. $f(x)=\sqrt{x+2}$ and $g(x)=3x^2$
2. $f(x)=|x|$ and $g(x)=4x-3$
3. $f(x)=\sqrt{x}$ and $g(x)=9x^2$
4. $f(x)=\frac{1}{x+1}$ and $g(x)=4x$
5. $f(x)=\frac{1-x}{3}$ and $g(x)=1-3x$

For 6–10, use $f(x) = \sqrt{x} - 1$ *and* $g(x) = x^2 + 1$ *to evaluate each expression, if possible.*

6. $(f \circ g)(3)$
7. $(f \circ g)(0)$
8. $(f \circ f)(9)$
9. $(g \circ f)(-4)$
10. $(g \circ f)(16)$

One-to-one functions

A function is **one-to-one** if no two different ordered pairs have the same second component. In other words, in a one-to-one function, each first component is paired with *one and only one* second component, and each second component is paired with *one and only one* first component. Symbolically, this definition means that if (c, y) and (d, y) are ordered pairs in a one-to-one function, then $c = d$.

PROBLEM Determine whether each function is one-to-one.

$$f = \{(-2,4),(0,0),(1,1),(2,4),(3,9)\}$$

$$g = \{(0,0),(1,1),(2,4),(3,9)\}$$

$$h = \{(x,y) \mid y = x^2\}$$

SOLUTION The function f is not one-to-one because it contains two different ordered pairs, namely, $(-2,4)$ and $(2,4)$, that have the same second component.

The function g is one-to-one because no two different ordered pairs in the function g have the same second component.

The function h is not one-to-one because it contains different ordered pairs that have the same second component. For instance, both $(-3, 9)$ and $(3, 9)$ are elements of h.

EXERCISE 4·6

Determine whether each function is one-to-one.

1. $f = \{(1,4),(2,5),(3,6),(4,7),(5,8)\}$
2. $g = \left\{(-3,\pi),(0,0),(1,3.14),(2,4),\left(3,\frac{22}{7}\right)\right\}$
3. $h = \left\{\left(-1,\frac{1}{2}\right),\left(2,\frac{3}{4}\right),(3,0.5),(4,0.75),(5,1.5)\right\}$
4. $r = \{(x,y) \mid y = \sqrt{x} + 1\}$
5. $\left\{(x,y) \mid y = \frac{5}{x^2}\right\}$

Inverses of functions

If f is a one-to-one function, the **inverse** of f is the function f^{-1} (read "f inverse") whose ordered pairs are obtained from f by interchanging the first and second components of each of the ordered pairs in f. Accordingly, the domain of f^{-1} is the range of f, and the range of f^{-1} is the domain of f.

Note: The notation f^{-1} does not indicate a reciprocal. The raised $^{-1}$ is not an exponent. Do *not* write $f^{-1} = \frac{1}{f}$.

Here is an example of a one-to-one function and its inverse:

$$g = \{(0,1),(1,3),(2,5),(3,7)\} \text{ is a one-to-one function;}$$

$$g^{-1} = \{(1,0),(3,1),(5,2),(7,3)\} \text{ is its inverse.}$$

As you can see, the functions g and g^{-1} undo what the other does. For instance, $g(1) = 3$ and $g^{-1}(3) = 1$. You express this idea as follows.

If f is a one-to-one function and y is in the range of f, then $f^{-1}(y)$ is the number x in the domain of f such that $f(x) = y$.

Thus, you have that for every x in the domain of f, $(f^{-1} \circ f)(x) = x$; and for every x in the domain of f^{-1}, $(f \circ f^{-1})(x) = x$.

If a function is one-to-one, its inverse exists. When a one-to-one function f is defined by an equation $y = f(x)$, there are two common methods that you can use to find an equation that defines f^{-1}.

Method 1

Set $(f \circ f^{-1})(x) = x$ and solve for $f^{-1}(x)$. Here is an example:

Suppose $f(x) = 2x + 1$, find $f^{-1}(x)$.

Set $(f \circ f^{-1})(x) = x$ and solve for $f^{-1}(x)$:

$$(f \circ f^{-1})(x) = x:$$

$$f(f^{-1}(x)) = x$$

$$2f^{-1}(x) + 1 = x$$

$$2f^{-1}(x) = x - 1$$

$$f^{-1}(x) = \frac{x-1}{2}$$

Method 2

First, interchange x and y in $y = f(x)$, and then solve $x = f(y)$ for y. Here is an example:

Suppose $y = f(x) = 2x + 1$, find $f^{-1}(x)$.

Interchange x and y in $y = 2x + 1$:

$$x = 2y + 1$$

Solve for y:

$$x = 2y + 1$$

$$x - 1 = 2y$$

$$\frac{x-1}{2} = y$$

Thus, $y = f^{-1}(x) = \frac{x-1}{2}$, the same as obtained in method 1.

If a function is not a one-to-one function, you might restrict the domain so that the function is one-to-one in the restricted domain. The inverse of the function will be defined in the restricted domain of the function. Here is an example.

Consider the function f defined by $f(x) = x^2$. You know from the previous section that f is not a one-to-one function because, for example, $(-3,9)$ and $(3,9)$ are ordered pairs in f. However, if the domain of f is restricted to the set of *nonnegative* real numbers, then f will be a one-to-one function. You will not have the problem that pairs such as $(-3,9)$ and $(3,9)$ are in f, because negative first components (e.g., -3) are not in the restricted domain of f. Thus, $f^{-1}(x) = \sqrt{x}$ will undo $f(x) = x^2$ by returning all the squares back to their *nonnegative* roots in the restricted domain of f.

EXERCISE

4·7

For 1–5, indicate whether the statement is true or false.

1. All functions have inverses that are functions.
2. The inverse of the function $\{(2,2),(3,3),(4,4)\}$ is the function $\left\{\left(2,\frac{1}{2}\right),\left(3,\frac{1}{3}\right),\left(4,\frac{1}{4}\right)\right\}$.
3. $f^{-1}(x) = \frac{1}{f(x)}$.
4. If $f(x) = (x+1)^2$, then $f^{-1}(x) = \sqrt{x+1}$.
5. The domain of f^{-1} equals the range of f.

For 6–12, find an equation that defines the inverse of the function defined by the given equation.

6. $f(x) = -x$
7. $h(x) = \frac{1}{x}$
8. $g(x) = x^3$
9. $f(x) = x - 8$
10. $h(x) = 5x$
11. $g(x) = \frac{x+2}{x-1}$
12. $f(x) = x^2 + 8$, where the domain of $f = \{x \mid x \geq 0\}$

For 13–15, evaluate $f^{-1}(x)$ *at the value x for the function f defined as indicated.*

13. $f(x)=\frac{5}{9}(x-32),\ x=100$

14. $f(x)=\frac{3}{4}x+5,\ x=11$

15. $f(x)=\frac{x^3-6}{2},\ x=-7$

Graphs of functions

·5·

Graphs of relations and the vertical line test

The **graph** of a relation f in $R \times R$ is the set of all ordered pairs (x, y) for which x is in the domain of f. You can sketch the graph by first plotting enough points (ordered pairs) to determine the shape of the graph and then drawing a smooth line or curve (or curves, possibly) through the points. Here is an example:

PROBLEM Graph the relation defined by $y^2 = x$ and state its domain and range.

SOLUTION Make a table of ordered pairs that satisfy $y^2 = x$, or $y = \pm\sqrt{x}$.

x	4	1	0	1	4
y	−2	−1	0	1	2

These ordered pairs indicate a graph that has the shape shown in Figure 5.1. The domain is $\{x | x \geq 0\}$. The range is all real numbers.

You can use the vertical line test to determine whether the graph shown in Figure 5.1 is a function.

The **vertical line test** specifies that a graph represents a function if and only if no vertical line intersects the graph in more than one point.

Notice that you could draw a vertical line that would intersect the graph of $y^2 = x$ in more than one point (see Figure 5.2, on the next page).

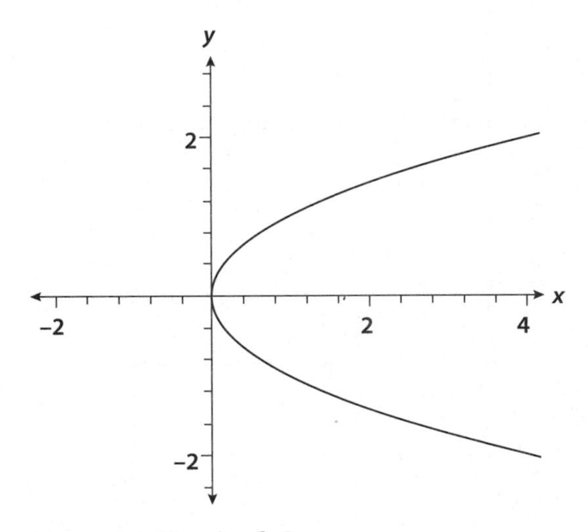

Figure 5.1 Graph of $y^2 = x$

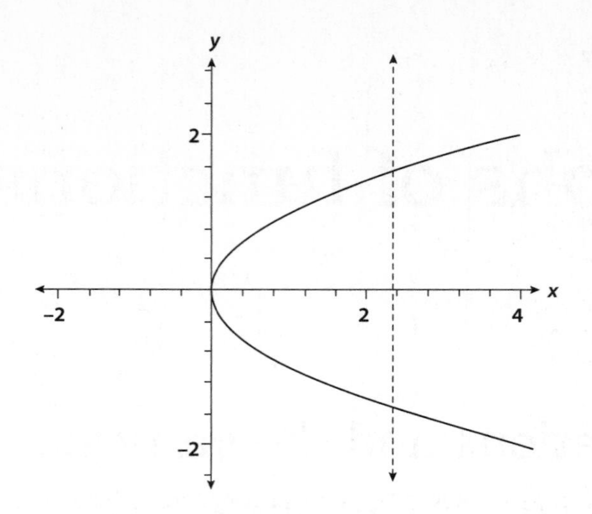

Figure 5.2 Example of vertical line test

This circumstance indicates that the graph of the relation defined by $y^2 = x$ is *not* a function.

Note: Needless to say, a graphing calculator is a helpful tool when you are exploring graphs of functions. Most graphing calculators require that you enter the equation of the graph in the form $y = f(x)$. This form excludes graphs of relations that are not functions. For such relations, break the equation into two parts, so that each part defines a function, and then graph the two parts on the same coordinate grid.

For 1–4, (a) graph the given relation, (b) state its domain and range, and (c) determine whether the relation is a function.

1. Graph the relation defined by $y = x^2$ and state its domain and range. Determine whether the relation is a function.
2. Graph the relation defined by $y = \sqrt{x}$ and state its domain and range. Determine whether the relation is a function.
3. Graph the relation defined by $y^2 = x - 2$ and state its domain and range. Determine whether the relation is a function.
4. Graph the relation defined by $y = x^3$ and state its domain and range. Determine whether the relation is a function.

For 5, use the given graph to answer the question.

5. Determine whether the graph shown is the graph of a function. Explain your reasoning.

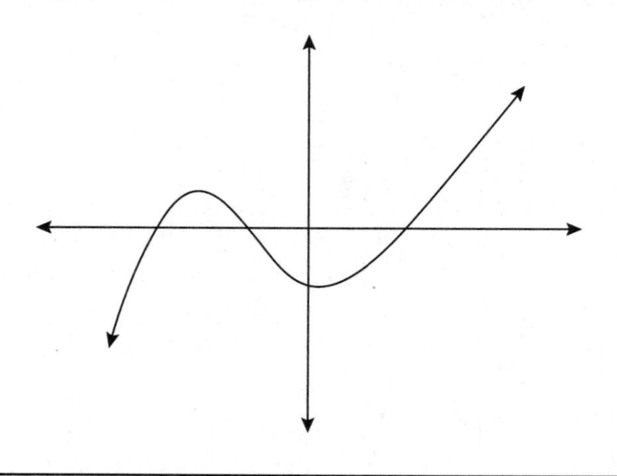

x- and y-intercepts

When you graph a function, you will find it helpful to determine the points (if any) at which the graph intersects the axes. As shown in Figure 5.3, an **x-intercept** is the x-coordinate of the point at which the graph intersects the x-axis, and the **y-intercept** is the y-coordinate of the point at which the graph intersects the y-axis.

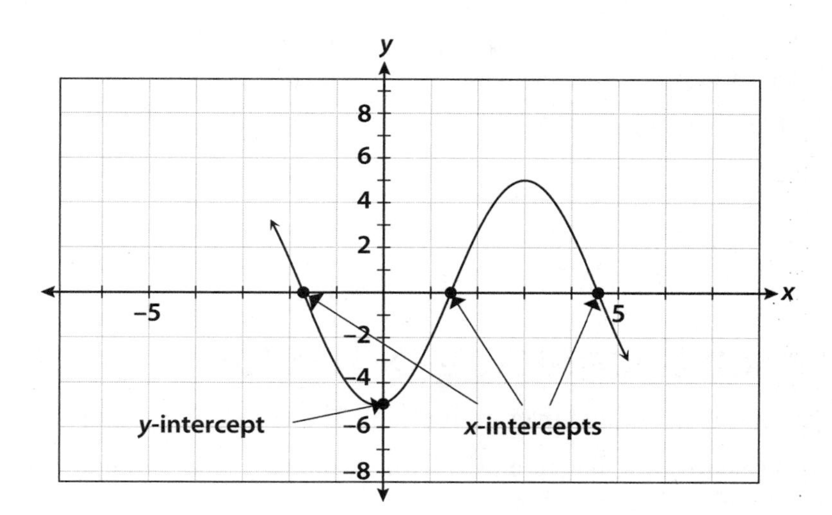

Figure 5.3 x- and y-intercepts

You have only *one* y-intercept when you graph a function because each x in the domain of the function corresponds to *one and only one* y in the range.

To determine the x-intercept(s), if any, for a function f, set $f(x) = 0$ and then solve for x. Similarly, provided that 0 is in the domain of f, to determine the y-intercept, if any, let $x = 0$, and then solve $f(0) = y$ for y. Here are examples:

PROBLEM Find the x- and y-intercepts for the graph of the function defined by the equation $f(x) = 2x + 6$. Give the coordinates where the graph crosses the axes.

SOLUTION To find the x-intercepts, set $f(x) = 0$. Solving $0 = 2x + 6$ for x yields $x = -3$. Thus, the x-intercept is -3. The graph crosses the x-axis at $(-3, 0)$.

To find the y-intercept, let $x = 0$. Solving $y = 2(0) + 6$ for y yields $y = 6$. Thus, the y-intercept is 6. The graph crosses the y-axis at $(0, 6)$.

PROBLEM Find the x- and y-intercepts for the graph of the function defined by the equation $f(x) = x^2 + x - 2$.

SOLUTION To find the x-intercepts, set $f(x) = 0$. To solve $0 = x^2 + x - 2$, factor the quadratic expression on the right to obtain $0 = (x + 2)(x - 1)$. Solving for x yields $x = -2$ and $x = 1$. Thus, the x-intercepts are -2 and 1. The graph crosses the x-axis at $(-2, 0)$ and $(1, 0)$. *Note*: See the section "Quadratic equations" in Chapter 8, "Quadratic functions," for a review of quadratic equations.

To find the y-intercept, let $x = 0$. Solving $y = 0^2 + 0 - 2$ for y yields $y = -2$. Thus, the y-intercept is -2. The graph crosses the y-axis at $(0, -2)$.

EXERCISE

5·2

For the graph of the function defined by the equation given, (a) find the x- and y-intercepts and (b) give the coordinates where the graph crosses the axes.

1. $f(x)=4x-1$
2. $2x-3y=5$
3. $g(x)=x^2-x-12$
4. $p(x)=(x-1)(x+5)(x+2)(x-4)$
5. $f(x)=\sqrt{2x-3}$
6. $f(x)=\sqrt{2x}-3$
7. $h(x)=|3x+5|$
8. $g(x)=\sqrt{x^2+1}$
9. $f(x)=(x-3)^2-1$
10. $g(x)=(x-1)^3+8$

Increasing or decreasing behavior

Suppose a function f is defined over an interval. Then:

- f is **increasing** on the interval if, for every pair of numbers x_1 and x_2 in the interval, $f(x_1)<f(x_2)$ whenever $x_1<x_2$
- f is **decreasing** on the interval if, for every pair of numbers x_1 and x_2 in the interval, $f(x_1)>f(x_2)$ whenever $x_1<x_2$
- f is **constant** on the interval if $f(x_1)=f(x_1)$ for every pair of numbers x_1 and x_2 in the interval

Thus, a function is increasing on an interval if its graph moves upward, from left to right, as the independent variable assumes values from left to right in the interval. A function is decreasing on an interval if its graph moves downward, from left to right, as the independent variable assumes values from left to right in the interval. A function is constant on an interval if the function value stays the same as the independent variable assumes values from left to right in the interval.

Here is an example:

PROBLEM For the graph of the function shown, list the intervals on which the function is decreasing, increasing, or constant.

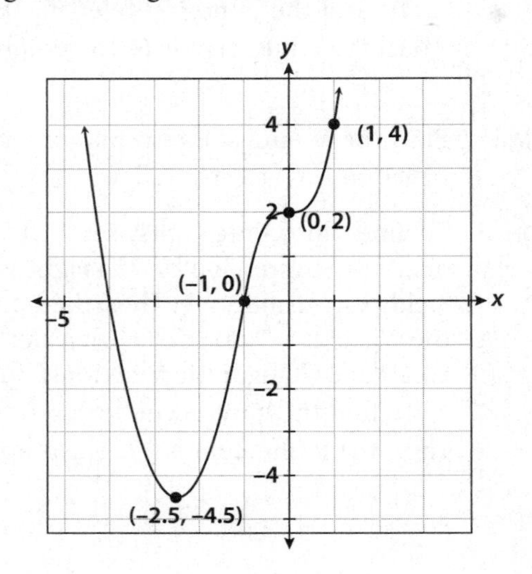

SOLUTION The function shown is decreasing in the interval $(-\infty, -2.5]$ and increasing in the interval $[-2.5, \infty)$. The function shows no constant behavior.
Note: When you are determining intervals to describe increasing and decreasing behavior, give x-intervals, not y-intervals.

PROBLEM The function f defined by $y = \sqrt{x}$ is increasing on $[0, \infty)$, true or false? Explain your reasoning.

SOLUTION True. The domain of the function is $\{x | x \geq 0\} = [0, \infty)$. The range of f is $\{y | y \geq 0\}$ because the principal square root of a number is always nonnegative. For any two numbers a and b in the interval $[0, \infty)$ for which $a < b$, you have $\sqrt{a} < \sqrt{b}$. Thus, the function f is increasing in $[0, \infty)$.

For 1–5, indicate whether the statement is true or false. Explain your reasoning.

1. If x_1 and x_2 are in an interval with $x_1 < x_2$, the function f is decreasing on the interval if $f(x_1) < f(x_2)$.
2. The function f defined by $y = 4x - 1$ is increasing on the interval $(-\infty, \infty)$.
3. The function f defined by $y = \sqrt{2x - 3}$ is increasing on $\left[\frac{3}{2}, \infty\right)$.
4. The function h defined by $h(x) = |3x + 5|$ is decreasing on $\left(-\infty, -\frac{5}{3}\right]$ and increasing on $\left[-\frac{5}{3}, \infty\right)$.
5. The function f defined by $f(x) = \sqrt{2x} - 3$ is increasing on $[-3, \infty)$.

For 6–10, use the graph of the function f to determine intervals where f is increasing, decreasing, or constant.

6. 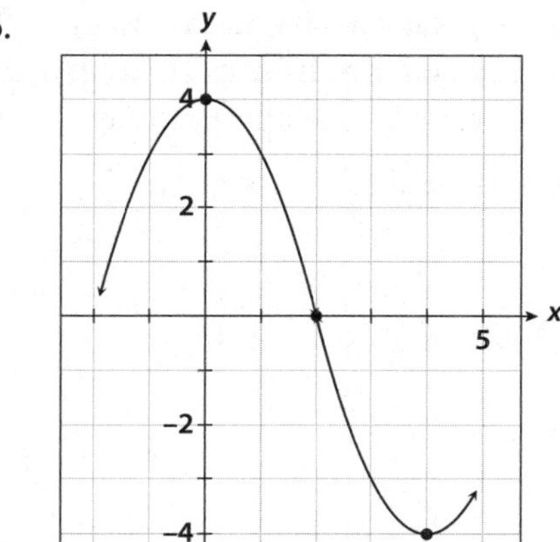

7.

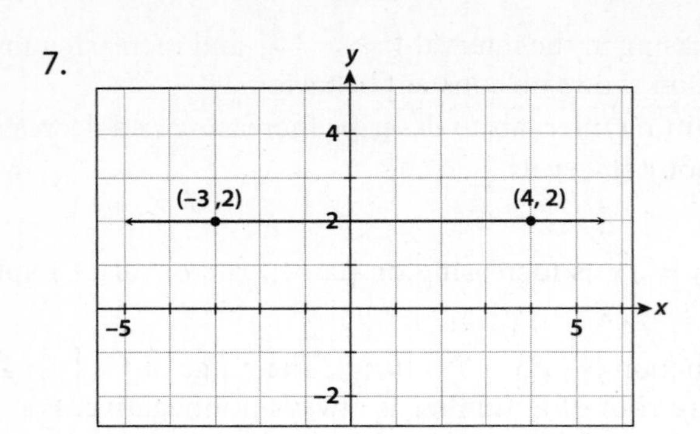

8.

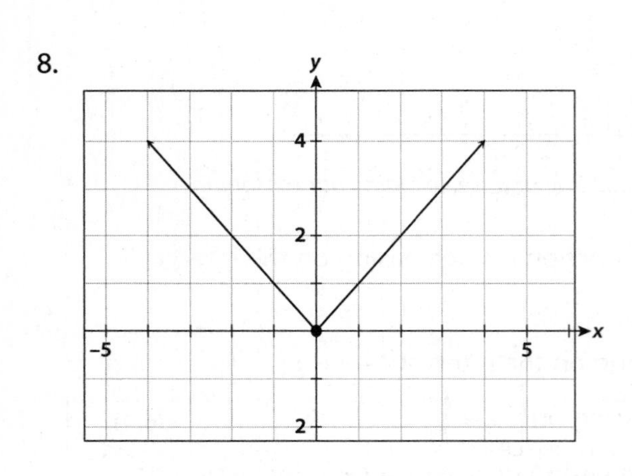

9.

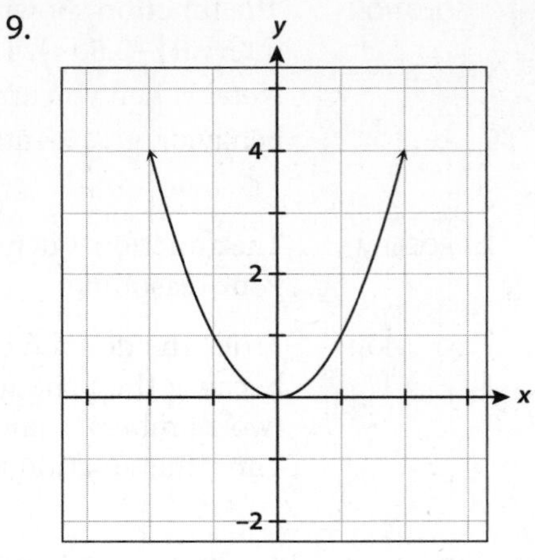

10.

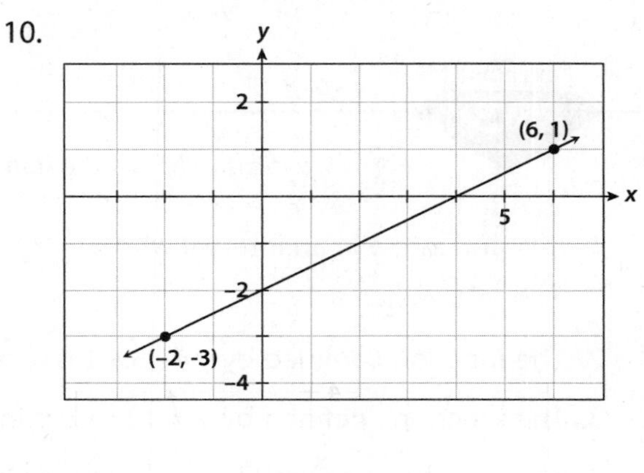

Extrema

Let c be in the domain of a function f. The number $f(c)$ is an **absolute minimum** of f if $f(c) \leq f(x)$ for all x in the domain of f. Similarly, $f(c)$ is an **absolute maximum** of f if $f(c) \geq f(x)$ for all x in the domain of f. The minimum and maximum values of a function are the **extreme values**, or **extrema** (plural of *extremum*), of the function.

The number $f(c)$ is a **relative minimum** of a function f if there exists an open interval containing c in which $f(c)$ is a minimum; similarly, the number $f(c)$ is a **relative maximum** of a function f if there exists an open interval containing c in which $f(c)$ is a maximum. If $f(c)$ is a relative minimum or maximum of f, it is called a **relative extremum** of f. Here is an example:

PROBLEM Use the graph shown to estimate any relative or absolute extrema.

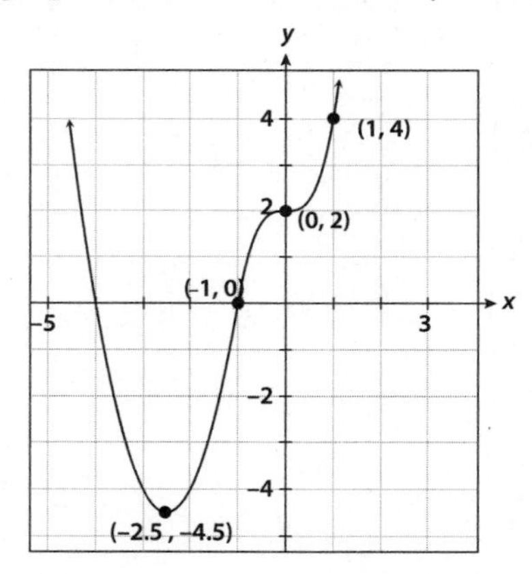

SOLUTION The point $(-2.5, -4.5)$ is the lowest point on the graph. Thus, there is an absolute minimum of -4.5.

EXERCISE
5·4

Use the graph shown to estimate any relative or absolute extrema.

1.
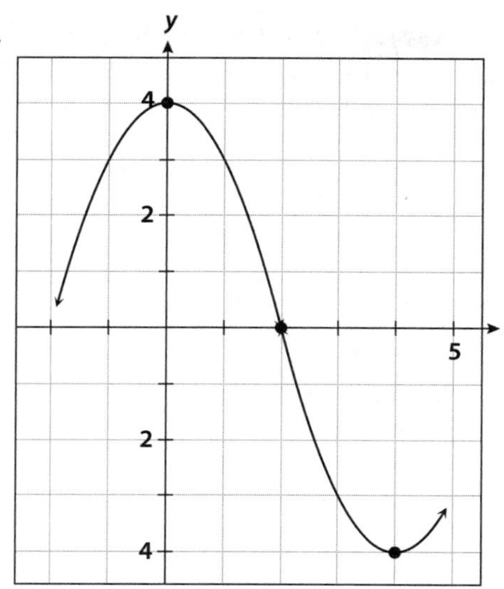

4.
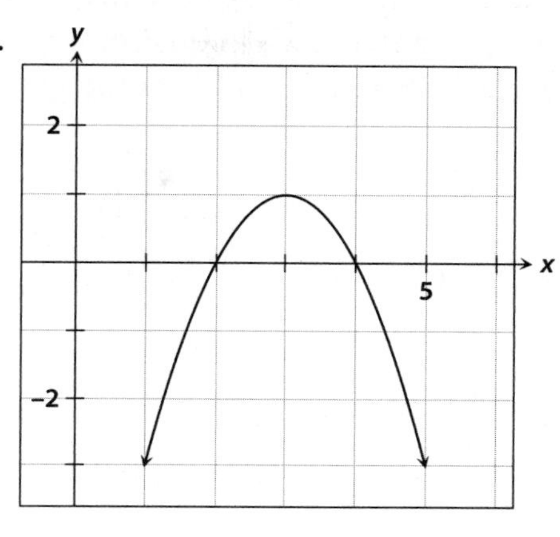

2.
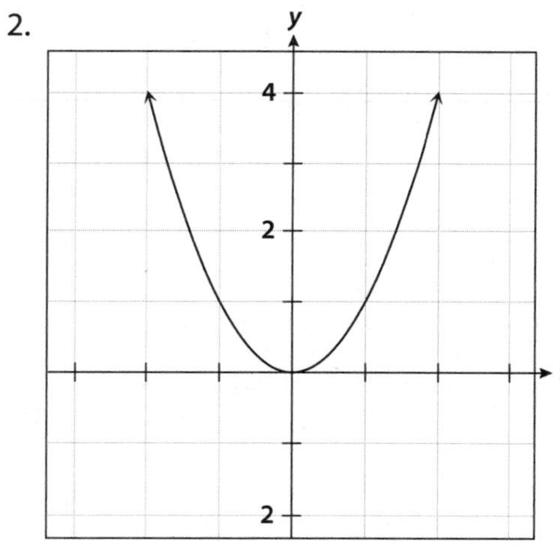

5.
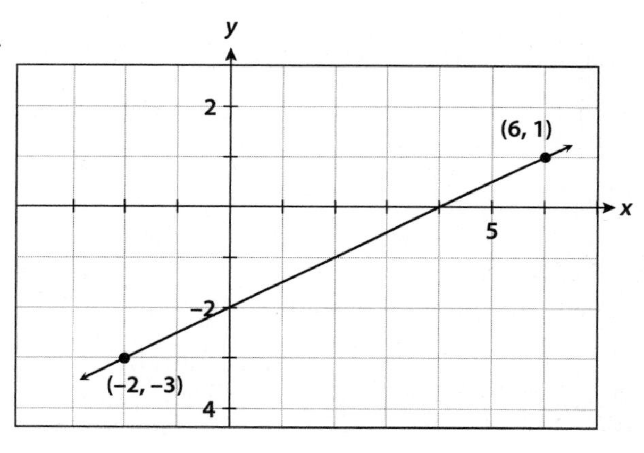

3.
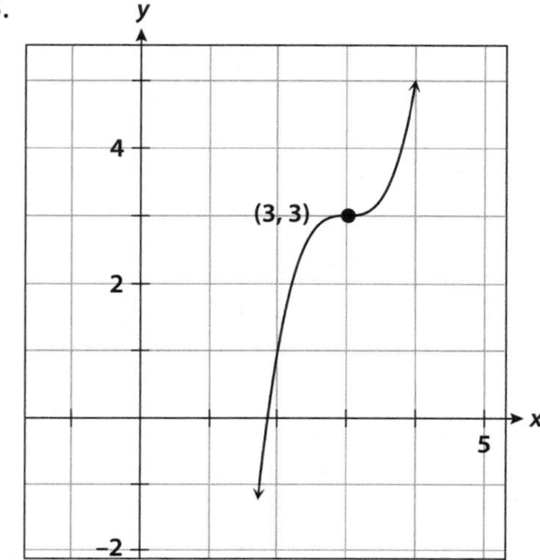

Vertical and horizontal asymptotes

An **asymptote** of the graph of a function f is a line that the graph gets closer and closer to in at least one direction along the line. The vertical line with equation $x = a$ is a **vertical asymptote** of the graph of f if $f(x)$ either increases or decreases without bound, as x approaches a from the left or right. The horizontal line with equation $y = b$ is a **horizontal asymptote** of the graph of f if $f(x)$ approaches b as x approaches positive or negative infinity. See Figure 5.4.

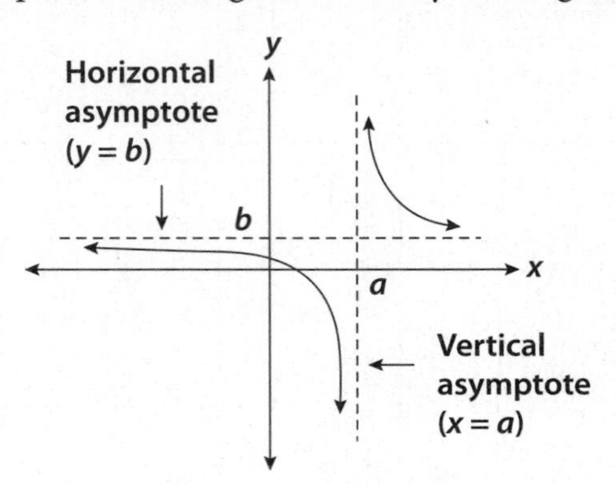

Figure 5.4 Vertical and horizontal asymptotes

Commonly, vertical and horizontal asymptotes are associated with rational functions (see Chapter 10, "Rational functions," for a fuller discussion of rational functions). Here are examples:

PROBLEM Find the vertical and horizontal asymptotes for the function defined by the equation $y = \dfrac{1}{x-2} + 4$.

SOLUTION You find vertical asymptotes by setting the denominator equal to zero and solving for x (provided the rational function is in simplified form). The denominator $x - 2$ equals zero when $x = 2$. Thus, the vertical asymptote is $x = 2$.

You find horizontal asymptotes by determining the value that $y = f(x)$ approaches as x approaches positive or negative infinity (again, provided the rational function is in simplified form). As x approaches ∞ or $-\infty$, $\dfrac{1}{x-2}$ approaches 0. Thus, as x approaches ∞ or $-\infty$, $\dfrac{1}{x-2} + 4$ approaches $0 + 4 = 4$, so $y = 4$ is a horizontal asymptote. The graph is shown in Figure 5.5.

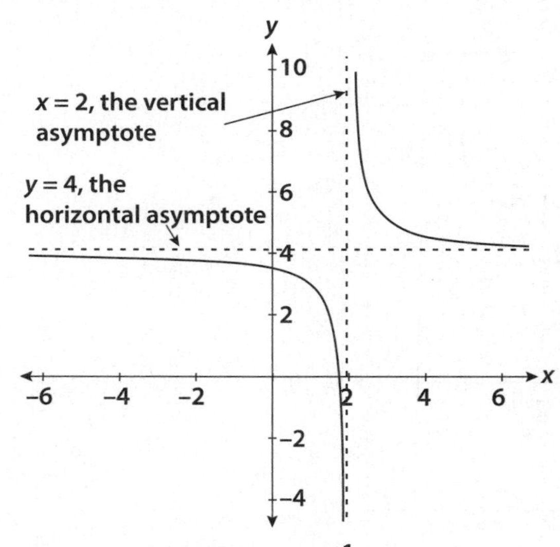

Figure 5.5 Graph of $y = \dfrac{1}{x-2} + 4$

EXERCISE 5·5

Find the vertical and horizontal asymptotes for the function defined by the equation given.

1. $y = \frac{1}{x}$
2. $f(x) = \frac{5}{x^2 - 1}$
3. $g(x) = \frac{x-2}{2x+6}$
4. $h(x) = \frac{x}{x^2 - 9}$
5. $y = 3 + \frac{5}{x^2 - 1}$

Average rate of change and difference quotient

If (x_1, y_1) and (x_2, y_2) are two distinct points on the graph of a function *f*, the **average rate of change** of *f* as *x* goes from x_1 to x_2 is given by

$$\frac{\Delta y}{\Delta x} = \frac{y_2 - y_1}{x_2 - x_1} = \frac{f(x_2) - f(x_1)}{x_2 - x_1}$$

where Δy is the change in *y*-values and Δx is the change in *x*-values. The average rate of change measures the "speed," *on average*, at which a function is changing over an interval of its domain.

PROBLEM Find the average rate of change of $f(x) = x^2$ (a) on the interval [0, 3] and (b) on the interval [1, 4].

SOLUTION a. $\frac{\Delta y}{\Delta x} = \frac{f(x_2) - f(x_1)}{x_2 - x_1} = \frac{f(3) - f(0)}{3 - 0} = \frac{3^2 - 0^2}{3} = \frac{9}{3} = 3$

b. $\frac{\Delta y}{\Delta x} = \frac{f(x_2) - f(x_1)}{x_2 - x_1} = \frac{f(4) - f(1)}{4 - 1} = \frac{4^2 - 1^2}{3} = \frac{15}{3} = 5$

Observe from this problem that even though the intervals [0, 3] and [1, 4] are each of length 3, the average rates of change on the two intervals are not equal. That is, the average rate of change for $f(x) = x^2$ is *not* constant.

The next problem is an example of a function that has a constant average rate of change.

PROBLEM Find the average rate of change of $f(x) = 2x$ (a) on the interval [0, 3] and (b) on the interval [1, 4].

SOLUTION a. $\frac{\Delta y}{\Delta x} = \frac{f(x_2) - f(x_1)}{x_2 - x_1} = \frac{f(3) - f(0)}{3 - 0} = \frac{2 \cdot 3 - 2 \cdot 0}{3} = \frac{6}{3} = 2$

b. $\frac{\Delta y}{\Delta x} = \frac{f(x_2) - f(x_1)}{x_2 - x_1} = \frac{f(4) - f(1)}{4 - 1} = \frac{2 \cdot 4 - 2 \cdot 1}{3} = \frac{6}{3} = 2$

If $(x + h, f(x + h))$ and $(x, f(x))$ are two ordered pairs in a function *f*, the **difference quotient** is the expression

$$\frac{f(x+h) - f(x)}{(x+h) - (x)} = \frac{f(x+h) - f(x)}{h}$$

where $h \neq 0$.

The difference quotient often is used in calculus to find a general expression for the average rate of change of a function. It is the average rate of change of f as x goes from x to $x+h$.

PROBLEM Find and simplify the difference quotient for (a) $f(x)=x^2$ and (b) $f(x)=2x$.

SOLUTION a. $\dfrac{f(x+h)-f(x)}{h}=\dfrac{(x+h)^2-(x)^2}{h}=\dfrac{x^2+2xh+h^2-x^2}{h}=\dfrac{2xh+h^2}{h}=2x+h$

b. $\dfrac{f(x+h)-f(x)}{h}=\dfrac{2(x+h)-2(x)}{h}=\dfrac{2x+2h-2x}{h}=\dfrac{2h}{h}=2$

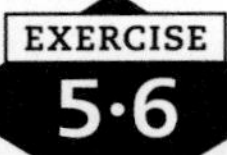

For 1–5, find the average rate of change for each function on the given interval.

1. $f(x)=-2x+5,\ [-5,5]$
2. $f(x)=x^2+2x,\ [-5,5]$
3. $f(x)=x^3,\ [2,5]$
4. $f(x)=\dfrac{1}{x},\ [5,10]$
5. $f(x)=\dfrac{1}{\sqrt{x}},\ [9,25]$

For 6–10, find and simplify the difference quotient for each function.

6. $f(x)=-2x+5$
7. $f(x)=x^2+2x$
8. $f(x)=x^3$
9. $f(x)=\dfrac{1}{x}$
10. $f(x)=\dfrac{1}{\sqrt{x}}$ *Hint*: Rationalize the numerator.

Function transformations and symmetry

Vertical and horizontal translations

A **translation**, or **shift**, is a geometric transformation of the graph of a function f that results in a new graph that is congruent to the graph of f, but for which every point P on the graph of f is "moved" the same distance and in the same direction along a straight line to a new point P'. Informally, a translation is a **slide** of the graph of a function in a horizontal or vertical direction.

You perform a **vertical shift** by adding or subtracting a positive constant k to or from $f(x)$. You perform a **horizontal shift** by adding or subtracting a positive constant h to or from the independent variable x. Table 6.1 contains a summary of vertical and horizontal shifts.

Table 6.1 Vertical and horizontal shifts

Type of translation (*h*, *k* both positive)	Effect on graph of *f*
$y = f(x) + k$	vertical shift: k units up
$y = f(x) - k$	vertical shift: k units down
$y = f(x + h)$	horizontal shift: h units to left
$y = f(x - h)$	horizontal shift: h units to right

PROBLEM The graph of the function g is the result of a vertical shift of 2 units down of the graph of the function f defined by $f(x) = x^2$. Write the equation for the graph of the function g.

SOLUTION To shift the graph of f down 2 units, subtract 2 from $f(x)$ to obtain $g(x) = f(x) - 2 = x^2 - 2$. The graph of g is shown in Figure 6.1, on the next page.

PROBLEM Given the cubic function defined by $f(x) = x^3$, which of the following best describes the graph of the function $g(x) = (x + 4)^3$? (a) Congruent to $f(x) = x^3$, but shifted up by 4 units; (b) congruent to $f(x) = x^3$, but shifted down by 4 units; (c) congruent to $f(x) = x^3$, but shifted right by 4 units; or (d) congruent to $f(x) = x^3$, but shifted left by 4 units.

SOLUTION (d). Adding a positive constant 4 to x results in a horizontal shift of 4 units to the left. The graph $g(x) = (x + 4)^3$ is congruent to $f(x) = x^3$, but shifted left by 4 units.

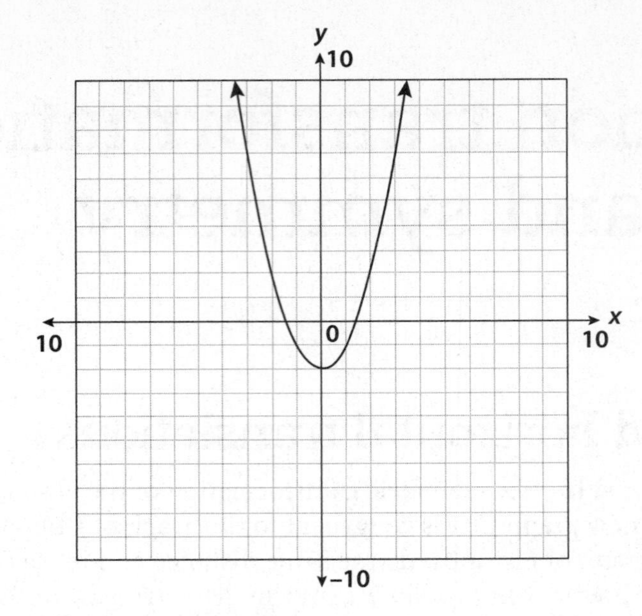

Figure 6.1 Graph of $g(x) = x^2 - 2$

EXERCISE
6·1

For 1–6, write the equation for the graph of the function g that results when the given transformation is applied to the function f. Do not simplify the equation.

1. A vertical shift of 5 units up of the graph defined by $f(x) = \sqrt{x}$.
2. A horizontal shift of 5 units to the right of the graph defined by $f(x) = \sqrt{x}$.
3. A vertical shift of $\frac{3}{4}$ unit down of the graph defined by $f(x) = |x|$.
4. A horizontal shift of $\frac{3}{4}$ unit to the left of the graph defined by $f(x) = |x|$.
5. A horizontal shift of 7 units to the right of the graph defined by $f(x) = \frac{1}{x}$.
6. A vertical shift of 7 units up of the graph defined by $f(x) = \frac{1}{x}$.

For 7–10, given that the function g is the result of a translation of the function f, which of the following best describes the graph of the function g: (a) Congruent to f(x), but shifted up by 3 units; (b) congruent to f(x), but shifted down by 3 units; (c) congruent to f(x), but shifted right by 3 units; or (d) congruent to f(x), but shifted left by 3 units?

7. $f(x) = \frac{1}{x},\ g(x) = \frac{1}{x} + 3$
8. $f(x) = x^2,\ g(x) = (x-3)^2$
9. $f(x) = |x|,\ g(x) = |x| - 3$
10. $f(x) = \sqrt{x},\ g(x) = \sqrt{x+3}$

Vertical and horizontal reflections

A **reflection** about a coordinate axis (the *x*-axis or the *y*-axis) is a geometric transformation of the graph of a function *f* that results in a new graph that is congruent to the graph of *f*, but for which every point *P* on the graph of *f* has an image *P′* that is symmetric to the given point with respect to the axis about which the reflection is made. The axis is the **line of reflection**. Informally, a reflection is a **flip** of the graph of the function across one of the axes, so that the new graph is a mirror image of the original.

A **reflection about the *x*-axis** of the graph defined by $y = f(x)$ is given by $y = -f(x)$. Figure 6.2 shows an example.

A **reflection about the *y*-axis** of the graph defined by $y = f(x)$ is given by $y = f(-x)$. Figure 6.3 shows an example.

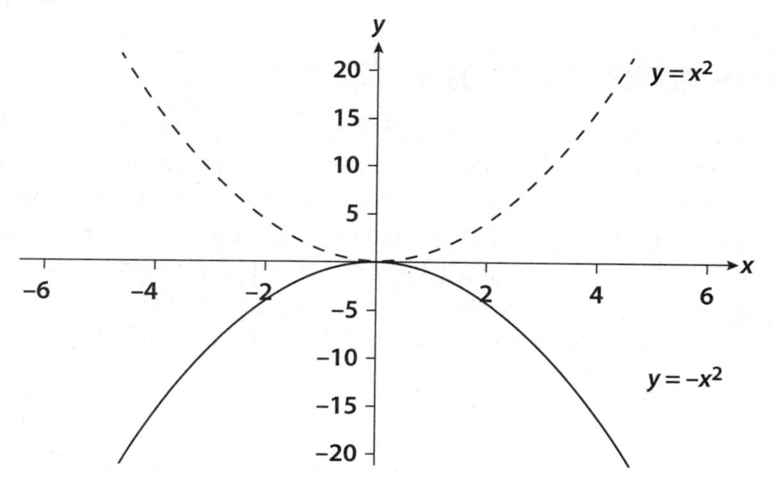

Figure 6.2 A reflection about the *x*-axis

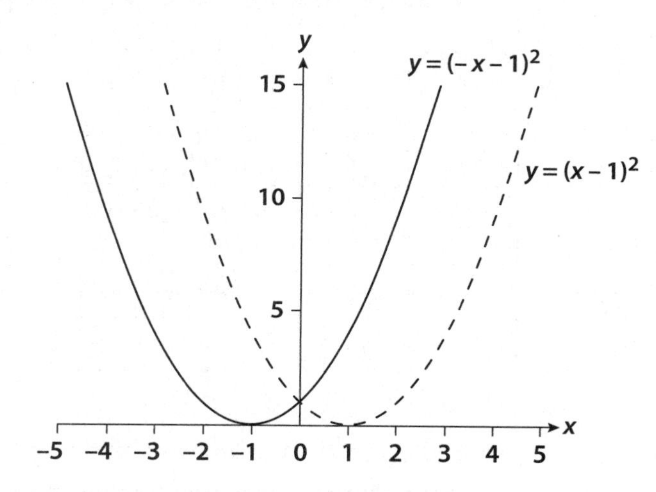

Figure 6.3 A reflection about the *y*-axis

PROBLEM Write an equation for a function *g* whose graph is a reflection about the *y*-axis of the graph defined by $y = \sqrt{x}$.

SOLUTION A reflection about the *y*-axis of the graph defined by $y = f(x)$ is given by $y = f(-x)$, so $g(x) = \sqrt{-x}$. *Note*: The domain of *g* is $(-\infty, 0]$.

EXERCISE 6·2

Write an equation for a function g *whose graph is congruent to the graph defined by* y = f(x) *and that satisfies the given condition. Do not simplify the equation.*

1. $f(x) = \frac{1}{x^3}$, a reflection about the *x*-axis
2. $f(x) = |x+5|$, a reflection about the *x*-axis

3. $f(x)=|x+5|$, a reflection about the y-axis

4. $f(x)=3x^2+2x-1$, a reflection about the x-axis

5. $f(x)=3x^2+2x-1$, a reflection about the y-axis

Stretches and compressions

A **dilation** is a geometric transformation of the graph of a function f that results in a new graph that is geometrically similar in shape to the graph of f, but for which the graph of f has undergone a vertical stretch or compression or a horizontal stretch or compression.

If $a > 1$, the graph defined by $y_2 = af(x)$ is a **vertical stretch** *away* from the x-axis of the graph defined by $y_1 = f(x)$; whereas if $0 < a < 1$, the graph defined by $y_2 = af(x)$ is a **vertical compression** *toward* the x-axis of the graph defined by $y_1 = f(x)$. In either case, if (x, y) is on the graph defined by $y_1 = f(x)$, then (x, ay) is on the graph defined by $y_2 = af(x)$. Figure 6.4 shows examples.

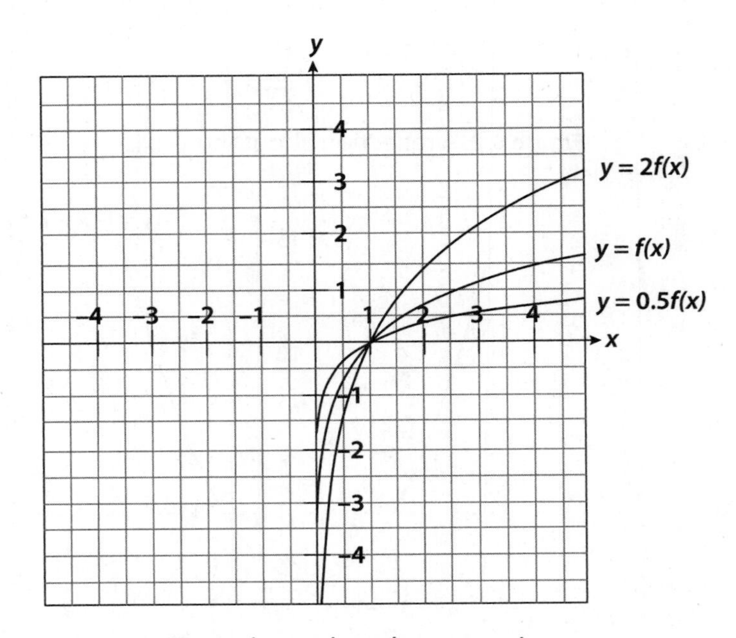

Figure 6.4 Vertical stretch and compression

If $b > 1$, the graph defined by $y_3 = f(bx)$ is a **horizontal compression** *toward* the y-axis of the graph defined by $y_1 = f(x)$; whereas if $0 < b < 1$, the graph defined by $y_3 = f(bx)$ is a **horizontal stretch** *away* from the y-axis of the graph defined by $y_1 = f(x)$. In either case, if (x, y) is on the graph defined by $y_1 = f(x)$, then $\left(\frac{x}{b}, y\right)$ is on the graph defined by $y_3 = f(bx)$. Figure 6.5 shows examples.

PROBLEM Given that the graph of the function g defined by $g(x)=\frac{1}{3}x^3$ is a dilation of the graph of the function f defined by $f(x)=x^3$, describe the dilation as (a) a vertical stretch, (b) a vertical compression, (c) a horizontal stretch, or (d) a horizontal compression.

SOLUTION (b). Given that $g(x)=\frac{1}{3}x^3=\frac{1}{3}f(x)$ and that $0<\frac{1}{3}<1$, the function g is a vertical compression of the function f.

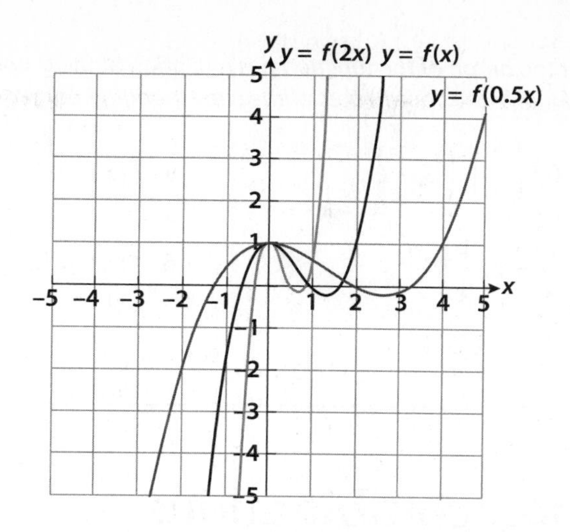

Figure 6.5 Horizontal stretch and compression

PROBLEM Write an equation for a function g whose graph is a dilation of the graph defined by $f(x) = x^3$, satisfying the condition that $g(x) = f(3x)$. Do not simplify the equation.

SOLUTION $g(x) = f(3x) = (3x)^3$

PROBLEM For the point $(-6, -216)$ on the graph defined by $f(x) = x^3$, give the coordinates of the corresponding point on the graph defined by $g(x) = f(3x)$.

SOLUTION If $(-6, -216)$ is on the graph defined by $f(x) = x^3$, then $\left(\frac{-6}{3}, -216\right) = (-2, -216)$ is its corresponding point on the graph defined by $g(x) = (3x)^3$.

EXERCISE
6·3

For 1–5, given that the graph of the function g is a dilation of the graph of the function f, describe the dilation as (a) a vertical stretch, (b) a vertical compression, (c) a horizontal stretch, or (d) a horizontal compression.

1. $f(x) = \frac{1}{x}$, $g(x) = 0.5\left(\frac{1}{x}\right)$
2. $f(x) = \frac{1}{x}$, $g(x) = \frac{2}{x}$
3. $f(x) = x^2$, $g(x) = 5x^2$
4. $f(x) = |x|$, $g(x) = \left|\frac{x}{4}\right|$
5. $f(x) = \sqrt{x}$, $g(x) = \sqrt{3x}$

For 6–10, write an equation for a function g whose graph is a dilation of the graph defined by y = f(x), satisfying the condition given. Do not simplify the equation.

6. $f(x) = \frac{1}{x} + 3$, $g(x) = f(5x)$
7. $f(x) = \sqrt{x+1}$, $g(x) = 5f(x)$
8. $f(x) = x^2$, $g(x) = f\left(\frac{1}{4}x\right)$
9. $f(x) = |x-6|$, $g(x) = f\left(\frac{2x}{3}\right)$
10. $f(x) = \sqrt{x} + 100$, $g(x) = 100f(x)$

For 11–15, suppose that the graph of the function g is a dilation of the graph of the function f; if the point (x, y) is on the graph of f, give the coordinates of the corresponding point on the graph of g.

11. $f(x)=\frac{1}{x}$, $g(x)=0.5\left(\frac{1}{x}\right)$, $\left(8,\frac{1}{8}\right)$

12. $f(x)=\frac{1}{x}$, $g(x)=\frac{2}{x}$, $\left(8,\frac{1}{8}\right)$

13. $f(x)=x^2$, $g(x)=5x^2$, $(-2,4)$

14. $f(x)=|x|$, $g(x)=\left|\frac{x}{4}\right|$, $(-4,4)$

15. $f(x)=\sqrt{x}$, $g(x)=\sqrt{3x}$, $(25,5)$

Transformation combinations

You can combine transformations of graphs of functions to obtain new graphs. If a transformation results in a graph that is congruent to the original graph, then the transformation is a **rigid** transformation. Therefore, translations are rigid transformations. If a transformation results in a graph that is geometrically similar to the original graph, but not congruent to it, then the transformation is a **nonrigid** transformation. Therefore, dilations are nonrigid transformations.

For example, you can obtain the graph defined by $y=-3(2x+5)^2-4$ by performing five transformations on the graph defined by $y=x^2$.

1. Horizontally compress the graph $y=x^2$ by a factor of $\frac{1}{2}$: $y=(2x)^2$.
2. Horizontally shift the graph $y=(2x)^2$ to the left 2.5 units: $y=(2x+5)^2$.
3. Vertically stretch the graph $y=(2x+5)^2$ by a factor of 3: $y=3(2x+5)^2$.
4. Reflect the graph $y=3(2x+5)^2$ about the x-axis: $y=-3(2x+5)^2$.
5. Vertically shift the graph $y=-3(2x+5)^2$ down 4 units: $y=-3(2x+5)^2-4$.

Notice that in step 2 the shift is 2.5 units to the left because $2x + 5 = 2(x + 2.5)$. Also, the vertical shift is performed last. *Always* do vertical shifts last because neglecting to do so might result in an incorrect graph.

PROBLEM Write an equation for a function g whose graph is the result of transformations on the graph $y=\sqrt{x}$ that satisfy the following conditions: reflected about the x-axis, shifted right 3 units, and shifted up 5 units.

SOLUTION $g(x)=-\sqrt{x-3}+5$

Write an equation for a function g whose graph is the result of transformations on the graph y = f(x) that satisfy the given conditions. Do not simplify the equation.

1. $f(x)=\frac{1}{x^3}$; reflected about the y-axis, shifted right 10 units, up 5 units, and vertically stretched by a factor of 20
2. $f(x)=|x|$; reflected about the x-axis, shifted left 4.5 units, up 9.25 units, and vertically compressed by a factor of 0.75

3. $f(x)=x^3$; shifted right 6 units, up 11 units, vertically stretched by a factor of 7, and horizontally compressed by a factor of $\frac{1}{2}$
4. $f(x)=3x^2+2x-1$; shifted left 5 units and up 10 units
5. $f(x)=\sqrt{x}$; shifted left 1 unit, down 5 units, horizontally stretched by a factor of 3, and vertically stretched by a factor of 1,000

Even and odd functions

A function f is an **even function** if for every x in the domain of f, $-x$ is in the domain of f and $f(-x)=f(x)$. Even functions are **symmetric about the y-axis**. Figure 6.6 shows an example of an even function.

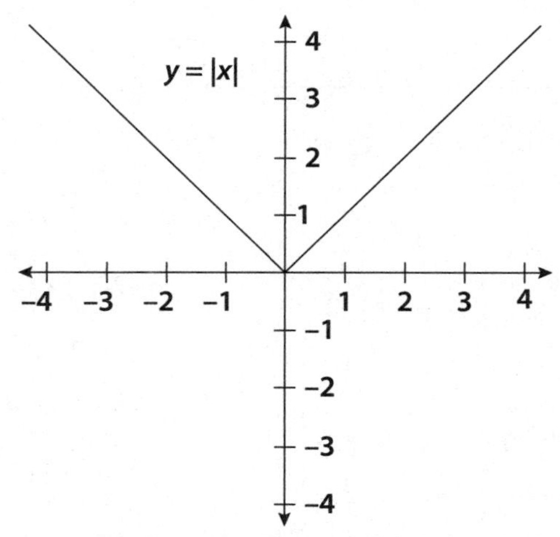

Figure 6.6 Graph of the even function $y=|x|$

A function f is an **odd function** if for every x in the domain of f, $-x$ is in the domain of f and $f(-x)=-f(x)$. Odd functions are **symmetric about the origin**. Figure 6.7 shows an example of an odd function.

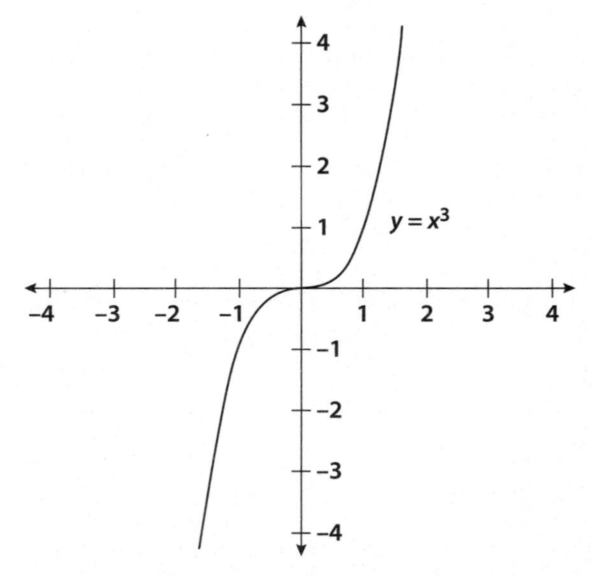

Figure 6.7 Graph of the odd function $y=x^3$

It is important to note that many functions are neither even nor odd. Their graphs show no symmetry with respect to either the y-axis or the origin.

PROBLEM Determine whether the function defined by the given equation is even, odd, or neither.

a. $f(x) = 3x + 5$

b. $g(x) = 2x^3 + x$

c. $h(x) = |3x| + 5$

SOLUTION a. Neither, because $f(-x) = 3(-x) + 5 = -3x + 5$, $f(-x) \neq f(x)$, and $f(-x) \neq -f(x)$.

b. Odd, because $g(-x) = 2(-x)^3 + (-x) = -2x^3 - x = -(2x^3 + x) = -g(x)$.

c. Even, because $h(-x) = |3(-x)| + 5 = |-3x| + 5 = |3x| + 5 = h(x)$.

Determine whether the function defined by the given equation is even, odd, or neither.

1. $f(x) = \dfrac{1}{x^3}$
2. $g(x) = |x + 5|$
3. $h(x) = 4x^2 - 1$
4. $f(x) = 3x^2 + 2x - 1$
5. $g(x) = 3x^6 + 2x^4 - 1$
6. $g(x) = 3x^5 + 2x^3 - x$
7. $f(x) = \sqrt{x}$
8. $g(x) = |x| - 25$
9. $f(x) = \sqrt{x^4 + 4}$
10. $h(x) = (x + 1)^3$

Linear functions

·7·

Definition of a linear function

Linear functions are defined by equations of the form $\boldsymbol{f(x) = mx + b}$ (or $y = mx + b$). The domain for all linear functions is R, the set of real numbers; and when $m \neq 0$, the range is R. When $m = 0$, the range is the set $\{b\}$, which contains the single value b. The graph of a linear function is always a nonvertical line with slope m and y-intercept b. The equation $y = mx + b$ is the **slope-intercept form** of the equation of a line.

PROBLEM The linear function f is defined by $f(x) = \frac{1}{2}x + 7$. State the domain and range of f. What is the slope m and y-intercept of the graph of f?

SOLUTION Both the domain and the range of f are all real numbers. The equation $f(x) = \frac{1}{2}x + 7$ is a linear equation in slope-intercept form, so $m = \frac{1}{2}$ and y-intercept = 7.

When $m \neq 0$, the x-intercept of the graph of a linear function f defined by $f(x) = mx + b$ is $-\frac{b}{m}$. For instance, the linear function f defined by $f(x) = \frac{1}{2}x + 7$ has x-intercept $= -\frac{b}{m} = -\frac{7}{\frac{1}{2}} = -14$.

Notice that when you evaluate the function $f(x) = \frac{1}{2}x + 7$ at $x = -14$ (the x-intercept), you obtain $f(14) = \frac{1}{2}(-14) + 7 = -7 + 7 = 0$. In general, for a linear function f defined by $f(x) = mx + b$, with $m \neq 0$, $f(x\text{-intercept}) = 0$. Thus, the x-intercept is a value for x that results in a zero value for $f(x)$. Therefore, for a linear function f, the x-intercept is a **zero** of the function f. When the slope of the graph of a linear function f is not zero, you have exactly one x-intercept for the graph and, thus, one zero for f.

Note: It is important to point out that the above discussion about the relationship between x-intercepts and zeros is specific to linear functions. To clarify the relationship in general: For any function f, x-intercepts (if any) of the graph of f are *always* zeros of f; however, only *real* zeros (if any) of f are x-intercepts of its graph. Some functions have zeros that are not real numbers, so these zeros do not correspond to x-intercepts because the graphs of their functions do not intersect the x-axis.

PROBLEM Find the zero for the linear function f defined by $f(x) = -2x + 6$.

SOLUTION The function f is a linear function; therefore, the x-intercept of its graph is the zero, so the zero $= -\frac{b}{m} = -\frac{6}{-2} = 3$.

Note: In the previous problem, you also can determine that f has the zero 3 by setting $f(x) = -2x + 6$ equal to zero and solving for x. That is, when $0 = -2x + 6$, you obtain $x = 3$.

EXERCISE

7·1

Fill in the blank(s) to make a true statement.

1. A(n) __________ function f is defined by an equation of the form $f(x) = mx + b$, where m is the __________ and b is the __________ of the graph of f.
2. If f is a linear function whose graph has nonzero slope, then the domain of f is __________, and the range of f is __________.
3. If f is a linear function whose graph has zero slope and y-intercept b, then the domain of f is __________, and the range of f is __________.
4. The graph of a linear function is always a(n) __________ line.
5. The equation $y = mx + b$ is the __________ form of the equation of a line.
6. When the slope of the graph of a function f __________ (is, is not) zero, you have exactly one zero for f.
7. An x-intercept for a function f corresponds to a(n) __________ of the graph of f.
8. The linear function f defined by $f(x) = -3x - 8$ has slope $m =$ __________ and y-intercept = __________. Its domain is __________, and its range is __________.
9. The linear function f defined by $f(x) = -2$ has slope $m =$ __________ and y-intercept = __________. Its domain is __________, and its range is __________.
10. The linear function f defined by $f(x) = -3x - 8$ has one zero at __________, and its graph has x-intercept = __________.

Graphs of linear functions

The graph of a linear function f defined by $f(x) = mx + b$ with $m \neq 0$ has exactly one y-intercept b and exactly one x-intercept $-\frac{b}{m}$. Thus, the points $(0, b)$ and $\left(\frac{-b}{m}, 0\right)$ are contained in the graph. If $m > 0$, f is increasing on its domain R; and if $m < 0$, f is decreasing on its domain R. Figure 7.1 shows the graph of the linear function $y = -\frac{5}{4}x + 5$ that has slope $-\frac{5}{4}$, y-intercept 5, and x-intercept 4.

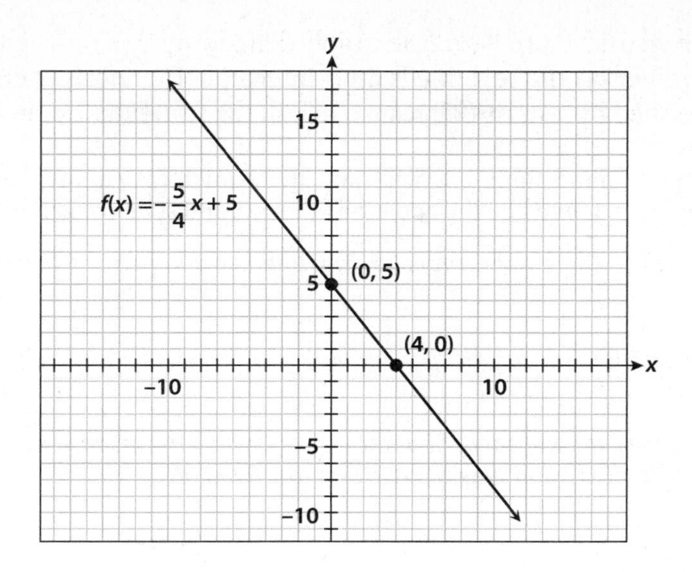

Figure 7.1 Graph of the linear function $f(x) = -\frac{5}{4}x + 5$

PROBLEM Find the y-intercept and the x-intercept (if any) for the graph of the function defined by the equation $f(x) = -2x + 8$, and state the zero or zeros, if any, of f.

SOLUTION The y-intercept is $b = 8$, x-intercept $= -\frac{b}{m} = -\frac{8}{-2} = 4$, and zero $= 4$.

The graph of a linear function f defined by $f(x) = 0 \cdot x + b = b$ with $m = 0$ and $b \neq 0$ is a horizontal line that is parallel to the x-axis and $|b|$ units away from it. When $b > 0$, the graph lies above the x-axis; and when $b < 0$, the graph lies below the x-axis. The graph has y-intercept b, but no x-intercept (because it does not cross the x-axis). Therefore, the function f defined by $f(x) = b$ has no zeros. Because $m = 0$, the function is constant on its domain R. Figure 7.2 shows the graph of the linear function $y = 4$ that has slope 0 and y-intercept 4.

PROBLEM Find the y-intercept and the x-intercept (if any) for the graph of the function defined by the equation $f(x) = 8$, and state the zero or zeros, if any, of f.

SOLUTION The y-intercept is $b = 8$, there is no x-intercept, and there are no zeros.

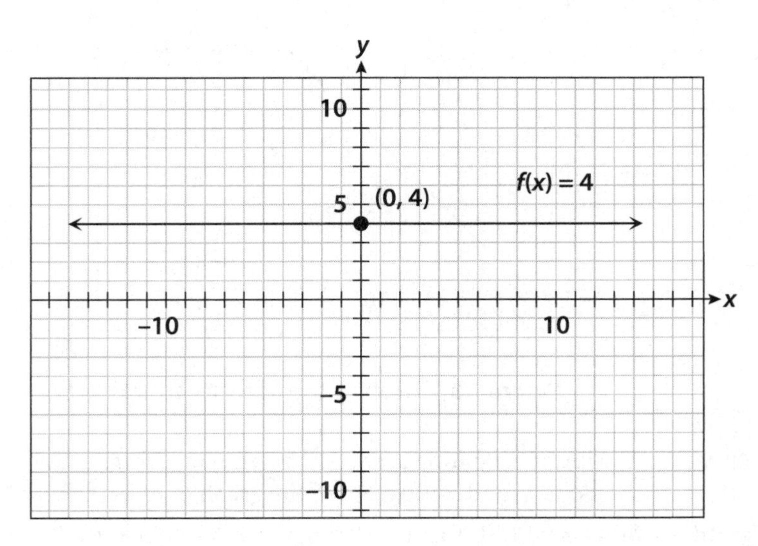

Figure 7.2 Graph of the linear function $f(x) = 4$

When both m and b are zero, the graph defined by $f(x)=0\cdot x+0=0$ coincides with the x-axis. Every number on the real line is an x-intercept. Therefore, every real number is a zero of the function f defined by $f(x)=0$. Because $m=0$, the function is constant on its domain R.

PROBLEM Find the y-intercept and the x-intercept (if any) for the graph of the function defined by the equation $f(x)=0$, and state the zero or zeros, if any, of f.

SOLUTION The y-intercept is $b=0$, x-intercepts $=\{x|x\in R\}$, zeros $=\{x|x\in R\}$.

Find the y-intercept and the x-intercept (if any) for the graph of the function defined by the equation $f(x)$, *and state the zero or zeros, if any, of* f.

1. $f(x)=3x+2$
2. $f(x)=-\frac{3}{4}x+9$
3. $f(x)=0$
4. $f(x)=100$
5. $f(x)=4x-5$

Two simple linear functions

The **identity function** is the linear function defined by the equation $f(x)=x$. The domain and range are both R. The graph of the function has slope $m=1$. The graph passes through the origin, so both the x- and y-intercepts are zero. The only zero is $x=0$. The identity function pairs each x-value to an identical y-value. Figure 7.3 shows the identity function.

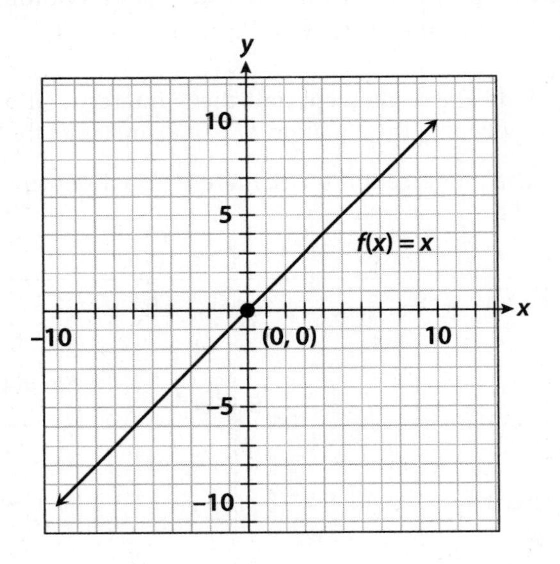

Figure 7.3 Graph of the identity function $f(x)=x$

PROBLEM The linear function f is defined by $f(x)=x$. State the domain and range of f. What is the slope m and y-intercept of the graph of f? Find the x-intercepts and zeros, if any.

SOLUTION Slope $m=1$, domain $=R$, range $=R$, y-intercept $=0$, x-intercept $=0$, zero $=0$.

Commonly, the identity function is denoted I, where $I(x)=x$. In the composition of functions, the role of I is analogous to the role of the identity 1 for the real numbers. That is, $f \circ f^{-1} = f^{-1} \circ f = I$ (provided f^{-1} is defined) and $f \circ I = I \circ f = f$ for all functions f. *Note*: See "Composition of functions" and "Inverses of functions" in Chapter 4, "Basic function concepts," for a discussion of function composition and inverses.

PROBLEM For the function f defined by $f(x)=x^3$, show that (a) $f \circ f^{-1} = f^{-1} \circ f = I$ and (b) $f \circ I = I \circ f = f$.

SOLUTION a. The inverse for the function f defined by $f(x)=x^3$ is the function f^{-1} defined by $f^{-1}(x)=\sqrt[3]{x}$; thus:

$$(f \circ f^{-1})(x) = f(f^{-1}(x)) = f(\sqrt[3]{x}) = (\sqrt[3]{x})^3 = x = I(x), \text{ so } f \circ f^{-1} = I$$
$$(f^{-1} \circ f)(x) = f^{-1}(f(x)) = f^{-1}(x^3) = (\sqrt[3]{x^3}) = x = I(x), \text{ so } f^{-1} \circ f = I$$

b. $(f \circ I)x = f(I(x)) = f(x), \text{ so } f \circ I = f$

$(I \circ f)x = I(f(x)) = f(x), \text{ so } I \circ f = f$

Constant functions are linear functions defined by equations of the form $f(x)=b$. The domain is R, and the range is the set $\{b\}$ containing the single element b. The graph of a constant function has slope 0 and y-intercept b. The graph can have either no x-intercepts or infinitely many: If $b \neq 0$, it has none; if $b=0$, every real number x is an x-intercept. The same is true of the zeros for a constant function.

When $b \neq 0$, the graph of a constant function is a horizontal line that is $|b|$ units above or below the x-axis; and when $b=0$, the graph is coincident with the x-axis. Figure 7.4 shows the constant function $f(x)=-6$.

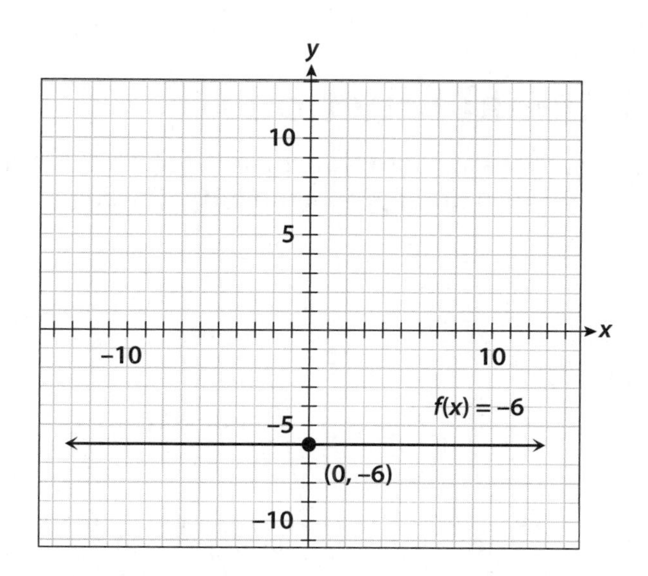

Figure 7.4 Graph of the constant function $f(x)=-6$

PROBLEM The linear function f is defined by $f(x)=7$. State the domain and range of f. What is the slope m and y-intercept of the graph of f? Find the x-intercepts and zeros, if any.

SOLUTION Slope $m = 1$, domain $= R$, range $= \{7\}$, y-intercept $= 7$, there are no x-intercepts and no zeros.

EXERCISE 7·3

Fill in the blank(s) to make a true statement.

1. The identity function is defined by the equation __________. Its domain is __________, and its range is __________.
2. If $\left(\frac{1}{2}, y\right) \in I$, then $y =$ __________.
3. The graph of the identity function has slope = __________. Its x-intercept is __________, and its y-intercept is __________. Its only zero is __________.
4. The graph of the identity function passes through the __________.
5. In the composition of functions, $f \circ f^{-1} = f^{-1} \circ f =$ __________ and $f \circ I = I \circ f =$ __________.
6. The linear function f defined by $f(x) = b$ is best described as a(n) __________ function.
7. The domain of a constant function is __________, and the range is __________.
8. The graph of a constant function defined by $f(x) = b$ has slope = __________ and y-intercept = __________.
9. For a constant function f defined by $f(x) = b$, if $b \neq 0$, then f has __________ (no, infinitely many) zeros.
10. The graph of the function f defined by $f(x) = -15$ is a __________ (horizontal, vertical) line that is __________ units __________ (above, below) the x-axis.

Directly proportional functions

Directly proportional functions are linear functions defined by equations of the form $f(x) = kx$, where $k \in R$ is the nonzero **constant of proportionality**. The domain and range are both R. The

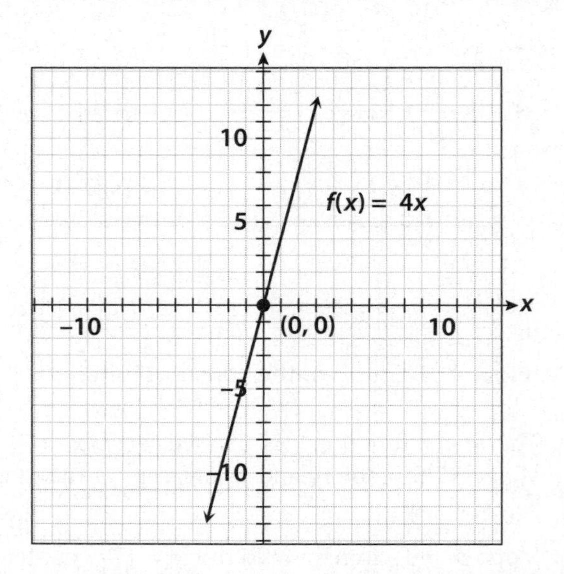

Figure 7.5 Graph of the directly proportional function $f(x) = 4x$

graph of the function has slope k and passes through the origin, so both the x- and y-intercepts are zero. The only zero is $x = 0$. Figure 7.5 shows the directly proportional function $f(x) = 4x$.

PROBLEM The constant of proportionality of a directly proportional function f is π. Write an equation that defines f, state its domain and range, and give the slope and intercepts of its graph.

SOLUTION The equation is $f(x) = \pi x$, domain = R, range = R, slope = π, x-intercept = 0, and y-intercept = 0.

PROBLEM Suppose f is a directly proportional function defined by $f(x) = kx$. If (2, 140) and (5, y) are ordered pairs in f, find y.

SOLUTION Given that (2, 140) is in f, you have $140 = k \cdot 2$; therefore, $k = \frac{140}{2} = 70$. Thus, $f(x) = 70x$ and $y = f(5) = 70(5) = 350$.

EXERCISE 7·4

Fill in the blank(s) to make a true statement.

1. Directly proportional functions are defined by equations of the form $f(x) = kx$, where $k \in R$ is the nonzero __________ (phrase). Its domain is __________, and its range is __________.
2. The graph of a directly proportional function defined by $f(x) = kx$ has slope = __________ and passes through the __________. The x-intercept is __________, and the y-intercept is __________.
3. The only zero of the directly proportional function is __________.
4. Suppose f is a directly proportional function defined by $f(x) = kx$. If $(14\pi, 7)$ is an ordered pair in f, the constant of proportionality for f is __________.
5. Suppose f is a directly proportional function defined by $f(x) = kx$. If $(-9, -3)$ and $(12, y)$ are ordered pairs in f, then y = __________.

Average rate of change and difference quotient of linear functions

The **average rate of change** of a linear function f defined by $f(x) = mx + b$ is constant over any interval on the x-axis and equals the slope m of the graph of f.

Note: See the section "Average rate of change and difference quotient" in Chapter 5, "Graphs of functions," for a general discussion of average rate of change.

PROBLEM Find the average rate of change of $f(x) = \frac{1}{2}x + 3$ on the interval $[-4, 4]$.

SOLUTION The average rate of change of $f(x) = \frac{1}{2}x + 3$ is constant on any interval and equals $\frac{1}{2}$, the slope of the graph of f.

PROBLEM Suppose f is a linear function whose graph contains the points (7,5) and (−4,−6). What is the average rate of change of f?

SOLUTION The average rate of change of f is the slope m, where:

$$m = \frac{y_2 - y_1}{x_2 - x_1} = \frac{(-6) - (5)}{(-4) - (7)} = \frac{-6 - 5}{-4 - 7} = \frac{-11}{-11} = 1$$

The **difference quotient** of a linear function f defined by $f(x) = mx + b$ equals the slope m of the graph of f.

Note: See the section "Average rate of change and difference quotient" in Chapter 5, "Graphs of functions," for a general discussion of the difference quotient.

PROBLEM Find the difference quotient for $f(x) = -5x + 2$.

SOLUTION The difference quotient for $f(x) = -5x + 2$ equals -5, the slope of the graph of f.

Fill in the blank(s) to make a true statement.

1. The average rate of change of a linear function f __________ (is, is not) constant and equals the __________ of the graph of f.
2. The difference quotient of a linear function f equals the __________ of the graph of f.
3. The average rate of change of $f(x) = \frac{2}{3}x + \frac{1}{2}$ is __________.
4. The difference quotient for $f(x) = 1.25x - 1000$ is __________.
5. If f is a linear function whose graph contains the points $(-2, 7)$ and $(3, -5)$, the average rate of change of f is __________.

Linear equations

The **standard form** of a linear equation is $\boldsymbol{Ax + By = C}$. If $B \neq 0$, you can rewrite this equation as an equivalent equation that has the form $y = mx + b$ by solving for y to obtain $y = -\frac{A}{B}x + \frac{C}{B}$. Thus, every linear equation with $B \neq 0$ defines a linear function. The graph is a nonvertical line with slope $-\frac{A}{B}$ and y-intercept $\frac{C}{B}$.

PROBLEM Suppose the equation $3x - 5y = 10$ defines a linear function. (a) Rewrite the equation in slope-intercept form, and (b) identify the slope m and y-intercept.

SOLUTION a. Solving $3x - 5y = 10$ for y yields $y = \frac{3}{5}x - 2$.

b. $m = \frac{3}{5}$ and y-intercept $= -2$.

Another form for a linear equation that you can use to define a linear function is the **point-slope form**: $\boldsymbol{y - y_1 = m(x - x_1)}$, where $m \neq 0$ and (x_1, y_1) is a point on the line. Solving this equation for y gives $y = mx + (y_1 - mx_1)$. This equation defines a linear function whose graph is a line with slope m and y-intercept $(y_1 - mx_1)$.

PROBLEM Suppose f is a linear function whose graph contains the point $(3, -4)$ and has slope $m = -2$. Write the equation $f(x) = mx + b$ that defines f.

SOLUTION Using the point-slope form,

$$y - y_1 = m(x - x_1);\ y - (-4) = -2(x - (3));\ y + 4 = -2x + 6,$$

which is $y = -2x + 2$ in slope-intercept form. Thus, $f(x) = -2x + 2$ defines f.

PROBLEM Suppose f is a linear function whose graph contains the points $(7,5)$ and $(-4,-6)$. Write the equation $f(x) = mx + b$ that defines f.

SOLUTION First, use the slope formula to find m, where:

$$m = \frac{y_2 - y_1}{x_2 - x_1} = \frac{(-6)-(5)}{(-4)-(7)} = \frac{-6-5}{-4-7} = \frac{-11}{-11} = 1$$

Note: See the section "Slope of a line" in Chapter 3, "The Cartesian coordinate system," for a discussion on finding the slope of a line.

Next, use the point-slope form with $m = 1$ and point $(7,5)$ to find the equations

$$y - y_1 = m(x - x_1); y - (5) = 1(x - (7)); y - 5 = x - 7$$

which is $y = x - 2$ in slope-intercept form. Thus, $f(x) = x - 2$ defines f.

Note: Instead of $(7,5)$, you can use $(-4,-6)$ in the point-slope form to find the equation.

The summary of linear equations in Table 7.1 will be helpful when you are working with linear functions and their graphs.

Table 7.1 Summary of linear equations

Slope-intercept form (functional form)	$y = mx + b$
Standard form	$Ax + By = C$
Point-slope form	$y - y_1 = m(x - x_1)$
Horizontal line	$y = b$ for any constant b
Vertical line (*not* a function)	$x = a$ for any constant a

For 1–5, (a) rewrite the equation in slope-intercept form and (b) identify the slope m *and y-intercept.*

1. $x + 2y = 6$
2. $3x - 4y = -8$
3. $5x + 6y = 16$
4. $5x - 5y = 8$
5. $-5y = 15$

For 6–10, suppose f *is a linear function whose graph contains the given point and has slope* m. *Write the equation* $y = mx + b$ *that defines* f.

6. $m = \frac{4}{5}, (-1, -3)$
7. $m = -1, (5, 5)$
8. $m = -\frac{1}{2}, (-3, 2)$
9. $m = -3, (-5, 2)$
10. $m = 0, (6, -5)$

For 11–15, suppose f *is a linear function whose graph contains the given points. Write the equation* $y = mx + b$ *that defines* f.

11. $(-2, -4), (-1, -3)$

12. $(0, 5), (8, 0)$

13. $(-5, 5), (5, -5)$

14. $(6, -4), (-1, -4)$

15. $(3, 4), (0, 0)$

Quadratic functions ·8·

Definition of a quadratic function

Quadratic functions are defined by equations of the form $\boldsymbol{f(x) = ax^2 + bx + c}$, $a \neq 0$. The domain is R and the range is a subset of R. The graph of a quadratic function is always a **parabola**, a U-shaped figure, that opens either upward (when $a > 0$) or downward (when $a < 0$) (see the section "Parabola" in Chapter 20, "Conics," for a discussion of parabolas). The graph of the quadratic function f defined by $f(x) = ax^2 + bx + c$ has a y-intercept at c and x-intercepts at the real zeros (if any) of f.

The zeros of f are the roots of the quadratic equation $ax^2 + bx + c = 0$ (see the section "Quadratic equations" in this chapter for a review of quadratic equations). The **discriminate** $b^2 - 4ac$ of the quadratic equation gives you three possibilities for real zeros of f and thereby the x-intercepts:

- If $b^2 - 4ac > 0$, you will have two real unequal zeros and, therefore, two x-intercepts.
- If $b^2 - 4ac = 0$, you will have exactly one real zero and, therefore, exactly one x-intercept.
- If $b^2 - 4ac < 0$, you will have no real zeros and, therefore, no x-intercepts.

PROBLEM For a–c, find the number of real zeros of the function and, thereby, the number of x-intercepts for the graph. See Figure 8.1 (on the next page).

a. $f(x) = x^2 - x - 6$

b. $g(x) = -x^2 + 8x - 16$

c. $h(x) = x^2 + 2x + 4$

SOLUTION a. $b^2 - 4ac = (-1)^2 - 4(1)(-6) = 1 + 24 = 25 > 0$, so f has two real zeros and thus two x-intercepts.

b. $b^2 - 4ac = (8)^2 - 4(-1)(-16) = 64 - 64 = 0$, so g has one real zero and thus one x-intercept.

c. $b^2 - 4ac = (2)^2 - 4(1)(4) = 4 - 16 = -12 < 0$, so h has no real zeros and thus no x-intercepts.

Figure 8.1 depicts the graphs of $f(x) = x^2 - x - 6$, $g(x) = -x^2 + 8x - 16$, and $h(x) = x^2 + 2x + 4$. Observe that $f(x)$ intersects the x-axis at two points because f has two real zeros; $g(x)$ intersects the x-axis at just one point because g has only one real zero; and $h(x)$ does not intersect the x-axis because h has no real zeros.

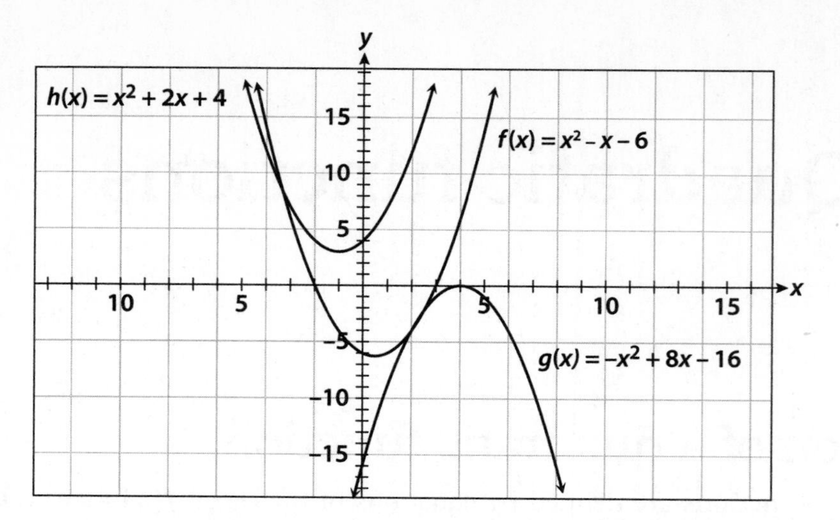

Figure 8.1 Graphs of $f(x) = x^2 - x - 6$, $g(x) = x^2 + 8x - 16$, and $h(x) = x^2 + 2x - 4$

PROBLEM For a–c, find the y-intercept, the zeros of the function, and the x-intercepts for the graph.

a. $f(x) = x^2 - x - 6$

b. $g(x) = -x^2 + 8x - 16$

c. $h(x) = x^2 + 2x + 4$

SOLUTION a. The y-intercept = -6. To find the zeros, set $f(x) = 0$. To solve $0 = x^2 - x - 6$, factor the quadratic expression on the right to obtain $0 = (x+2)(x-3)$. Solving for x yields the zeros: $x = -2$ and $x = 3$. Thus, the x-intercepts are -2 and 3.

b. The y-intercept $= -16$. To find the zeros, set $f(x) = 0$. To solve $0 = -x^2 + 8x - 16$, factor the quadratic expression on the right to obtain $0 = -(x-4)^2$. Solving for x yields one zero: $x = 4$. Thus, the x-intercept is 4.

c. The y-intercept = 4. To find the zeros, set $f(x) = 0$. To solve $0 = x^2 + 2x + 4$, use the quadratic formula to obtain $x = \dfrac{-b \pm \sqrt{b^2 - 4ac}}{2a} = \dfrac{-2 \pm \sqrt{(2)^2 - 4(1)(4)}}{2(1)}$ $= \dfrac{-2 \pm \sqrt{-12}}{2} = \dfrac{-2 \pm 2i\sqrt{3}}{2} = -1 \pm i\sqrt{3}$, two complex zeros. Since the domain of f is R, then f has no real zeros and thus no x-intercepts.

The form $f(x) = ax^2 + bx + c,\ a \neq 0$, is the **general form** for the equation of a quadratic function. The **vertex** (or **standard**) **form** for a quadratic function is $f(x) = a(x-h)^2 + k$, for $a \neq 0$. You can convert from the general form to vertex form by completing the square.

Note: Recall that when you have $x^2 + mx$, you complete the square by adding $\left(\frac{1}{2} \cdot m\right)^2$ to obtain $x^2 + mx + \frac{1}{4}m^2 = \left(x + \frac{m}{2}\right)^2$, which is a perfect square.

PROBLEM Rewrite $f(x) = x^2 - 10x + 28$ in vertex form.

SOLUTION $f(x) = x^2 - 10x + 28$

$f(x) = (x^2 - 10x) + 28$ Group the x terms.

$f(x) = (x^2 - 10x + 25) + 28 - 25$ Add $\left(\frac{10}{2}\right)^2 = 25$ inside the parentheses to complete the square. Subtract 25 outside the parentheses to keep the function the same.

$f(x)=(x-5)^2+3$ Factor the perfect square and simplify.

Thus, $f(x)=(x-5)^2+3$ is the standard form for $f(x)=x^2-10x+28$.

PROBLEM Rewrite $g(x)=2x^2-20x+46$ in vertex form.

SOLUTION $g(x)=2x^2-20x+46$

$g(x)=(2x^2-20x)+46$ Group the x terms.

$g(x)=2(x^2-10x)+46$ Factor 2 out of the x terms, so that the coefficient of x^2 is 1.

$g(x)=2(x^2-10x+25)+46-50$ Add $\left(\frac{10}{2}\right)^2=25$ inside the parentheses to complete the square. Subtract 50 outside the parentheses to keep the function the same.

Note: Because of the factor of 2 preceding the parentheses, you subtract $2\cdot 25=50$ outside the parentheses so that you add and subtract the same value.

$g(x)=2(x-5)^2-4$ Factor the perfect square and simplify.

Thus, $g(x)=2(x-5)^2-4$ is the vertex form for $g(x)=2x^2-20x+46$.

You can convert from vertex form to standard form by performing the indicated operations and simplifying. For example:

$$g(x)=2(x-5)^2-4=2(x^2-10x+25)-4=2x^2-20x+50-4=2x^2-20x+46$$

Note: Quadratic equations of the form $x=ay^2+by+c$, for $a\neq 0$, do *not* define functions of x. Their graphs are parabolas that open to the right or left. Clearly, the graphs of these relations do not pass the vertical line test for functions. See the section "Parabola" in Chapter 20, "Conics," for a discussion of parabolic relations.

EXERCISE

8·1

For 1–5, fill in the blank(s) to make a true statement.

1. The general form for a quadratic function is __________.
2. The vertex form for a quadratic function is __________.
3. The domain of a quadratic function is __________, and the range is a(n) __________ of *R*.
4. The graph of a quadratic function is always a(n) __________.
5. The graph of a quadratic function always intersects the __________ (*x*-axis, *y*-axis).

For 6–10, find (a) the number of real zeros, (b) the y-intercept, and (c) the x-intercepts.

6. $f(x)=x^2-10x+28$
7. $g(x)=2x^2-20x+46$
8. $h(x)=x^2-10x+25$
9. $g(x)=-2x^2+3x-1$
10. $f(x)=x^2+1$

For 11–15, rewrite the equation for the quadratic function in vertex form.

11. $f(x) = x^2 - x - 6$
12. $g(x) = -x^2 + 8x - 16$
13. $h(x) = x^2 - 10x + 25$
14. $g(x) = -2x^2 + 3x - 1$
15. $h(x) = x^2 + 2x + 4$

Graphs of quadratic functions

The graph of a quadratic function defined by the general form $f(x) = ax^2 + bx + c$ or the vertex form $f(x) = a(x-h)^2 + k$, with leading coefficient $a \neq 0$, is always a parabola. If $a > 0$, the parabola opens upward and has a lowest point; and if $a < 0$, the parabola opens downward and has a highest point. The **vertex** is the highest point of a parabola that opens downward or the lowest point of a parabola that opens upward. For the general form $f(x) = ax^2 + bx + c$, the vertex of the parabola is $\left(-\frac{b}{2a}, f\left(-\frac{b}{2b}\right)\right)$; and for the vertex form $f(x) = a(x-h)^2 + k$ the vertex is $(\boldsymbol{h}, \boldsymbol{k})$.

The parabola is symmetric about a vertical line through its vertex. This line, with equation $x = -\frac{b}{2a} = h$, is the **axis of symmetry** for the parabola (see Figure 8.2).

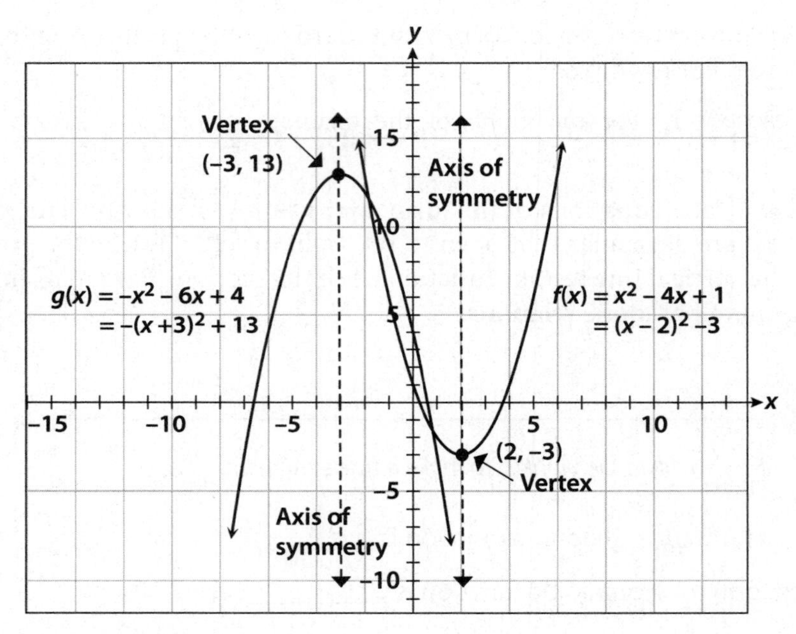

Figure 8.2 Axes of Symmetry

PROBLEM For the graph of the quadratic function $f(x) = x^2 + 8x - 2$, (a) state whether the parabola opens upward or downward, (b) find the vertex, and (c) find the axis of symmetry.

SOLUTION
a. Since $a = 1 > 0$, the parabola opens upward.

b. Vertex $= \left(-\frac{b}{2a}, f\left(-\frac{b}{2a}\right)\right)$, $-\frac{b}{2a} = -\frac{8}{2 \cdot 1} = -4$; $f\left(-\frac{b}{2a}\right) = f(-4)$
$= (-4)^2 + 8(-4) - 2 = 16 - 32 - 2 = -18$; vertex $= (-4, -18)$.

c. The line $x = -4$ is the axis of symmetry.

PROBLEM For the graph of the quadratic function $g(x)=-2(x-3)^2+5$, (a) state whether the parabola opens upward or downward, (b) find the vertex, and (c) find the axis of symmetry.

SOLUTION
a. $a=-2<0$, so the parabola opens downward.
b. Vertex $=(h,k)=(3,5)$.
c. The axis of symmetry is $x=3$.

As you already know, R is the domain of a quadratic function f. When the vertex (h, k) is the highest point of the graph of f, as it is when $\boldsymbol{a<0}$, then k, the **y-coordinate** of the vertex, is the **absolute maximum value** of f. Thus, if $\boldsymbol{a<0}$, the **range** of f is $(-\infty,\boldsymbol{k}]$. The function is increasing on $(-\infty,h]$ and decreasing on $[h,\infty)$. When the vertex is the lowest point of the graph of f, as it is when $\boldsymbol{a>0}$, then k, the **y-coordinate** of the vertex, is the **absolute minimum value** of f. Thus, if $\boldsymbol{a>0}$, the **range** of f is $[\boldsymbol{k},\infty)$. The function is decreasing on $(-\infty,h]$ and increasing on $[h,\infty)$.

PROBLEM For the graph of the quadratic function $f(x)=x^2+8x-2$, (a) find the maximum or minimum value, (b) state the range, and (c) state the intervals on which f is increasing or decreasing. See Figure 8.3.

SOLUTION
a. The vertex is $\left(-\frac{b}{2a}, f\left(-\frac{b}{2a}\right)\right)=(-4,-18)$. Since $a=1>0$, the y-coordinate, -18, of the vertex is an absolute minimum.
b. The range is $[-18,\infty)$.
c. The function f is decreasing on $(-\infty,-4]$ and increasing on $[-4,\infty)$.

PROBLEM For the graph of the quadratic function $g(x)=-2(x-3)^2+5$, (a) find the maximum or minimum value, (b) state the range, and (c) state the intervals on which f is increasing or decreasing. See Figure 8.3.

SOLUTION
a. The vertex is $(h,k)=(3,5)$. The coefficient a is $-2<0$, so the y-coordinate, 5, of the vertex is an absolute maximum.
b. The range is $(-\infty,5]$.
c. The function f is increasing on $(-\infty,3]$ and decreasing on $[3,\infty)$.

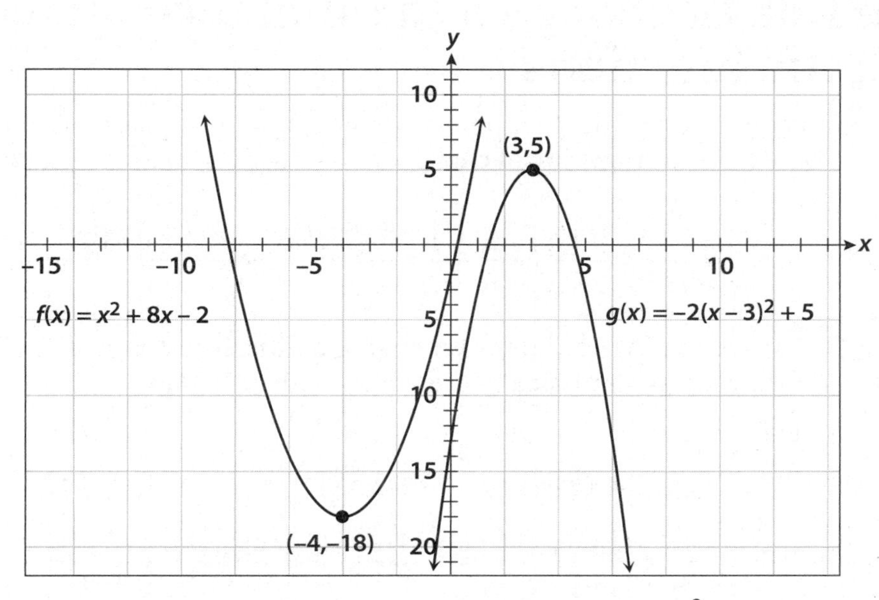

Figure 8.3 Graphs of $f(x)=x^2+8x-2$ and $g(x)=-2(x-3)^2+5$

EXERCISE

8·2

For 1–5, fill in the blank(s) to make a true statement.

1. If $a > 0$, the parabola that is the graph $f(x) = ax^2 + bx + c$ opens __________ (downward, upward) and has a(n) __________ point; and if $a < 0$, the parabola opens __________ (downward, upward) and has a(n) __________ point.
2. For $f(x) = ax^2 + bx + c$, the vertex of the parabola is __________; and for $f(x) = a(x-h)^2 + k$ the vertex is __________.
3. A parabola that is the graph of a quadratic function is symmetric about its ____________________ (phrase).
4. When the leading coefficient of a quadratic function is negative, the y-coordinate of the vertex is the absolute __________ (maximum, minimum) value of f.
5. When the leading coefficient of a quadratic function is positive, the y-coordinate of the vertex is the absolute __________ (maximum, minimum) value of f.

In 6–10, for the graph of the given quadratic function, (a) state whether the parabola opens upward or downward, (b) find the vertex, (c) find the axis of symmetry, (d) find the maximum or minimum value, (e) state the range, and (f) state the intervals on which the function is increasing or decreasing.

6. $f(x) = x^2 - 10x + 28$
7. $g(x) = 2x^2 - 20x + 46$
8. $h(x) = x^2 - 10x + 25$
9. $g(x) = -2(x+6)^2 - 1$
10. $f(x) = x^2 + 1$

Average rate of change and difference quotient of quadratic functions

If (x_1, y_1) and (x_2, y_2) are two distinct points on the graph of a quadratic function f defined by $f(x) = ax^2 + bx + c$, the **average rate of change** of f as x goes from x_1 to x_2 is given by:

$$\frac{\Delta y}{\Delta x} = \frac{f(x_2) - f(x_1)}{x_2 - x_1} = \frac{(ax_2^2 + bx_2 + c) - (ax_1^2 + bx_1 + c)}{x_2 - x_1}$$

Note: See the section "Average rate of change and difference quotient" in Chapter 5, "Graphs of functions," for a general discussion of the average rate of change.

PROBLEM Find the average rate of change of $f(x) = x^2 - x - 6$ on the interval $[1, 4]$.

SOLUTION $$\frac{\Delta y}{\Delta x} = \frac{f(x_2) - f(x_1)}{x_2 - x_1} = \frac{f(4) - f(1)}{4-1} = \frac{((4)^2 - (4) - 6) - ((1)^2 - (1) - 6)}{4-1} = \frac{12}{3} = 4$$

The **difference quotient** of a quadratic function f defined by $f(x) = ax^2 + bx + c$ is

$$\frac{f(x+h) - f(x)}{(x+h) - (x)} = \frac{f(x+h) - f(x)}{h}$$

where $h \neq 0$.

Note: See the section "Average rate of change and difference quotient" in Chapter 5, "Graphs of functions," for a general discussion of the difference quotient.

PROBLEM Find and simplify the difference quotient for $f(x) = x^2 - x - 6$.

SOLUTION

$$\frac{f(x+h) - f(x)}{h} = \frac{\left[(x+h)^2 - (x+h) - 6\right] - \left[x^2 - x - 6\right]}{h}$$

$$= \frac{x^2 + 2xh + h^2 - x - h - 6 - x^2 + x + 6}{h}$$

$$= \frac{2xh + h^2 - h}{h} = 2x + h - 1$$

EXERCISE 8·3

For 1–5, find the average rate of change for each function on the given interval.

1. $f(x) = -2(x-3)^2 + 5$, $[3, 5]$
2. $f(x) = -2(x-3)^2 + 5$, $[-5, 3]$
3. $f(x) = x^2 + 2x$, $[-5, 5]$
4. $g(x) = x^2 + 8x - 2$, $[-8, -4]$
5. $g(x) = x^2 + 8x - 2$, $[-4, 8]$

For 6–10, find and simplify the difference quotient for each function.

6. $f(x) = x^2 + 2x$
7. $f(x) = x^2 + 2x + 1$
8. $f(x) = x^2 + 2x + 100$
9. $f(x) = -2(x-3)^2 + 5$
10. $g(x) = x^2 + 8x - 2$

Quadratic equations

The zeros of a quadratic function f defined by $f(x) = ax^2 + bx + c$ are the roots of the quadratic equation $ax^2 + bx + c = 0$. Three common methods for solving quadratic equations are (1) by factoring, (2) by completing the square, and (3) by using the quadratic formula. Here are examples of each method:

PROBLEM Solve $x^2 + 2x - 3 = 0$ by factoring.

SOLUTION

$x^2 + 2x - 3 = 0$	
$(x+3)(x-1) = 0$	Factor the quadratic expression.
$(x+3) = 0$ or $(x-1) = 0$	Set each factor each to zero.
$x = -3$ or $x = 1$	Solve for x.

PROBLEM Solve $2x^2 + 16x - 4 = 0$ by completing the square.

SOLUTION $2x^2 + 16x - 4 = 0$

$2x^2 + 16x = 4$	Get variable terms on one side and all other terms on other side.
$x^2 + 8x = 2$	If the leading coefficient is not 1, divide by it.
$x^2 + 8x + 16 = 2 + 16$	Add $\left(\frac{8}{2}\right)^2 = 16$ to both sides.
$(x+4)^2 = 18$	Factor the perfect square and simplify.
$(x+4) = \pm\sqrt{18}$	Take the square root of both sides.
$x = -4 \pm \sqrt{18} = -4 \pm 3\sqrt{2}$	Solve for x and simplify.

PROBLEM Solve $x^2 + 8x - 2 = 0$ by using the quadratic formula.

SOLUTION $x^2 + 8x - 2 = 0$

$$x = \frac{-b \pm \sqrt{b^2 - 4ac}}{2a} = \frac{-(8) \pm \sqrt{(8)^2 - 4(1)(-2)}}{2(1)} = \frac{-8 \pm \sqrt{72}}{2} = \frac{-8 \pm 6\sqrt{2}}{2} = -4 \pm 3\sqrt{2}$$

Note: For most situations, unless the quadratic expression is a simple one that you can easily factor, you should use the quadratic formula to find the roots of a quadratic equation. It is the most reliable and efficient way to find the roots of a quadratic equation.

EXERCISE 8·4

Solve by any method.

1. $x^2 - 10x + 28 = 0$
2. $2x^2 - 20x + 46 = 0$
3. $x^2 - 10x + 25 = 0$
4. $-2(x+6)^2 - 1 = 0$
5. $x^2 + 1 = 0$

Polynomial functions

·9·

Definition of a polynomial function

Polynomial functions are defined by equations of the form $\boldsymbol{p(x) = a_n x^n + a_{n-1}x^{n-1} + a_{n-2}x^{n-2} + \ldots + a_2x^2 + a_1x + a_0}$, where $a_n \neq 0$ and n is a *nonnegative* integer. The **leading coefficient** is $\boldsymbol{a_n}$, the coefficient of x^n, the highest power of x. The **degree** of $p(x)$ is $\boldsymbol{n}$, the highest exponent of x. The degree of a constant polynomial defined by $p(x) = c$ is zero, where c is a nonzero constant. The degree of the zero polynomial defined by $p(x) = 0$ is undefined. The domain of a polynomial function is R. If n is *odd*, the range is R; and if n is *even*, the range is a subset of R. Linear functions and quadratic functions are particular types of polynomial functions. The zeros of p are the roots of the equation $p(x) = 0$. That is, r is a zero of p if and only if $p(r) = 0$. If r is a real zero of p, then r is an x-intercept of the graph of p. The y-intercept of the graph of p is $p(0)$.

PROBLEM Given the polynomial function defined by $p(x) = 0.1(x+1)(x+2)(x-3)(x+5)(x-6)$ (see Figure 9.1), (a) determine the degree of $p(x)$, (b) state the domain and range of the graph of p, (c) determine the real zeros of p, (d) find x-intercepts of the graph of p, and (e) find the y-intercept of the graph of p.

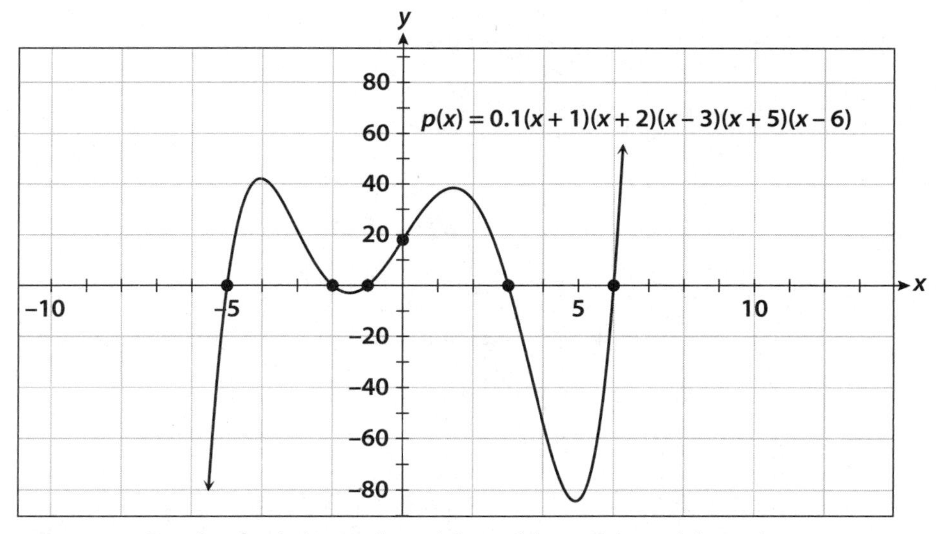

Figure 9.1 Graph of $p(x) = 0.1(x+1)(x+2)(x-3)(x+5)(x-6)$

SOLUTION a. When you expand $p(x)=0.1(x+1)(x+2)(x-3)(x+5)(x-6)$, the term with the highest exponent in the expansion is the product of 0.1 and the first terms of the five binomial factors of $p(x)$. Thus, the term with the highest exponent in $p(x)$ is $0.1x^5$, so the degree of the polynomial is 5.

b. The domain is R and the range is R (because n is odd).

c. Setting $p(r)=0$ and solving for x yield $x=-1, -2, 3, -5$, or 6. Thus, the zeros of p are $-1, -2, 3, -5$ and 6.

d. The x-intercepts are $-1, -2, 3, -5$ and 6 because these values correspond to the real zeros of p. The graph of p intersects the x-axis at $-1, -2, 3, -5$ and 6 (see Figure 9.1).

e. The y-intercept is $p(0)=0.1(0+1)(0+2)(0-3)(0+5)(0-6)=0.1(1)(2)(-3)(5)(-6)=18$. The graph of p intersects the y-axis at 18 (see Figure 9.1).

PROBLEM Given the polynomial function defined by $p(x)=x^4-1$ (see Figure 9.2), (a) determine the degree of $p(x)$, (b) state the domain and range of the graph of p, (c) determine the real zeros of p, (d) find x-intercepts of the graph of p, and (e) find the y-intercept of the graph of p.

SOLUTION a. The degree of $p(x)=x^4-1$ is 4.

b. The domain is R. Given that x^4 is always nonnegative, then $x^4-1\geq -1$; thus, the range is $[-1,\infty)$ (see Figure 9.2).

c. Setting $p(r)=0$, you have $x^4-1=0$. Factoring gives:

$$x^4-1=(x^2+1)(x^2-1)=(x^2+1)(x+1)(x-1)=0$$

Since the factor x^2+1 has no real roots, solving $(x^2+1)(x+1)(x-1)=0$ for x yields two real zeros, -1 and 1.

d. The x-intercepts are -1 and 1 because these values correspond to the real zeros of p. The graph of p intersects the x-axis at -1 and 1 (see Figure 9.2).

e. The y-intercept is $p(0)=0^4-1=-1$. The graph of p intersects the y-axis at -1 (see Figure 9.2).

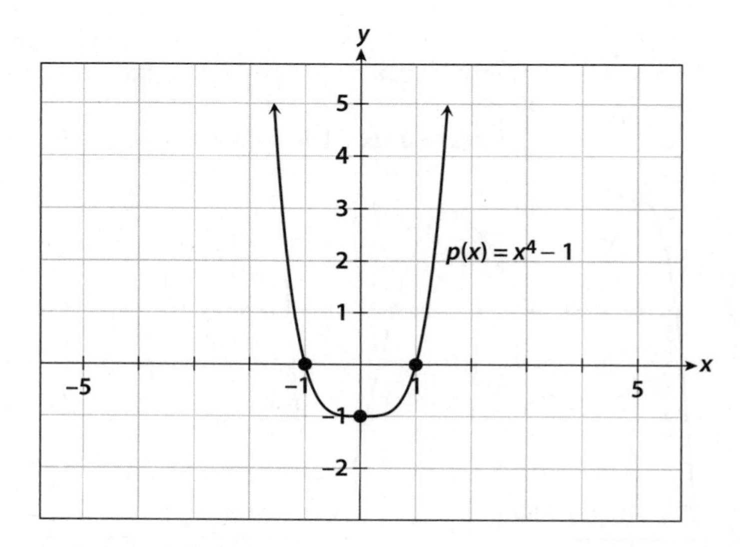

Figure 9.2 Graph of $p(x)=x^4-1$

EXERCISE

9·1

For 1–5, fill in the blank(s) to make a true statement.

1. The leading coefficient of a polynomial function is the coefficient of the term that contains the __________ power of the variable.
2. The degree of a polynomial function defined by $p(x)$ is the highest __________ of x. The degree of a constant polynomial defined by $p(x) = c\ (c \neq 0)$ for $x \in R$ is __________. The degree of a constant polynomial defined by $p(x) = 0$ for $x \in R$ is __________.
3. The domain of a polynomial function is __________. If the degree of the polynomial is __________, the range is R; and if the degree is __________, the range is a subset of R.
4. The number r is a zero of a polynomial function p if and only if __________. If r is a __________ zero of p, the graph of p intersects the x-axis at r.
5. The y-intercept of the graph of a polynomial function p is __________.

For 6–10, given the polynomial function, (a) determine the degree of p(x), *(b) state the domain and range of the graph of* p, *(c) determine the real zeros of* p, *(d) find the* x*-intercepts of the graph of* p, *and (e) find the* y*-intercept of the graph of* p.

6. $p(x) = 2(x-1)(x+2)(x-2)(x+3)(x-4)$
7. $f(x) = (x+4)(x^2-5)(x^2-9)$
8. $g(x) = x^4 - 81$
9. $g(x) = -2x^2 + 3x - 1$
10. $f(x) = 3x + 5$

Graphs of polynomial functions

The graph of a polynomial function p is a continuous smooth curve (or line) without breaks of any kind; furthermore, it has no cusps (sharp corners). The domain is R. The y-intercept of the graph is $p(0)$. The x-intercepts correspond to the real zeros (if any) of p.

Recall that graphs of polynomials of degrees 0 and 1 were discussed in Chapter 7, and those of degree 2 were discussed in Chapter 8. As the degree increases, the graphs of polynomial functions become more complex.

The graph of a polynomial function will have a **turning point** (x, y) whenever the graph changes from increasing to decreasing or from decreasing to increasing. The y-value of a turning point is either a relative maximum or relative minimum value for the function. An nth-degree polynomial will have at most $n-1$ turning points. *Note*: A turning point is an ordered pair (x, y) that identifies a point on the graph where the graph changes from increasing to decreasing or conversely. A maximum or minimum value is a value of the function, not a point on the graph.

PROBLEM Given the polynomial function $p(x) = 0.1(x+1)(x+2)(x-3)(x+5)(x-6)$ shown in Figure 9.3, on the next page, identify (a) turning points and (b) relative or absolute extrema.

SOLUTION a. Turning points: $(-4.06, 42.08)$, $(-1.50, -2.95)$, $(1.43, 38.45)$, and $(4.93, -84.27)$.

b. Relative maxima: 42.08, 38.45; relative minima: -2.95 and -84.27; no absolute extrema. *Note*: *Maxima* and *minima* are the plurals of *maximum* and *minimum*, respectively.

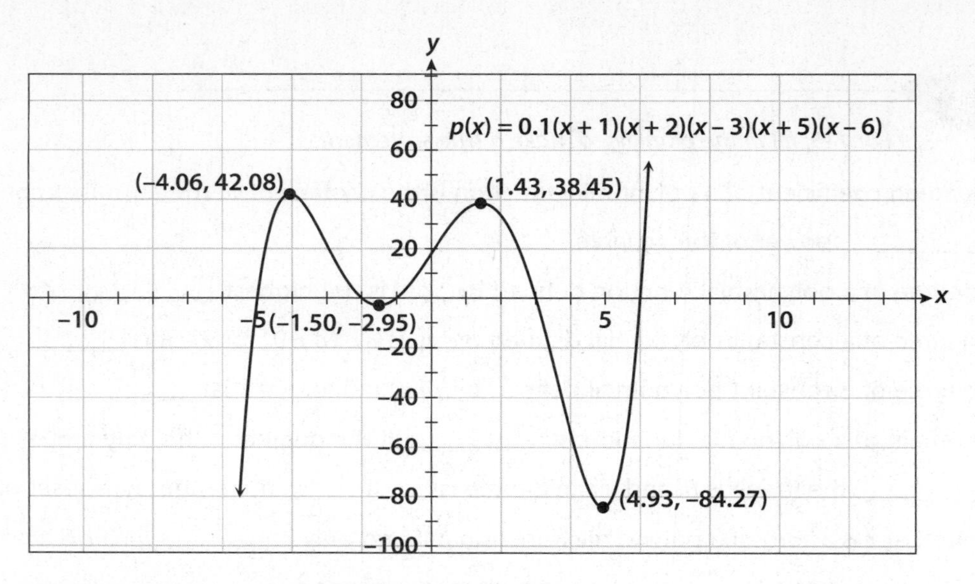

Figure 9.3 Turning points and extrema

EXERCISE
9·2

For 1–5, fill in the blank to make a true statement.

1. The graph of a polynomial function p is a(n) __________ smooth curve.
2. The graph of a polynomial function will have a(n) __________ (two words) whenever the graph changes from increasing to decreasing or from decreasing to increasing.
3. How many turning points does a constant function have? Answer: __________
4. How many turning points does a linear function have? Answer: __________
5. How many turning points does a quadratic function have? Answer: __________

In 6–10, for the graph shown, identify (a) turning points and (b) relative or absolute extrema.

6.

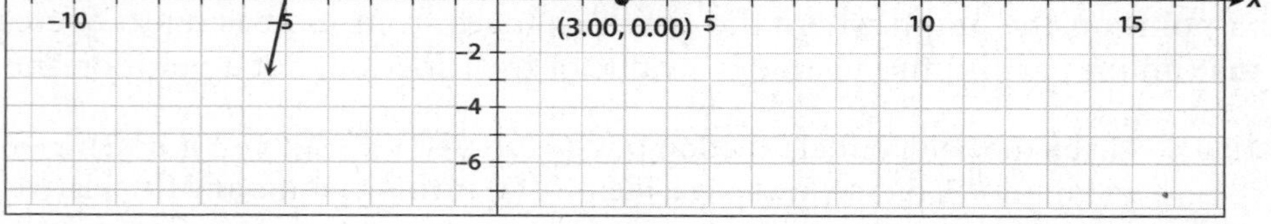

$p(x) = 0.1(x^3 - x^2 - 21x + 45)$

(−2.39, 7.58)

(3.00, 0.00)

7.

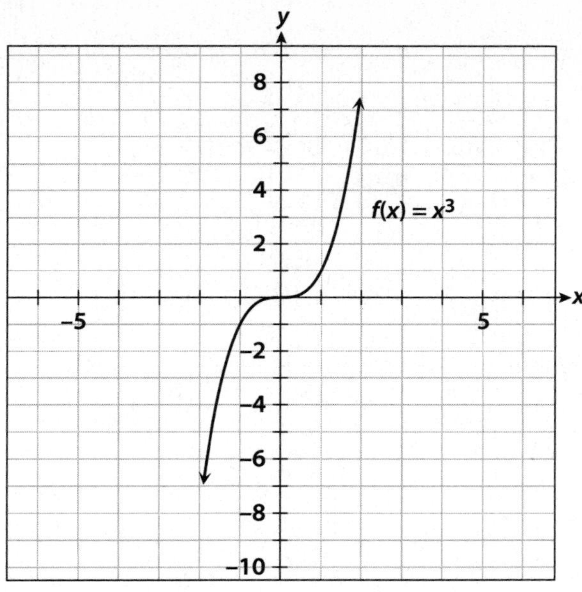

y
8
6
4
2
$f(x) = x^3$
x
−5
5
−2
−4
−6
−8
−10

8.

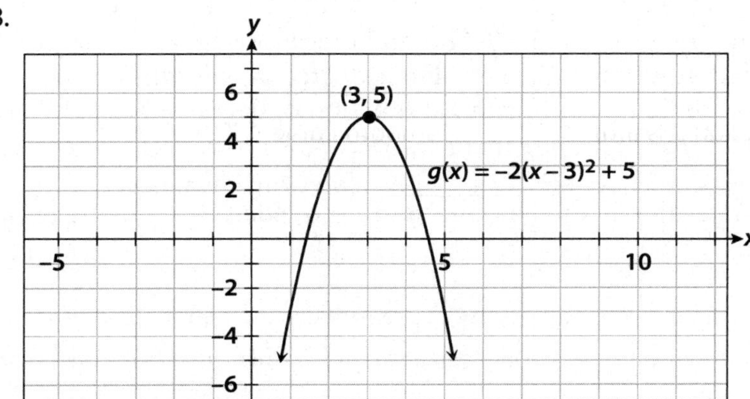

y
6
4
2
(3, 5)
$g(x) = -2(x - 3)^2 + 5$
x
−5
5
10
−2
−4
−6

9.

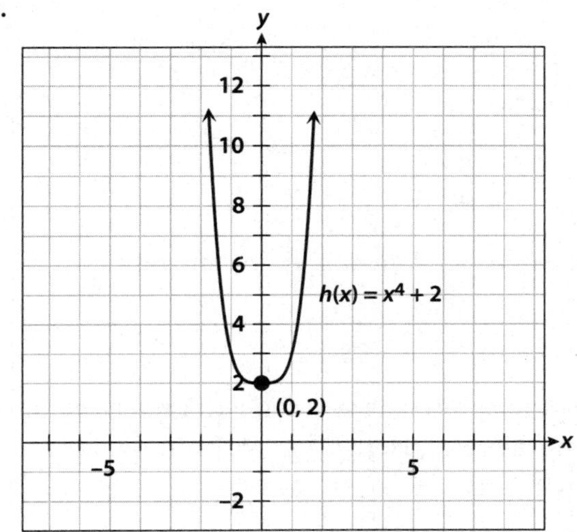

y
12
10
8
6
4
2
$h(x) = x^4 + 2$
(0, 2)
x
−5
5
−2

10.

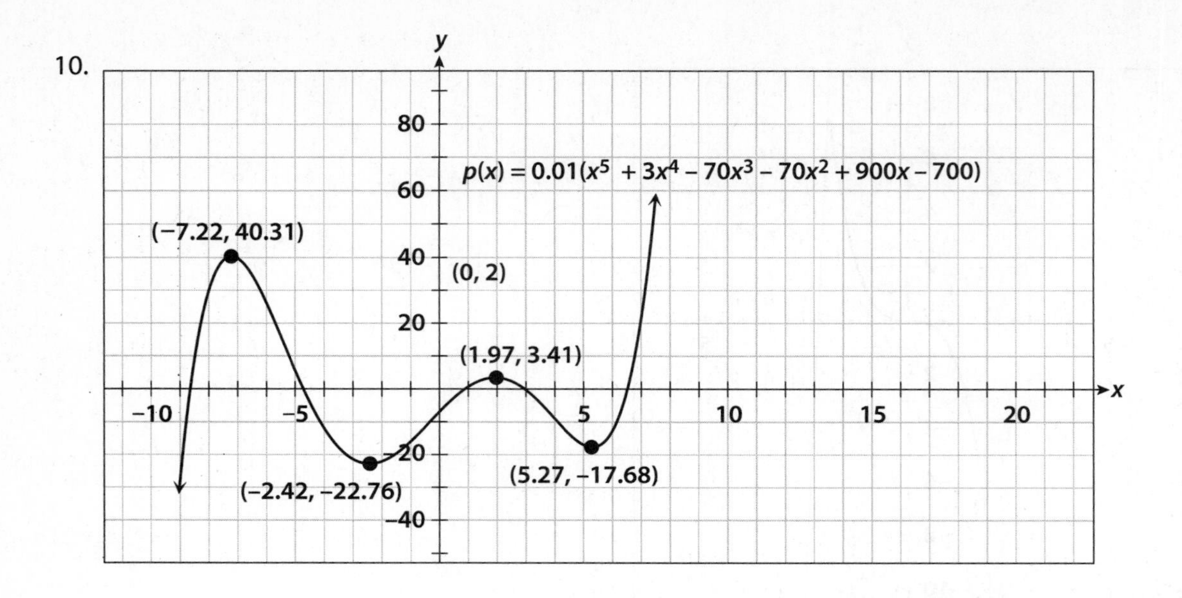

Remainder theorem and factor theorem

The zeros of p are the roots of the equation $p(x)=0$. You find the zeros of polynomials of degree 2 or less by solving linear or quadratic equations. For higher-degree polynomials, finding values for x that satisfy $p(x)=0$ is often difficult. A polynomial function of degree n has at most n real zeros. Two theorems that will assist you in finding these values are the remainder theorem and the factor theorem.

The **remainder theorem** states that if a polynomial $p(x)$ is divided by $x-a$, the remainder is $p(a)$. Here is an example:

PROBLEM Given $p(x)=2x^3+x^2-13x+6$, use the remainder theorem to find $p(4)$.

SOLUTION By the remainder theorem, $p(4)$ is the remainder when $p(x)$ is divided by $x-4$:

$$\begin{array}{r l}
 & 2x^2+9x+23 \\
x-4 & \overline{)\,2x^3+\;x^2-13x+6} \\
 & \underline{2x^3-8x^2} \\
 & \quad 9x^2-13x \\
 & \quad \underline{9x^2-36x} \\
 & \qquad 23x+\;6 \\
 & \qquad \underline{23x-92} \\
 & \qquad\quad \text{R } 98
\end{array}$$

The remainder (R) is 98, so $p(4)=98$. You can verify this result by evaluating $p(x)$ at 4:

$$p(4)=2\cdot 4^3+4^2-13\cdot 4+6=98$$

You can shorten the division process by using **synthetic division** (see Appendix A for an explanation of synthetic division). Here is the synthetic division version of $p(x)=2x^3+x^2-13x+6$ divided by $x-4$:

$$\begin{array}{r|rrrr l}
4 & 2 & 1 & -13 & 6 & \\
 & & 8 & 36 & 92 & \\
\hline
 & 2 & 9 & 23 & \boxed{98} & \text{Remainder}
\end{array}$$

Thus, the quotient is $2x^2+9x+23$ R 98.

The **factor theorem** states that $x-c$ is a factor of $p(x)$ if and only if $p(c)=0$. Thus, you can factor $p(x)$ by finding the zeros of p; and, conversely, you can determine the zeros of p by factoring $p(x)$.

PROBLEM Given the polynomial function p defined by $p(x)=(x+5)^2(x^2-7x+12)$, find the real zeros of p.

SOLUTION Factoring, you have $p(x)=(x+5)^2(x-3)(x-4)$. Because the factor $(x+5)$ occurs twice, its corresponding zero, -5, is a zero of multiplicity 2. Thus, the zeros of p are -5 (with multiplicity 2), 3, and 4.

Note: In general, a zero r of a polynomial function p has **multiplicity k** if $(x-r)^k$ is a factor of $p(x)$ and $(x-r)^{k+1}$ is not a factor of $p(x)$.

PROBLEM The zeros of a polynomial function p of degree 5 are $-4, \pm 3$, and $\pm\sqrt{5}$. If $p(x)$ has leading coefficient $a \neq 0$, express $p(x)$ in factored form.

SOLUTION Given that $-4, \pm 3$, and $\pm\sqrt{5}$ are zeros, you know that $p(-4)=0$, $p(3)=0$, $p(-3)=0$, $p(\sqrt{5})=0$, and $p(-\sqrt{5})=0$. Thus, by the factor theorem,

$$(x-(-4)), (x-3), (x-(-3)), (x-\sqrt{5}), \text{ and } \left(x-\left(-\sqrt{5}\right)\right) \text{ are factors of } p(x).$$

Because $p(x)$ has degree 5, there are no other factors. Hence:

$$p(x)=a(x+4)(x-3)(x+3)(x-\sqrt{5})(x-\sqrt{5})$$

For 1–5, fill in the blank to make a true statement.

1. The remainder theorem states that if a polynomial $p(x)$ is divided by $x-a$, the remainder is ________.
2. The factor theorem states that $x-c$ is a factor of $p(x)$ if and only if ________.
3. A zero r of a polynomial function p has multiplicity k if ________ is a factor of $p(x)$ and ________ is not a factor of $p(x)$.
4. A polynomial function of degree n has at most ________ real zeros.
5. A polynomial function of degree 5 has at most ________ real zeros.
6. Given $p(x)=2x^3+x^2-13x+6$, use the remainder theorem to find $p(2)$.
7. Given $p(x)=2x^3+x^2-13x+6$, use the remainder theorem to find $p(-2)$.
8. Find the real zeros of f defined by $f(x)=(x-4)(x^2-x-6)$.
9. The zeros of a polynomial function p of degree 4, with leading nonzero coefficient c, are 5, -2, and $\pm\sqrt{3}$. Express $p(x)$ in factored form.
10. Suppose $g(x)=2x^3-6x^2-2x+6$ has zeros ± 1 and 3. Express $g(x)$ in factored form.

Rational root theorem and Descartes's rule of signs

The **rational root theorem** states that if $p(x) = a_n x^n + a_{n-1}x^{n-1} + a_{n-2}x^{n-2} + \ldots + a_2 x^2 + a_1 x + a_0$ is a polynomial function with **integral** coefficients ($a_n \neq 0$ and $a_0 \neq 0$) and $\frac{p}{q}$ is a **rational zero** of $p(x)$ in simplified form, then p is a factor of a_0 and q is a factor of a_n.

You can use this theorem to obtain some possible rational roots of $p(x)$ that you then can test by using the remainder or factor theorem. Here is an example:

PROBLEM Find the real zeros for p defined by $p(x) = 2x^3 + x^2 - 13x + 6$.

SOLUTION Given that $p(x)$ has degree 3, it has at most three zeros. Possible numerators of a rational zero are factors of 6: $\pm 1, \pm 2, \pm 3$, and ± 6. Possible denominators are factors of 2: ± 1 and ± 2. Thus, possible rational zeros of $p(x)$ are $\pm 1, \pm 2, \pm 3$, $\pm 6, \pm\frac{1}{2}$, and $\pm\frac{3}{2}$.

Use the factor theorem to test ± 1:

$p(1) = 2\cdot 1^3 + 1^2 - 13\cdot 1 + 6 = -4 \neq 0$, so 1 is not a zero.

$p(-1) = 2\cdot(-1)^3 + (-1)^2 - 13\cdot(-1) + 6 = 18 \neq 0$, so -1 is not a zero.

Use the remainder theorem to test the other possible zeros:

$$\text{Test } x = 2: \quad \begin{array}{r|rrrr} 2 & 2 & 1 & -13 & 6 \\ & & 4 & 10 & -6 \\ \hline & 2 & 5 & -3 & 0 \end{array}$$

0; so 2 is a zero of p.

Hence, by the factor theorem, $(x-2)$ is a factor of $p(x)$. Using the coefficients from the synthetic division, you have $p(x) = 2x^3 + x^2 - 13x + 6 = (x-2)(2x^2 + 5x - 3)$. Factoring completely gives $p(x) = (x-2)(2x-1)(x+3)$. Thus, the zeros of p are 2, $\frac{1}{2}$, and -3.

When the terms of a polynomial $p(x)$ are written in descending (or ascending) order of powers of x, a **variation of sign** occurs if the coefficients of two consecutive terms have opposite signs (with missing terms being ignored). Here are examples:

$p(x) = 2x^3 + x^2 - 13x + 6$ has two variations in sign.

$g(x) = 3x^5 - x^4 + 5x^3 - 2x^2 + 4x + 10$ has four variations in sign.

If $p(x) = 0$ is a polynomial equation with real coefficients, **Descartes's rule of signs** states:

1. The maximum number of positive real roots of $p(x) = 0$ either equals the number of variations in sign of $p(x)$ or differs from it by an even number.
2. The maximum number of negative real roots of $p(x) = 0$ either equals the number of variations in sign of $p(-x)$ or differs from it by an even number.

PROBLEM Discuss the positive or negative nature of the roots of $p(x) = 2x^3 + x^2 - 13x + 6$.

SOLUTION The polynomial $p(x) = 2x^3 + x^2 - 13x + 6$ has two variations in sign. Thus, according to Descartes's rule of signs, $p(x)$ has either two or zero positive real roots;

$p(-x) = -2x^3 + x^2 + 13x + 6$ has one variation in sign. Thus, according to Descartes's rule of signs, $p(x)$ has one negative real root.

EXERCISE

9·4

For 1–5, fill in the blank to make a true statement.

1. The rational root theorem applies only to polynomials with __________ coefficients.
2. The number of __________ real roots of $p(x) = 0$ either equals the number of variations in sign of $p(x)$ or differs from it by an even number.
3. The number of negative real roots of $p(x) = 0$ either equals the number of variations in sign of __________ or differs from it by an even number.
4. The polynomial $p(x) = 2x^6 - 4x^5 + 2x^2 - x + 4$ has __________ variations in sign.
5. The polynomial $p(x) = 3x^5 - 4x^4 - x^2 + 5x - 1$ has __________ variations in sign.

For 6–10, discuss the positive or negative nature of the roots of the given polynomial equation.

6. $6x^4 + 7x^3 - 9x^2 - 7x + 3 = 0$
7. $2x^2 + 3x - 4 = 0$
8. $x^3 - 1 = 0$
9. $2x^3 - 3x^2 - 11x + 6 = 0$
10. $x^5 + 4x^4 - 4x^3 - 16x^2 + 3x + 12 = 0$

For 11–15, find the real zeros for the function defined by the given polynomial function.

11. $p(x) = 6x^4 + 7x^3 - 9x^2 - 7x + 3$
12. $f(x) = 2x^2 + 3x - 2$
13. $f(x) = x^3 - 1$
14. $g(x) = 2x^3 - 3x^2 - 11x + 6$
15. $f(x) = x^5 + 4x^4 - 4x^3 - 16x^2 + 3x + 12$

Fundamental theorem of algebra

Recall that the zeros of a polynomial function p defined by $p(x)$ are the roots of the equation $p(x) = 0$. Thus far, you have considered only real roots of $p(x) = 0$; however, some polynomial equations have zeros that are not real numbers. Here is an example:

PROBLEM Find the roots of the polynomial equation $x^2 + 4 = 0$.

SOLUTION $x^2 + 4 = 0$

$x^2 = -4$ Isolate the variable term.

$x = \pm\sqrt{-4} = \pm 2i$ Take the square root of both sides.

Hence, $p(x)$ has two roots, $2i$ and $-2i$, neither of which is a real number.

Thus, extending the discussion of polynomial functions to the venue of all complex numbers is a logical next step.

The **fundamental theorem of algebra** states that, over the complex numbers, every polynomial function of degree $n \geq 1$ has at least one root. Using this theorem, you can show that if you allow complex roots and count a root again each time it occurs more than once, every polynomial function of degree n has exactly n roots. Thus, every linear function has exactly one root, every quadratic function has exactly two roots, and so on.

Note: Keep in mind that the n roots of an nth-degree polynomial are not necessarily distinct. For instance, 5 is a root of multiplicity 2 for the second-degree quadratic function equation $f(x) = x^2 - 10x + 25$.

PROBLEM Find the zeros of the polynomial function f defined by $f(x) = x^3 - 1$.

SOLUTION The zeros of f are the roots of $x^3 - 1 = 0$. Since $f(x)$ has degree 3, the equation $x^3 - 1 = 0$ has exactly three roots. You find the roots as follows:

$$x^3 - 1 = 0$$

$$(x-1)(x^2 + x + 1) = 0 \quad \text{Factor.}$$

$$x - 1 = 0 \text{ and } (x^2 + x + 1) = 0 \quad \text{Set each factor equal to zero.}$$

$$x = 1 \text{ and } x = \frac{-1 \pm i\sqrt{3}}{2} \quad \text{Solve.}$$

Hence, 1, $-\frac{1}{2} + \frac{\sqrt{3}}{2}i$, and $-\frac{1}{2} - \frac{\sqrt{3}}{2}i$ are the zeros of f.

Note: Remember to use the $a + bi$ form.

In this problem, even though f has three zeros, it has only *one* real zero, namely, 1, so its graph will intersect the x-axis only once—at $x = 1$ (see Figure 9.4).

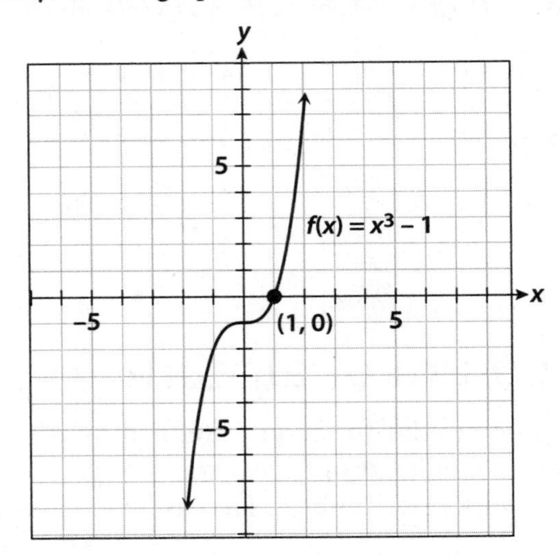

Figure 9.4 Graph of $f(x) = x^3 - 1$

The fundamental theorem of algebra guarantees that for every polynomial $p(x)$ of degree $n \geq 1$ there exist complex roots $r_1, r_2, \ldots, r_n$, so that you can factor $p(x)$ completely as $p(x) = a_n(x - r_1)(x - r_2)\ldots(x - r_n)$, where a_n is the leading coefficient of $p(x)$. Here is an example.

For the polynomial $f(x) = x^3 - 1$, there exist three roots, 1, $-\frac{1}{2} + \frac{\sqrt{3}}{2}i$, and $-\frac{1}{2} - \frac{\sqrt{3}}{2}i$, so that you can factor $f(x)$ completely as

$$f(x) = x^3 - 1 = (x-1)\left[x - \left(-\frac{1}{2} + \frac{\sqrt{3}}{2}i\right)\right]\left[x - \left(-\frac{1}{2} - \frac{\sqrt{3}}{2}i\right)\right]$$

Suppose $p(x)$ is a polynomial function with *real* coefficients. If $a+bi\ (b\neq 0)$ is a complex root of $p(x)$, then its **complex conjugate** a-bi is also a root of $p(x)$. In other words, nonreal roots of a polynomial with real coefficients come in pairs. See the section "Multiplication of complex numbers and complex conjugates" in Chapter 2, "Complex numbers," for a discussion of complex conjugates.

PROBLEM Find a polynomial $p(x)$ with real coefficients and with the least degree that has 3 and $2-i$ as roots.

SOLUTION The polynomial has real coefficients, so besides 3 and $2-i$, you have $2+i$ as a root. Each root corresponds to a factor of the polynomial, yielding:

$$p(x)=(x-3)[x-(2-i)][x-(2+i)]$$

$$p(x)=(x-3)[(x-2)+i][(x-2)-i] \quad \text{Regroup.}$$

$$p(x)=(x-3)\left[(x-2)^2-i^2\right] \quad \text{Multiply last two factors.}$$

$$p(x)=x^3-7x^2+17x-15 \quad \text{Simplify.}$$

This polynomial has the required roots and the least degree.

For 1–5, fill in the blank(s) to make a true statement.

1. The fundamental theorem of algebra states that, over the __________ numbers, every polynomial of degree $n\geq 1$ has at least __________ root(s).
2. The fundamental theorem of algebra guarantees that every polynomial of degree $n\geq 1$ has exactly __________ root(s).
3. Every quadratic equation has exactly __________ root(s).
4. Every polynomial equation of degree 100 has exactly __________ root(s).
5. If $p(x)$ is a polynomial with real coefficients and $5+3i$ is a root of $p(x)=0$, then __________ is also a root of $p(x)=0$.

For 6–10, (a) find all the zeros of p *and (b) factor* p(x) *completely.*

6. $p(x)=(x+4)(x^2-5)(x^2-36)$
7. $p(x)=x^6-1$
8. $p(x)=-2x^2+3x-1$
9. $p(x)=(3x+5)(x^3+8)(x^3-8)$
10. $p(x)$ is the simplest polynomial that has real coefficients and degree 5; p has zeros $\frac{1}{2}$, $1-i$, and $1+2i$.

Behavior of polynomial functions near $\pm\infty$

To determine the behavior of a polynomial function defined by the equation $p(x) = a_n x^n + a_{n-1}x^{n-1} + a_{n-2}x^{n-2} + \ldots + a_2x^2 + a_1x + a_0$ as x approaches ∞ or $-\infty$, factor out $a_n x^n$, the term with highest degree; then $p(x)$ behaves as $a_n x^n$ does.

PROBLEM Suppose $p(x) = 3x^4 - 50{,}000x^3 + 20x^2 - 5x + 17$. (a) As x approaches ∞, is $p(x)$ positive or negative? (b) As x approaches $-\infty$, is $p(x)$ positive or negative?

SOLUTION Factoring out the term with highest degree, you have

$$p(x) = 3x^4\left(1 - \frac{50{,}000}{3x} + \frac{20}{3x^2} - \frac{5}{3x^3} + \frac{17}{3x^4}\right)$$

As x approaches ∞ or $-\infty$, the quantity in parentheses above approaches 1. Thus, $p(x)$ behaves as $3x^4$ does.

a. As x approaches ∞, $x > 0$, so $3x^4 > 0$; therefore, $p(x)$ is positive.

b. As x approaches $-\infty$, $x < 0$, so $3x^4 > 0$; therefore, $p(x)$ is positive.

Note: Determining the behavior of functions as x nears $\pm\infty$ is a skill used in calculus.

EXERCISE 9·6

For the given polynomial: (a) As x approaches ∞, is the polynomial positive or negative? (b) As x approaches $-\infty$, is the polynomial positive or negative?

1. $p(x) = 6x^4 + 7x^3 - 9x^2 - 7x + 3$
2. $f(x) = 2x^2 + 3x - 4$
3. $f(x) = 8x^3 + 9999x^2$
4. $g(x) = 2x^3 - 3x^2 - 11x + 6$
5. $f(x) = x^5 + 4x^4 - 4x^3 - 16x^2 + 3x + 12$

Rational functions

·10·

Definition of rational functions

Rational functions are defined by equations of the form $f(x)=\frac{p(x)}{q(x)}$, where $p(x)$ and $q(x)$ are polynomials, provided that $q(x)$ is not the zero polynomial. The domain is the set $\{x|x\in R \text{ for which } q(x)\neq 0\}$. The range is a subset of R. The zeros occur at x values for which $p(x) = 0$, when $f(x)$ is in simplified form. The y-intercept is $f(0)$; and when $f(x)=\frac{p(x)}{q(x)}$ is in simplified form (that is, when the numerator and denominator polynomials have no common factors), the x-intercepts occur at values for which $p(x) = 0$.

PROBLEM Given the rational function f defined by $f(x) = \frac{2x-1}{x^2-1}$ (see Figure 10.1), (a) state the domain, (b) find the real zeros, and (c) find the intercepts.

SOLUTION a. The denominator, $x^2 - 1$, is zero when $x = \pm 1$, so the domain is $\{x|x\neq 1, x\neq -1\}$.

b. The numerator, $2x - 1$, is zero when $x=\frac{1}{2}$, so $\frac{1}{2}$ is a real zero of f.

c. The numerator, $2x - 1$, is zero when $x=\frac{1}{2}$, so there is one x-intercept at $\frac{1}{2}$; the y-intercept is $f(0)=\frac{2\cdot 0-1}{0^2-1}=1$.

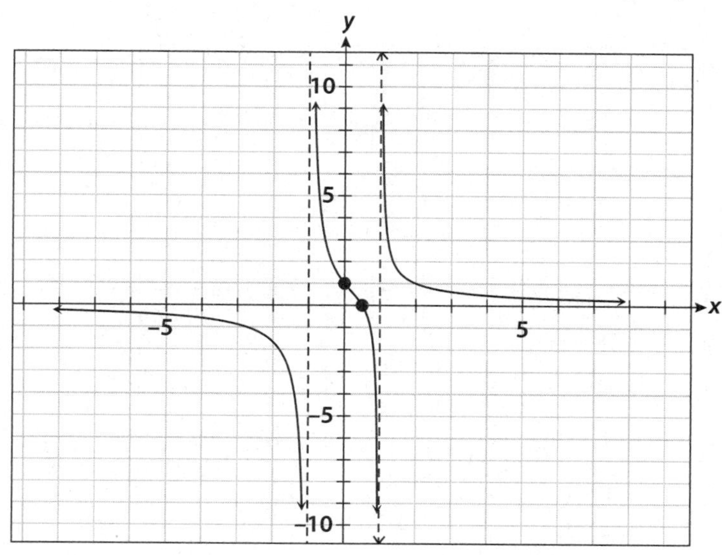

Figure 10.1 Graph of $f(x)=\frac{2x-1}{x^2-1}$

PROBLEM Given the rational function f defined by $f(x) = \frac{x+1.4}{x^2+1}$ (see Figure 10.2), (a) state the domain, (b) find the real zeros, and (c) find the intercepts.

SOLUTION
a. The denominator, $x^2 + 1$, is never zero over the real numbers, so the domain is R.

b. The numerator, $x + 1.4$, is zero when $x = -1.4$, so -1.4 is a real zero of f.

c. The numerator, $x + 1.4$, is zero when $x = -1.4$, so there is one x-intercept at -1.4; the y-intercept is $f(0) = \frac{0+1.4}{0^2+1} = 1.4$.

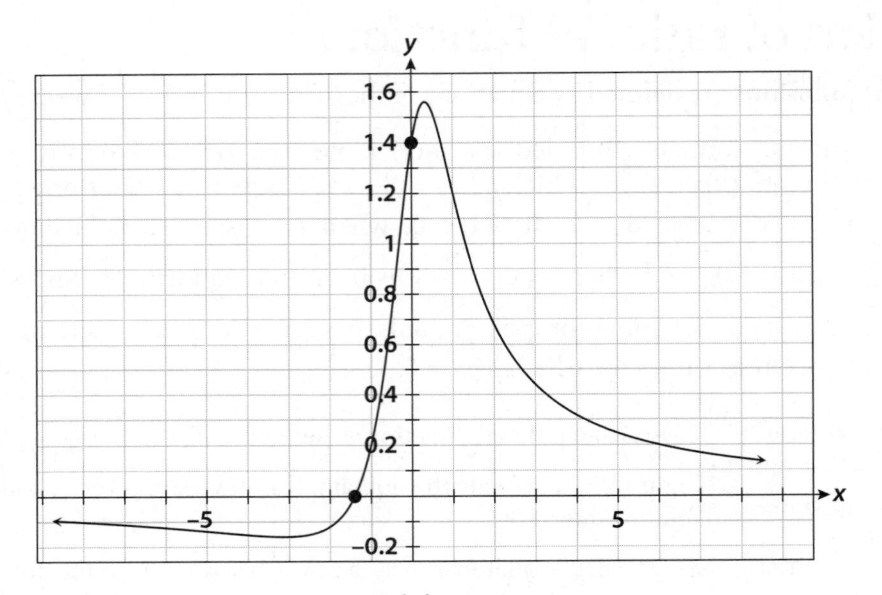

Figure 10.2 Graph of $f(x) = \frac{x+1.4}{x^2+1}$

PROBLEM Given the rational function f defined by $f(x) = \frac{x+1}{x^2-1}$ (see Figure 10.3), (a) state the domain, (b) find the real zeros, and (c) find the intercepts.

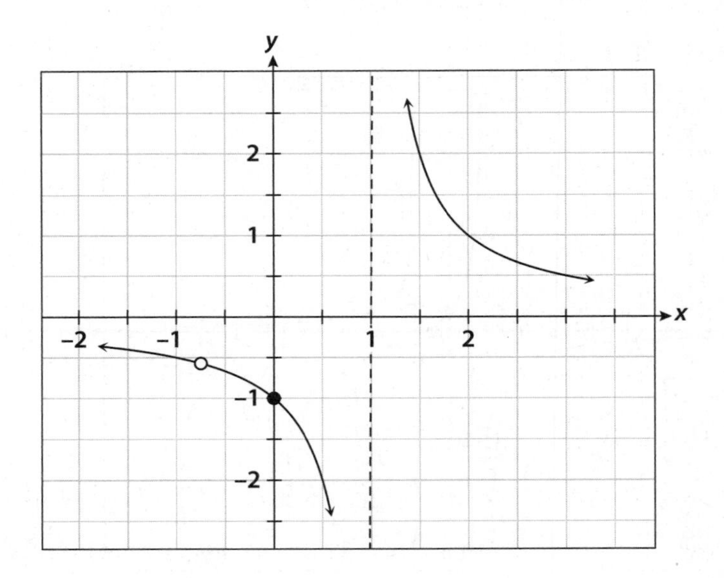

Figure 10.3 Graph of $f(x) = \frac{x+1}{x^2-1}$

SOLUTION

a. The denominator, $x^2 - 1$, is zero when $x = \pm 1$; so the domain is $\{x | x \neq 1, x \neq -1\}$.

b. The numerator, $x + 1$, is zero when $x = -1$, but -1 is *not* a real zero of f because -1 is not in the domain of f.

c. In simplified form, $f(x) = \frac{x+1}{(x+1)(x-1)} = \frac{1}{x-1}$; the numerator, 1, of the simplified form is never zero, so there are no x-intercepts. The y-intercept is $f(0) = \frac{0+1}{0^2-1} = -1$.

Note: Always simplify before determining x-intercepts of a rational function.

EXERCISE

10·1

For 1–5, fill in the blank(s) to make a true statement.

1. Rational functions are defined by equations of the form __________, where $p(x)$ and $q(x)$ are __________, provided that __________ is not the zero polynomial.
2. The domain of a rational function excludes any values for which the denominator polynomial is __________.
3. When the numerator and denominator polynomials have no common factors, the zeros of a rational function occur at domain values for which the __________ (denominator, numerator) polynomial is zero.
4. The y-intercept of a rational function f is __________.
5. When the numerator and denominator polynomials have no common factors, the x-intercepts occur at values for which the __________ (denominator, numerator) polynomial is zero.

For 6–10, for the given rational function (a) state the domain, (b) find the real zeros, and (c) find the intercepts.

6. $f(x) = \frac{1}{x}$
7. $g(x) = \frac{x^2 - 4}{x^2 + x - 12}$
8. $h(x) = \frac{x^2 - 4}{x^2 - x - 6}$
9. $f(x) = \frac{3x^3 - 24}{2x^2 + 4x + 8}$
10. $h(x) = \frac{x^4 - 81}{x^2 + x - 12}$

Graphs of rational functions

To graph a rational function defined by $f(x) = \frac{p(x)}{q(x)}$, first factor $p(x)$ and $q(x)$ to identify possible "holes" in the graph of f. Upon inspection, if $p(x)$ and $q(x)$ have a common factor, $(x - s)$, that will divide out completely from the denominator when $f(x)$ is simplified, then the graph will have a hole at $(s, f(s))$, where $f(s)$ is calculated after $f(x)$ is simplified. Following is an example.

Suppose $f(x)=\dfrac{x+1}{x^2-1}=\dfrac{x+1}{(x+1)(x-1)}$. By inspection, you can see that the common factor, $(x+1)$, will divide out completely from the denominator when $f(x)$ is simplified to $f(x)=\dfrac{1}{x-1}$. Therefore, the graph of f has a hole at $\left(-1,-\dfrac{1}{2}\right)$. See Figure 10.4.

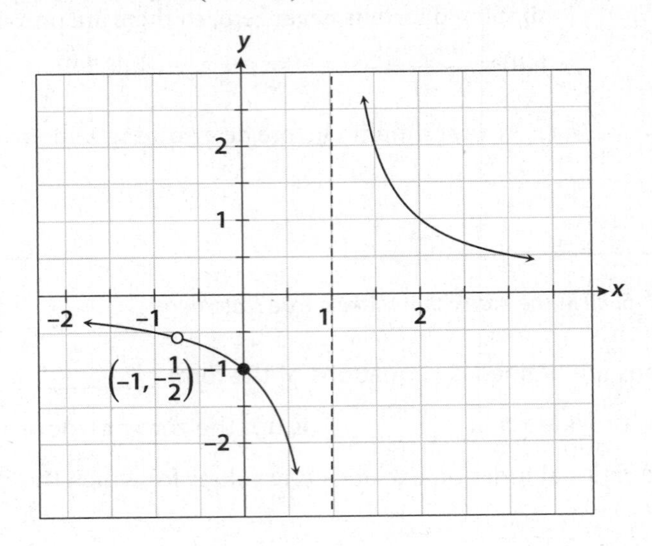

Figure 10.4 Graph of $f(x)=\dfrac{x+1}{x^2-1}$

After identifying any holes, next you use the simplified form of $f(x)$ to determine asymptotes of the graph (see the section "Vertical and horizontal asymptotes" in Chapter 5, "Graphs of functions," for a discussion of asymptotes). Assuming $f(x)=\dfrac{p(x)}{q(x)}$ is in simplified form and $q(x)$ has degree greater than 1, the graph of f might have one or more vertical asymptotes, a horizontal asymptote, or an oblique asymptote (explained below).

Vertical asymptotes: Vertical asymptotes occur at values of x for which $q(x)=0$. For instance, $f(x)=\dfrac{1}{x-1}$ has a vertical asymptote at $x=1$.

Horizontal asymptote: A rational function will have at most one horizontal asymptote. The following guidelines will help you identify a horizontal asymptote:

- If the degree of $p(x)$ is less than the degree of $q(x)$, then the x-axis is a horizontal asymptote. For example, the x-axis is a horizontal asymptote of $f(x)=\dfrac{1}{x-1}$.
- If the degree of $p(x)$ equals the degree of $q(x)$, then the graph will have a horizontal asymptote at $y_n=\dfrac{a_n}{b_n}$, where a_n is the leading coefficient of $p(x)$ and b_n is the leading coefficient of $q(x)$. For example, $y=1$ is a horizontal asymptote of $g(x)=\dfrac{x^2-4}{x^2+x-12}$.
- If the degree of $p(x)$ exceeds the degree of $q(x)$ by more than 1, the graph will *not* have a horizontal asymptote. For example, $g(x)=\dfrac{x^4-64}{x^2-2x-15}$ has no horizontal asymptote.

Oblique asymptote: A rational function will have at most one oblique asymptote. If the degree of $p(x)$ exceeds the degree of $q(x)$ by *exactly* 1, the graph will have an **oblique asymptote**. To find the equation of the oblique asymptote, use division to rewrite $f(x)=\dfrac{p(x)}{q(x)}$ as quotient plus

$\frac{\text{remainder}}{q(x)}$. The line with equation y = quotient is an oblique asymptote. For example, suppose $f(x)=\frac{x^2+3}{x-1}=x+1+\frac{4}{x-1}$; then $y = x + 1$ is an oblique asymptote of the graph of f.

PROBLEM Given the rational function f defined by $f(x) = \frac{x+1}{x^2-1}$ (see Figure 10.5), (a) state the domain, (b) determine holes, (c) determine asymptotes, and (d) find the intercepts.

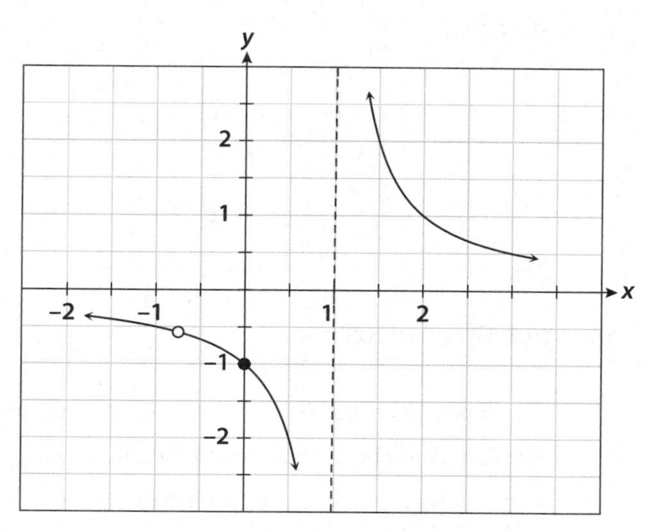

Figure 10.5 Graph of $f(x) = \frac{x+1}{x^2-1}$

SOLUTION

a. The denominator, $x^2 - 1$, is zero when $x=\pm 1$, so the domain is $\{x|x \neq 1, x \neq -1\}$.
b. In factored form, $f(x)=\frac{x+1}{x^2-1}=\frac{x+1}{(x+1)(x-1)}$. By inspection, you can see that the common factor, $x + 1$, will divide out completely from the denominator when $f(x)$ is simplified to $f(x)=\frac{1}{x-1}$. Therefore, the graph of f has a hole at $\left(-1,-\frac{1}{2}\right)$ because $x=-1$ is excluded from the domain of f.
c. *Vertical asymptotes*: The denominator of the simplified form $f(x)=\frac{1}{x-1}$ is zero when $x = 1$, so $x = 1$ is a vertical asymptote.

   Horizontal asymptote: In the simplified form $f(x)=\frac{1}{x-1}$, the degree of the numerator is less than the degree of the denominator; so the x-axis is a horizontal asymptote.

   Oblique asymptote: There is none.
d. The numerator, 1, of the simplified form is never zero, so there are no x-intercepts. The y-intercept is $f(0) = \frac{1}{0-1}=-1$.

PROBLEM Given the rational function f defined by $f(x) = \frac{2x-1}{x^2-1}$ (see Figure 10.6, on the next page), (a) state the domain, (b) determine holes, (c) determine asymptotes, and (d) find the intercepts.

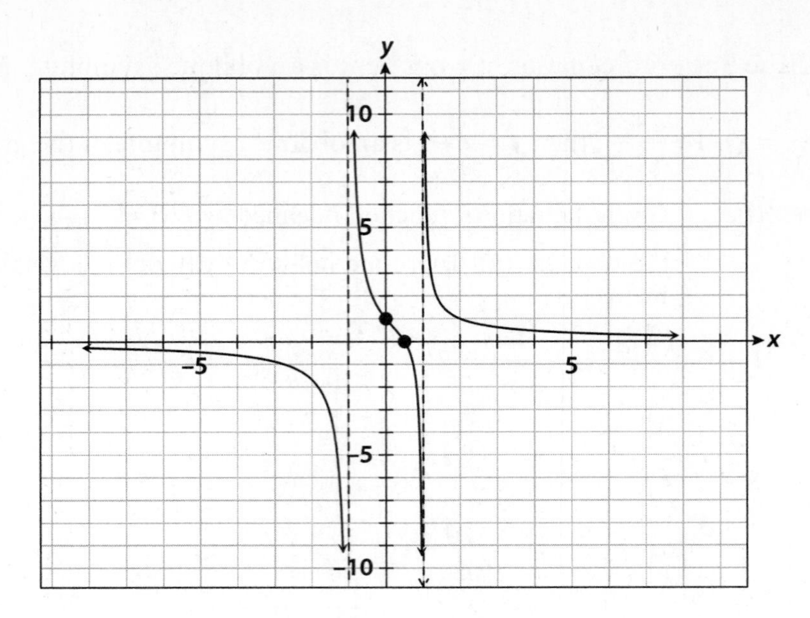

Figure 10.6 Graph of $f(x) = \dfrac{2x-1}{x^2-1}$

SOLUTION

a. The denominator, $x^2 - 1$, is zero when $x = \pm 1$, so the domain is $\{x | x \neq 1, x \neq -1\}$.

b. The function $f(x)$ is in simplified form, so the graph has no holes.

c. *Vertical asymptotes*: The denominator, $x^2 - 1$, is zero when $x = \pm 1$; so $x = 1$ and $x = -1$ are vertical asymptotes.

   Horizontal asymptote: The degree of the numerator is less than the degree of the denominator; so the x-axis is a horizontal asymptote.

   Oblique asymptote: There is none.

d. The numerator is zero when $x = \frac{1}{2}$, so the x-intercept is $\frac{1}{2}$; the y-intercept is $f(0) = \dfrac{2 \cdot 0 - 1}{0^2 - 1} = 1.$

PROBLEM

Given the rational function f defined by $f(x) = \dfrac{x^2+3}{x-1}$ (see Figure 10.7), (a) state the domain, (b) determine holes, (c) determine asymptotes, and (d) find the intercepts.

SOLUTION

a. The denominator, $x - 1$, is zero when $x = 1$, so the domain is $\{x | x \neq 1\}$.

b. The function $f(x)$ is in simplified form, so the graph has no holes.

c. *Vertical asymptotes*: The denominator, $x - 1$, is zero when $x = 1$, so $x = 1$ is a vertical asymptote.

   Horizontal asymptote: There is none.

   Oblique asymptote: $f(x) = \dfrac{x^2+3}{x-1} = x + 1 + \dfrac{4}{x-1}$, so $y = x + 1$ is an oblique asymptote.

d. The numerator, $x^2 + 3$, is never zero, so there are no x-intercepts. The y-intercept is $f(0) = \dfrac{0+3}{0-1} = -3$.

Note: You might want to use a graphing calculator to explore rational functions. Enter the function, using parentheses around both the numerator and denominator polynomials. Use trial and error and the Zoom feature to find a good viewing window; otherwise, you might be misled by the graph displayed.

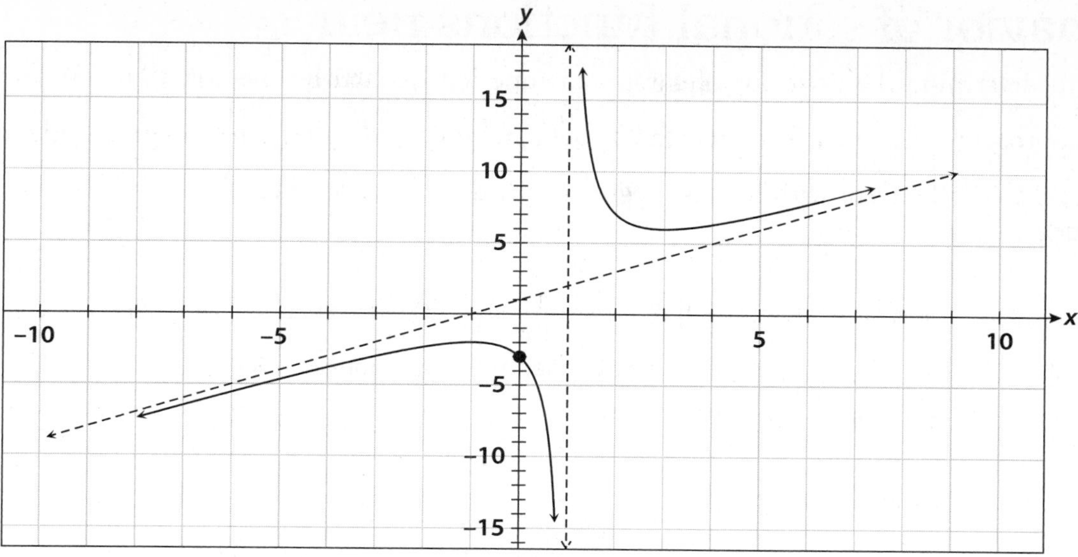

Figure 10.7 Graph of $f(x)=\dfrac{x^2+3}{x-1}$

EXERCISE
10·2

For 1–5, fill in the blank(s) to make a true statement.

1. Suppose $f(x)=\dfrac{x-2}{x^2-4}$. Will its graph have a hole at $x=2$? __________ (Yes, No). Will its graph have an asymptote at $x=2$? __________ (Yes, No).
2. Suppose $f(x)=\dfrac{(x-2)^2}{x^2-4}$. Will its graph have a hole at $x=2$? __________ (Yes, No). Will its graph have an asymptote at $x=2$? __________ (Yes, No).
3. Suppose $f(x)=\dfrac{x-2}{(x-2)(x^2-4)}$. Will its graph have a hole at $x=2$? __________ (Yes, No). Will its graph have an asymptote at $x=2$? __________ (Yes, No).
4. Suppose $f(x)=\dfrac{p(x)}{q(x)}$. If the degree of $p(x)$ is __________ (greater than, less than) the degree of $q(x)$, then the x-axis is a horizontal asymptote. If the degree of $p(x)$ equals the degree of $q(x)$, then the graph will have a horizontal asymptote at __________, where a_n is the leading coefficient of $p(x)$ and b_n is the leading coefficient of $q(x)$.
5. For a rational function, if the degree of the numerator polynomial exceeds the degree of the numerator polynomial by exactly 1, the graph of the function will have a(n) __________ asymptote.

In 6–10, for the given the rational function equation, (a) state the domain, (b) determine holes, (c) determine asymptotes, and (d) find the intercepts.

6. $f(x)=\dfrac{1}{x}$
7. $g(x)=\dfrac{x^2-4}{x^2+x-12}$
8. $h(x)=\dfrac{x^2-4}{x^2-x-6}$
9. $f(x)=\dfrac{3x^3-24}{2x^2+4x+8}$
10. $h(x)=\dfrac{x^4-81}{x^2+x-12}$

Behavior of rational functions near $\pm\infty$

To determine the behavior of a rational function defined by the equation $f(x) = \frac{p(x)}{q(x)}$ as x approaches ∞ or $-\infty$, separately factor out $a_n x^n$, the term with highest degree in the numerator, and $b_m x^m$, the term with highest degree in the denominator; then $f(x) = \frac{p(x)}{q(x)}$ behaves as $\frac{a_n x^n}{b_m x^m}$ does.

PROBLEM Suppose $f(x) = \frac{x^5 + 6x^3 - 7}{5x^6 + 6x^2 - 11}$. (a) Discuss the behavior of $f(x)$ as x approaches ∞. (b) Discuss the behavior of $f(x)$ as x approaches $-\infty$.

SOLUTION Factoring out the terms with highest degree, you have:

$$f(x) = \frac{x^5 + 6x^3 - 7}{5x^6 + 6x^2 - 11} = \frac{x^5\left(1 + \frac{6}{x^2} - \frac{7}{x^5}\right)}{5x^6\left(1 + \frac{6}{5x^4} - \frac{11}{5x^6}\right)}$$

As x approaches ∞, the quantities in parentheses above approach 1. Thus, $f(x)$ behaves as $\frac{x^5}{5x^6} = \frac{1}{5x}$ does.

a. As x approaches ∞, $x > 0$, so $\frac{1}{5x}$ is positive, but very close to zero.

b. As x approaches ∞, $x < 0$, so $\frac{1}{5x}$ is negative, but very close to zero.

PROBLEM Suppose $f(x) = \frac{7x^4 + 6x^2 - 3x}{-3x^3 - 7x + 5}$. (a) Discuss the behavior of $f(x)$ as x approaches ∞. (b) Discuss the behavior of $f(x)$ as x approaches $-\infty$.

SOLUTION Factoring out the terms with highest degree, you have:

$$f(x) = \frac{7x^4 + 6x^2 - 3x}{-3x^3 - 7x + 5} = \frac{7x^4\left(1 + \frac{6}{7x^2} - \frac{3}{7x^3}\right)}{-3x^3\left(1 + \frac{7}{3x^2} - \frac{5}{3x^3}\right)}$$

As x approaches ∞ or $-\infty$, the quantities in parentheses above approach 1. Thus, $f(x)$ behaves as $\frac{7x^4}{-3x^3} = -\frac{7x}{3}$ does.

a. As x approaches ∞, x > 0, so $-\frac{7x}{3}$ approaches $-\infty$.

b. As x approaches $-\infty$, $x < 0$, so $-\frac{7x}{3}$ approaches ∞.

EXERCISE

10·3

For the given rational function equation, (a) discuss the behavior of f(x) *as* x *approaches* ∞ *and (b) discuss the behavior of* f(x) *as* x *approaches* −∞.

1. $f(x) = \dfrac{x-4}{x^2-5x+6}$
2. $g(x) = \dfrac{x^2-4x+7}{x^2+x-12}$
3. $h(x) = \dfrac{2x^3+8x-5}{-5x^2+6}$
4. $f(x) = \dfrac{3x^7-5x^6+24}{2x^7+4x+8}$
5. $h(x) = \dfrac{8}{x-12}$

Exponential and logarithmic functions

Definition of exponential functions

Exponential functions are defined by equations of the form $f(x) = b^x$ $(b \neq 1, b > 0)$, where b, a constant, is the **base** of the exponential function. The domain is R, or $(-\infty, \infty)$, and the range is $(-0, \infty)$, which is to say that $b^x > 0$ for every real number x. The function is always greater than zero, so there are no zeros. The y-intercept is 1.

The **natural exponential function** is defined by $f(x) = e^x$, where the base e is the irrational number whose rational decimal approximation is 2.718281828 (to nine digits).

The **base-10 exponential function** is defined by $f(x) = 10^x$.

You evaluate exponential functions using your knowledge of exponents (see the section "Exponentiation" in Chapter 1, "Real numbers," for a review of exponents).

PROBLEM Suppose $f(x) = 2^x$, $g(x) = 1000^x$, and $h(x) = \left(\frac{1}{4}\right)^x$. Find the following function values:

a. $f(3)$

b. $f(-3)$

c. $g\left(\frac{2}{3}\right)$

d. $g\left(-\frac{2}{3}\right)$

e. $h\left(\frac{1}{2}\right)$

f. $h\left(-\frac{1}{2}\right)$

SOLUTION a. $f(3) = 2^3 = 8$

b. $f(-3) = 2^{-3} = \frac{1}{2^3} = \frac{1}{8}$

c. $g\left(\frac{2}{3}\right) = 1000^{\frac{2}{3}} = \left(\sqrt[3]{1000}\right)^2 = 10^2 = 100$

d. $g\left(-\frac{2}{3}\right) = 1000^{-\frac{2}{3}} = \frac{1}{1000^{\frac{2}{3}}} = \frac{1}{\left(\sqrt[3]{1000}\right)^2} = \frac{1}{10^2} = \frac{1}{100}$

e. $h\left(\frac{1}{2}\right)=\left(\frac{1}{4}\right)^{\frac{1}{2}}=\sqrt{\frac{1}{4}}=\frac{1}{2}$

f. $h\left(-\frac{1}{2}\right)=\left(\frac{1}{4}\right)^{-\frac{1}{2}}=\frac{1}{\sqrt{\frac{1}{4}}}=2$

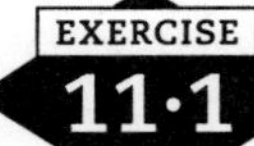

EXERCISE 11·1

For 1–5, fill in the blank(s) to make a true statement.

1. In the exponential function defined by equation $f(x)=b^x$, the two restrictions on the base b are __________ and __________.
2. The domain of an exponential function is __________, and the range is __________.
3. The graph of an exponential function does not cross the x-axis, so there are no real __________.
4. The y-intercept of an exponential function is __________.
5. The natural exponential function is defined by __________.

For 6–10, find the function value for the given exponential function.

6. $f(x)=3^x$; $f(4)$
7. $g(x)=64^x$; $g\left(\frac{4}{3}\right)$
8. $f(x)=\left(\frac{1}{2}\right)^x$; $f(-5)$
9. $h(x)=\left(\frac{4}{9}\right)^x$; $h\left(\frac{3}{2}\right)$
10. $g(x)=0.25^x$; $g\left(\frac{1}{2}\right)$

Graphs of exponential functions

The graph of $f(x)=b^x$ $(b\neq 1, b>0)$ is a smooth, continuous curve. The graph passes through the points (0, 1) and (1, b) and is located in the first and second quadrants only. The domain is R, and the range is $(-0, \infty)$. The y-intercept is 1. It has no x-intercepts. The x-axis is a horizontal asymptote.

Furthermore, for the function f defined by $f(x)=b^x$, the following hold:

- If $b>1$, the function is increasing. As x approaches ∞, $f(x)=b^x$ approaches ∞. As x approaches $-\infty$, $f(x)=b^x$ approaches 0 but never reaches 0.
- If $0<b<1$, the function is decreasing. As x approaches ∞, $f(x)=b^x$ approaches 0 but never reaches 0. As x approaches $-\infty$, $f(x)=b^x$ approaches ∞.

PROBLEM Given the exponential function f defined by $f(x)=2^x$ (see Figure 11.1, on the next page), (a) state the domain and range, (b) find zeros, (c) determine asymptotes, (d) find the intercepts, (e) discuss increasing and decreasing behavior, and (f) discuss behavior as x approaches $\pm\infty$.

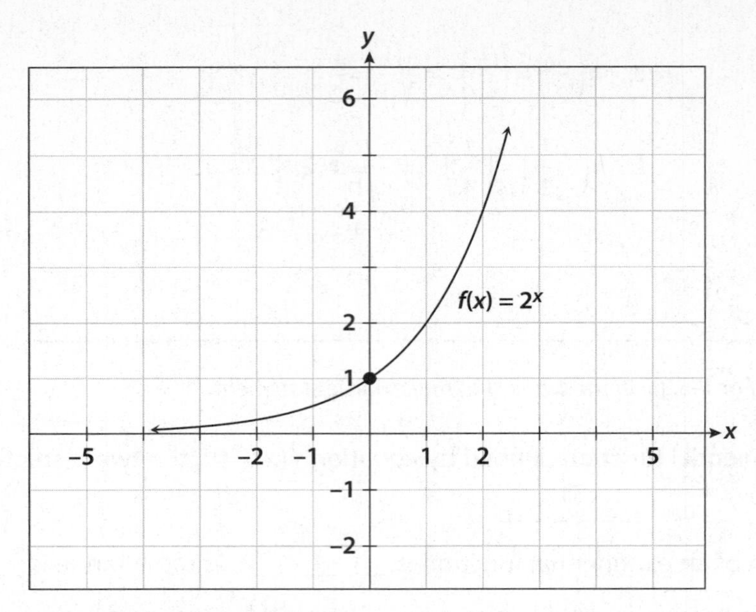

Figure 11.1 Graph of $f(x) = 2^x$

SOLUTION

a. The domain is R, and the range is $(0, \infty)$.

b. The function $f(x) = 2^x$ is always greater than 0, so it has no zeros.

c. The x-axis is a horizontal asymptote.

d. The y-intercept is 1. The graph has no x-intercepts.

e. $b = 2 > 1$; therefore, f is increasing on $(-\infty, \infty)$.

f. $b = 2 > 1$; therefore, as x approaches ∞, $f(x) = 2^x$ approaches ∞, and as x approaches $-\infty$, $f(x) = 2^x$ approaches 0 but never reaches 0.

PROBLEM

Given the exponential function f defined by $f(x) = \left(\frac{1}{2}\right)^x$ (see Figure 11.2), (a) state the domain and range, (b) find zeros, (c) determine asymptotes,

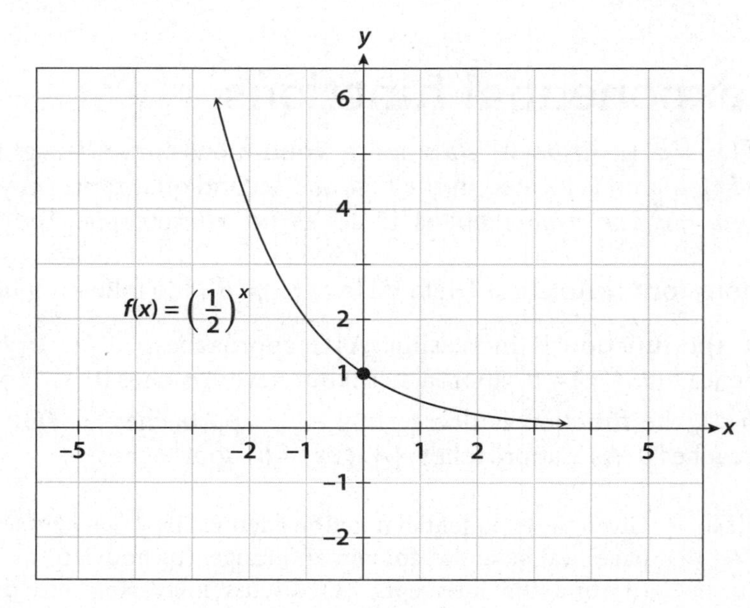

Figure 11.2 Graph of $f(x) = \left(\frac{1}{2}\right)^x$

(d) find the intercepts, (e) discuss increasing and decreasing behavior, and (f) discuss behavior as x approaches $\pm\infty$.

SOLUTION

a. The domain is R, and the range is $(0, \infty)$.

b. $f(x) = \left(\frac{1}{2}\right)^x$ is always greater than 0, so it has no zeros.

c. The x-axis is a horizontal asymptote.

d. The y-intercept is 1. The graph has no x-intercepts.

e. $0 < b = \frac{1}{2} < 1$; therefore, f is decreasing on $(-\infty, \infty)$.

f. $0 < b = \frac{1}{2} < 1$; therefore, as x approaches ∞, $f(x) = \left(\frac{1}{2}\right)^x$ approaches 0 but never reaches 0, and as x approaches $-\infty$, $f(x) = \left(\frac{1}{2}\right)^x$ approaches ∞.

For 1–5, fill in the blank(s) to make a true statement.

1. The graph of $f(x) = b^x$ is a smooth, __________ curve.
2. The graph of $f(x) = 10^x$ passes through the points __________ and __________.
3. The domain of f defined by $f(x) = \left(\frac{1}{3}\right)^x$ is __________, and the range is __________.
4. The y-intercept of f defined by $f(x) = e^x$ is __________.
5. The graph of f defined by $f(x) = 3^x$ has __________ x-intercepts.

For 6–10, for the given exponential function, (a) state the domain and range, (b) find zeros, (c) determine asymptotes, (d) find the intercepts, (e) discuss increasing and decreasing behavior, and (f) discuss behavior as x approaches $\pm\infty$.

6. $f(x) = 5^x$
7. $g(x) = 64^x$
8. $f(x) = \left(\frac{1}{2}\right)^x$
9. $h(x) = \left(\frac{4}{9}\right)^x$
10. $g(x) = 0.25^x$

Properties of exponential functions

Exponential functions have the following properties.

Properties of exponential functions

For the function f defined by $f(x) = b^x$ $(b \neq 1, b > 0)$:

$$f(x) = b^x > 0, \text{ for all real numbers}$$
$$f(0) = b^0 = 1$$
$$f(1) = b^1 = b$$

$$f(-x) = b^x = \frac{1}{b^x}$$

$$f(u) \cdot f(v) = b^u \cdot b^v = b^{u+v} = f(u+v)$$

$$\frac{f(u)}{f(v)} = \frac{b^u}{b^v} = b^{u-v} = f(u-v)$$

$$(f(x))^p = (b^x)^p = b^{xp} = f(xp)$$

One-to-one property: $f(u) = f(v)$ if and only if $u = v$; that is, $b^u = b^v$ if and only if $u = v$.

PROBLEM Suppose $f(x) = e^x$. Use the properties of exponential functions to find each of the following:

a. $f(0)$

b. $f(2) \cdot f(3)$

c. $f(-3)^4$

SOLUTION a. $f(0) = e^0 = 1$

b. $f(2) \cdot f(3) = e^2 \cdot e^3 = e^5$

c. $(f(-3))^4 = (e^{-3})^4 = e^{-12} = \frac{1}{e^{12}}$

Note: Do not confuse exponential functions with polynomial functions. The exponents in exponential functions are *variables*, whereas the exponents in polynomial functions are *constants*. For instance, $f(x) = 5^x$ defines an exponential function and $g(x) = x^5$ defines a polynomial function.

EXERCISE 11·3

Use the properties of exponential functions in each of the following.

1. Suppose $f(x) = e^x$. Find $\frac{f(7)}{f(2)}$.
2. Suppose $f(x) = 2^x$. Find $f(0) \cdot f(1)$.
3. Suppose $f(x) = 10^x$. Find $f(-x)$.
4. Suppose $f(x) = \left(\frac{4}{9}\right)^x$. Find (−1).
5. Suppose $f(x) = e^x$. If $e^u = e^4$, then $u =$ ________.

Definition of logarithmic functions

Logarithmic functions are defined by equations of the form $f(x) = \log_b x$, where $\log_b x = y$ if and only if $x = b^y$ ($b \neq 1$, $b > 0$). The constant b is the **base** of the logarithmic function. The domain is $(0, \infty)$, and the range is R. The function has one zero at $x = 1$. The graph of the function does not cross the y-axis, so it does not have a y-intercept.

The **natural logarithmic function** is defined by $\boldsymbol{f(x) = \log_{10} x}$. This function is denoted $\boldsymbol{f(x) = \ln x}$.

The **common logarithmic function** is defined by $\boldsymbol{f(x) = \log_{10} x}$. This function is sometimes represented as $f(x) = \log x$; however, avoid this notation when confusion might occur.

Note: Common logarithms were once used extensively in computations, particularly for problems involving exponents. Nowadays, it is more efficient to use calculators or computers to perform these calculations.

Logarithms are ways to write exponents. If $\log_b x = k$, then k is the *exponent* that is used on b to get x; that is, $x = b^k$. Notice that the domain is restricted to positive real numbers; this restriction is necessary because b is a positive number. Therefore, there is no exponent k for which $x = b^k$ is not positive. Consequently, the logarithms of negative numbers or zero are undefined. Here are examples of logarithms:

- $\log_2 8 = 3$ because 3 is the exponent such that $2^3 = 8$.
- $\log_5\left(\frac{1}{25}\right) = -2$ because -2 is the exponent such that $5^{-2} = \frac{1}{25}$.
- $\log_{10} 100 = 2$ because 2 is the exponent such that $10^2 = 100$.
- $\log_4 0.25 = -1$ because -1 is the exponent such that $4^{-1} = 0.25$.
- $\ln e^5 = 5$ because 5 is the exponent such that $e^5 = e^5$.
- $\log_4 0 =$ undefined because the logarithm of zero is undefined.
- $\ln(-e) =$ undefined because the logarithmic function is undefined for negative numbers.

From these examples, you can surmise that the logarithmic function is the *inverse* of the exponential function, and conversely. To confirm this relationship, let $f(x) = \log_b x$ and $g(x) = b^x$ $(b \neq 1, b > 0)$, and then show $(f \circ g)(x) = x$ and $(g \circ f)(x) = x$:

- $(f \circ g)(x) = f(g(x)) = \log_b b^x = x$, because x is the exponent such that $b^x = b^x$.
- $(g \circ f)(x) = g(f(x)) = b^{f(x)} = b^{\log_b x} = x$, because $\log_b x$ is the exponent you use on b to get x.

Figure 11.3, on the next page, shows the graphs of the logarithmic function defined by $f(x) = \ln x$ and its mutually inverse exponential function defined by $f(x) = e^x$.

PROBLEM For the given function, state the function that defines its inverse.

a. Function f defined by $f(x) = \log_2 x$
b. Function g defined by $g(x) = 3^x$

SOLUTION

a. f^{-1} defined by $f^{-1}(x) = 2^x$
b. g^{-1} defined by $g^{-1}(x) = \log_3 x$

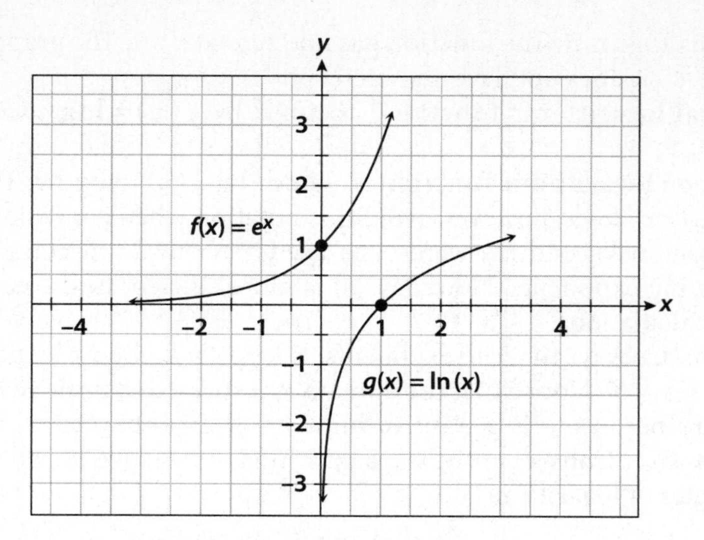

Figure 11.3 Inverse functions e^x and $\ln x$

EXERCISE
11·4

For 1–5, fill in the blank(s) to make a true statement.

1. In the logarithmic function defined by $f(x) = \log_b x$, the two restrictions on the base b are __________ and __________.
2. The domain of a logarithmic function is __________, and the range is __________.
3. The graph of a logarithmic function does not cross the y-axis, so there is no __________.
4. The logarithmic function is undefined for __________ numbers and __________.
5. The natural logarithmic function is defined by __________.

For 6–10, find the function value for the given logarithmic function.

6. $f(x) = \log_3 x;\ f(81)$
7. $g(x) = \log_{64} x;\ g(256)$ *Hint*: Consider fractional exponents.
8. $f(x) = \log_{1/2} x;\ f(32)$
9. $h(x) = \log_{4/9} x;\ h\left(\frac{8}{27}\right)$
10. $g(x) = \log_{0.25} x;\ g(0.5)$

For 11–15, for the given function, state the function that defines its inverse.

11. Function f defined by $f(x) = \log_8 x$
12. Function g defined by $g(x) = \log_{1/4} x$
13. Function h defined by $h(x) = 10^x$
14. Function g defined by $g(x) = (1.05)^x$
15. Function f defined by $f(x) = \ln x$

Graphs of logarithmic functions

The graph of $f(x) = \log_b x$ $(b \neq 1, b > 0)$ is a smooth, continuous curve. The graph passes through $(1,0)$ and $(b,1)$ and is located in the first and fourth quadrants only. The domain is $(0, \infty)$, and the range is R. The x-intercept is 1. It has no y-intercepts. The y-axis is a vertical asymptote.

Furthermore, for the function f defined by $f(x) = \log_b x$, the following hold:

- If $b > 1$, the function is increasing. As x approaches ∞, $f(x) = \log_b x$ approaches ∞. As x approaches 0, $f(x) = \log_b x$ approaches $-\infty$.
- If $0 < b < 1$, the function is decreasing. As x approaches 0, $f(x) = \log_b x$ approaches ∞. As x approaches ∞, $f(x) = \log_b x$ approaches $-\infty$.

PROBLEM Given the logarithmic function f defined by $f(x) = \log_2 x$ (see Figure 11.4), (a) state the domain and range, (b) find zeros, (c) determine asymptotes, (d) find the intercepts, (e) discuss increasing and decreasing behavior, and (f) discuss behavior as x approaches 0 or ∞.

SOLUTION
a. The domain is $(0, \infty)$, and the range is R.
b. The function f has one zero, namely, 1.
c. The y-axis is a vertical asymptote.
d. The x-intercept is 1. The graph has no y-intercept.
e. $b = 2 > 1$; therefore, f is increasing on $(0, \infty)$.
f. $b = 2 > 1$; therefore, as x approaches ∞, $f(x) = \log_2 x$ approaches ∞, and as x approaches 0, $f(x) = \log_2 x$ approaches $-\infty$.

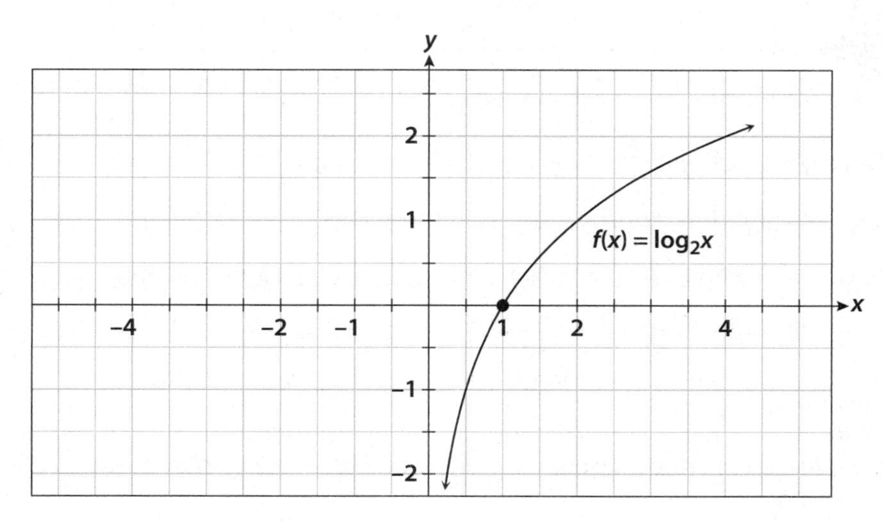

Figure 11.4 Graph of $f(x) = \log_2 x$

PROBLEM Given the logarithmic function f defined by $f(x) = \log_{1/2} x$ (see Figure 11.5, on the next page), (a) state the domain and range, (b) find zeros, (c) determine asymptotes, (d) find the intercepts, (e) discuss increasing and decreasing behavior, and (f) discuss behavior as x approaches 0 or ∞.

SOLUTION
a. The domain is $(0, \infty)$, and the range is R.
b. The function f has one zero, namely, 1.
c. The y-axis is a vertical asymptote.

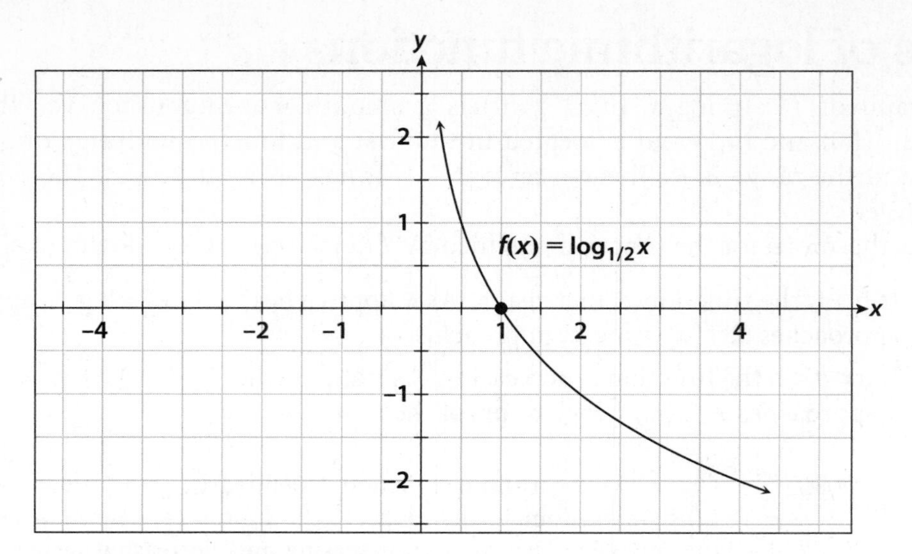

Figure 11.5 Graph of $f(x) = \log_{1/2} x$

d. The x-intercept is 1. The graph has no y-intercept.

e. $0 < b = \frac{1}{2} < 1$; therefore, f is decreasing on $(0, \infty)$.

f. $0 < b = \frac{1}{2} < 1$; therefore, as x approaches ∞, $f(x) = \log_{1/2} x$ approaches $-\infty$, and as x approaches 0, $f(x) = \log_{1/2} x$ approaches ∞.

For 1–5, fill in the blank(s) to make a true statement.

1. The graph of $f(x) = \log_b x$ is a smooth, __________ curve.
2. The graph of $f(x) = \log_{10} x$ passes through the points __________ and __________.
3. The domain of f defined by $f(x) = \log_{1/3} x$ is __________, and the range is __________.
4. The x-intercept of f defined by $f(x) = \ln x$ is __________.
5. The graph of f defined by $f(x) = \log_3 x$ has __________ y-intercepts.

For 6–10, for the given logarithmic function, (a) state the domain and range, (b) find zeros, (c) determine asymptotes, (d) find the intercepts, (e) discuss increasing and decreasing behavior, and (f) discuss behavior as x approaches 0 or ∞.

6. $f(x) = \log_8 x$
7. $g(x) = \log_{1/4} x$
8. $h(x) = 10^x$
9. $g(x) = (1.05)^x$
10. $f(x) = \ln x$

Properties of logarithmic functions

Logarithmic functions have the following properties.

Properties of logarithmic functions

For the function f defined by $f(x) = \log_b x$ $(b \neq 1, b > 0)$:

$$f(1) = \log_b 1 = 0$$

$$f(b) = \log_b b = 1$$

$$f(b^x) = \log_b b^x = x$$

$$f\left(\frac{1}{u}\right) = \log_b\left(\frac{1}{u}\right) = -\log_b u$$

$$f(uv) = \log_b(uv) = \log_b u + \log_b v$$

$$f\left(\frac{u}{v}\right) = \log_b\left(\frac{u}{v}\right) = \log_b u - \log_b v$$

$$f(u^p) = \log_b(u^p) = p\log_b u$$

Change-of-base formula: $f(x) = \log_b x = \dfrac{\log_a x}{\log_a b} = \dfrac{\ln x}{\ln b} = \dfrac{\log_{10} x}{\log_{10} b}$ $(a \neq 1, a > 0)$

One-to-one property: $f(u) = f(v)$ if and only if $u = v$; that is, $\log_b u = \log_b v$ if and only if $u = v$.

Note: In view of the fact that logarithms are exponents, think about the rules for exponents when you are working with logarithms.

PROBLEM Suppose $f(x) = \ln x$. Use the properties of logarithmic functions to find each of the following:

a. $f(1)$

b. $f\left(\frac{1}{e}\right)$

c. $f(e^4)$

SOLUTION a. $f(1) = \ln 1 = 0$

b. $f\left(\frac{1}{e}\right) = -\ln e = -1$

c. $f(e^4) = \ln(e^4) = 4$

PROBLEM Suppose $f(x) = \log_2 x$. Use the properties of logarithmic functions to find each of the following:

a. $f(2)$

b. $f(8 \cdot 64)$

c. $f(32^6)$

SOLUTION a. $f(2) = \log_2 2 = 1$

b. $f(8 \cdot 64) = \log_2 (8 \cdot 64) = \log_2 8 + \log_2 64 = 3 + 6 = 9$

c. $f(32^6) = \log_2 (32^6) = 6 \log_2 32 = 6 \cdot 5 = 30$

EXERCISE 11·6

For 1–5, use the properties of logarithmic functions in each of the following.

1. Suppose $f(x) = \ln x$. Find $f(e^{100})$.
2. Suppose $g(x) = \log_2 x$. Find $g(64^{12})$.
3. Suppose $h(x) = \log_{10} x$. Find $h\left(\dfrac{1000}{0.00001}\right)$.
4. Suppose $g(x) = \log_3 x$. Find $g(27 \cdot 81)$.
5. Suppose $f(x) = \ln x$. If $\ln u = \ln 200$, then $u =$ __________.

For 6–10, use the property that $f(x) = \log_b x = \dfrac{\ln x}{\ln b}$ to express each of the following logarithms in terms of the natural logarithm function; then evaluate each, using the LN *key on your calculator (round to two decimal places, if needed).*

6. $\log_8 (512)$
7. $\log_{1/4} (0.015625)$
8. $\log_5 (15{,}625)$
9. $\log_{1.05} (2.5)$
10. $\log_2 (200)$

Exponential and logarithmic equations

Recall that $\log_b x = k$ if and only if $x = b^k$. You can use this inverse relationship to solve exponential and logarithmic equations.

To solve logarithmic equations of the form $\log_b x = k$, exponentiate both sides, using b as the base of the exponent. Here is an example:

PROBLEM Solve $\log_3 x = 4$.

SOLUTION

$\log_3 x = 4$

$3^{\log_3 x} = 3^4$ Exponentiate both sides.

$x = 81$ Simplify.

When the equation is more complex, isolate the logarithmic function, and then proceed as in the previous problem. Here are examples:

PROBLEM Solve each of the following logarithmic equations. Round to two decimal places, when needed.

a. $3\log_2 x = 15$

b. $\log_{10}(2x+1) = 3$

c. $\log_{1.05} x = 14.21$

SOLUTION a.

$3\log_2 x = 15$

$\frac{3\log_2 x}{3} = \frac{15}{3}$ Divide both sides by 3.

$\log_2 x = 5$ Simplify.

$2^{\log_2 x} = 2^5$ Exponentiate both sides.

$x = 32$ Simplify.

b. $\log_{10}(2x+100) = 3$

$10^{\log_{10}(2x+100)} = 10^3$ Exponentiate both sides.

$2x + 100 = 1000$ Simplify.

$2x = 1000 - 100$ Subtract 100 from both sides.

$2x = 900$ Simplify.

$\frac{2x}{2} = \frac{900}{2}$ Divide both sides by 2.

$x = 450$ Simplify.

c.

$\log_{1.05} x = 14.21$

$1.05^{\log_{1.05} x} = 1.05^{14.21}$ Exponentiate both sides.

$x \approx 2.00$ Simplify.

To solve exponential equations of the form $b^x = k$, take the logarithm of both sides, using b as the base of the logarithm. If necessary, use the change-of-base formula so that you can evaluate the logarithm. Here is an example:

PROBLEM Solve $5^x = 15{,}625$.

SOLUTION

$5^x = 15{,}625$

$\log_5 5^x = \log_5(15{,}625)$ Take the logarithm of both sides.

$x = \frac{\ln 15{,}625}{\ln 5} = 6$ Use the change-of-base formula and simplify.

When the equation is more complex, isolate the exponential function, and then proceed as in the previous problem. Here are examples:

PROBLEM Solve each of the following exponential equations. Round to two decimal places, when needed.

a. $10^{2x-1} = 100{,}000$

b. $3e^{5x} = 21{,}000$

c. $500(1.05)^x = 1{,}000$

SOLUTION a.

$$10^{2x-1} = 100{,}000$$

$\log_{10} 10^{2x-1} = \log_{10}(100{,}000)$ Take the logarithm of both sides.

$2x - 1 = 5$ Simplify.

$2x = 5 + 1$ Add 1 to both sides.

$2x = 6$ Simplify.

$\frac{2x}{2} = \frac{6}{2}$ Divide both sides by 2.

$x = 3$ Simplify.

b.

$$3e^{5x} = 21{,}000$$

$\frac{3e^{5x}}{3} = \frac{21{,}000}{3}$ Divide both sides by 3.

$e^{5x} = 7{,}000$ Simplify.

$\ln e^{5x} = \ln 7{,}000$ Take the logarithm of both sides.

$x \approx 1.77$ Simplify.

c. $500(1.05)^x = 1{,}000$

$\frac{500(1.05)^x}{500} = \frac{1{,}000}{500}$ Divide both sides by 500.

$1.05^x = 2$ Simplify.

$\log_{1.05} 1.05^x = \log_{1.05} 2$ Take the logarithm of both sides.

$x = \frac{\ln 2}{\ln 1.05} \approx 14.21$ Use the change-of-base formula and simplify.

EXERCISE 11·7

Solve. Round to two decimal places, when needed.

1. $5\log_2(3x-1) = 160$
2. $\ln 8x = 3.5$
3. $2{,}000(1.005)^x = 6{,}000$
4. $10.5e^{.05x} = 21$
5. $e^{5x+1} = e^{3x-4}$

Additional common functions

·12·

Piecewise functions

A **piecewise function** is a function defined by different equations on different parts (usually intervals) of its domain. Here is an example of a piecewise function (also see Figure 12.1):

$$f(x)=\begin{cases} x^2 & \text{if } x \le 0 \\ \sqrt{x} & \text{if } x > 0 \end{cases}$$

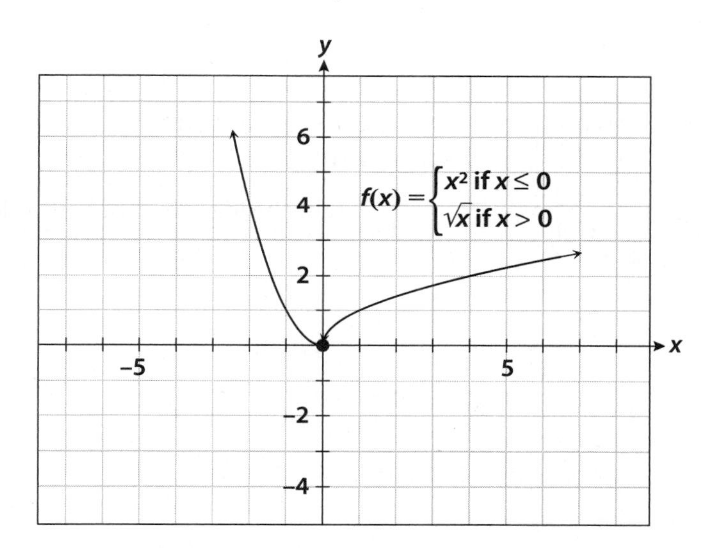

Figure 12.1 Piecewise function

PROBLEM Suppose $f(x)=\begin{cases} x^2 & \text{if } x \le 0 \\ \sqrt{x} & \text{if } x > 0 \end{cases}$. Find each of the following function values:

a. $f(-9)$
b. $f(9)$
c. $f(0)$

SOLUTION a. $f(-9) = f(-9)^2 = 81$

b. $f(9) = \sqrt{9} = 3$

c. $f(0) = 0^2 = 0$

You can use piecewise functions to "repair" holes in graphs of a function that otherwise would be continuous. For example, as shown in Figure 12.2, on the next page, the graph of the function defined by $f(x)=\frac{x^2-1}{x-1}=\frac{(x+1)(x-1)}{(x-1)}$ has a "hole" at the point $(1,2)$.

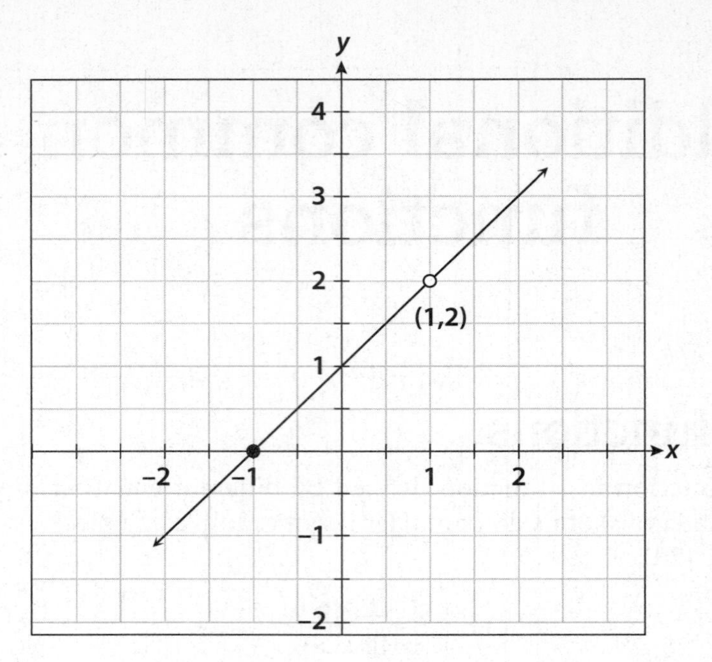

Figure 12.2 Graph of $f(x)=\dfrac{x^2-1}{x-1}$

The function f is discontinuous at $x = 1$ although it is continuous at all other values of x. However, the discontinuity is a **removable discontinuity**. The graph of the function defined by $g(x)=\begin{cases}\dfrac{x^2-1}{x-1} & \text{if } x\neq 1\\ 2 & \text{if } x=1\end{cases}$ is the function g with the discontinuity removed (see Figure 12.3).

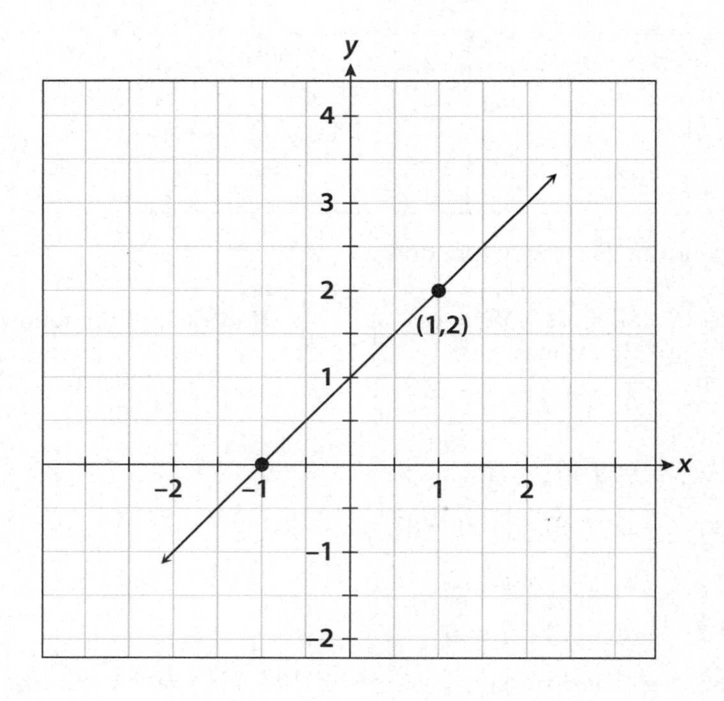

Figure 12.3 Graph of $g(x)=\begin{cases}\dfrac{x^2-1}{x-1} & \text{if } x\neq 1\\ 2 & \text{if } x=1\end{cases}$

EXERCISE

12·1

Suppose $f(x)=\begin{cases} 2x+5 & \text{if } x<0 \\ 3 & \text{if } x=0 \\ \sqrt{2x+5} & \text{if } x>0 \end{cases}$. *Find each of the following function values.*

1. $f(-2)$
2. $f\left(-\frac{1}{2}\right)$
3. $f(0)$
4. $f(2)$
5. $f(3)$

Greatest integer function

The **greatest integer function** (also called the **step function**) is defined by $f(x)=[x]$, where the brackets denote finding the greatest integer n such that $n \geq x$. The domain is R, and the range is the integers. The zeros lie in the interval [0, 1). The y-intercept is 0. The x-intercepts lie in the interval [0, 1). The graph of the function is constant between the integers, but "jumps" at each integer. Figure 12.4 shows the greatest integer function.

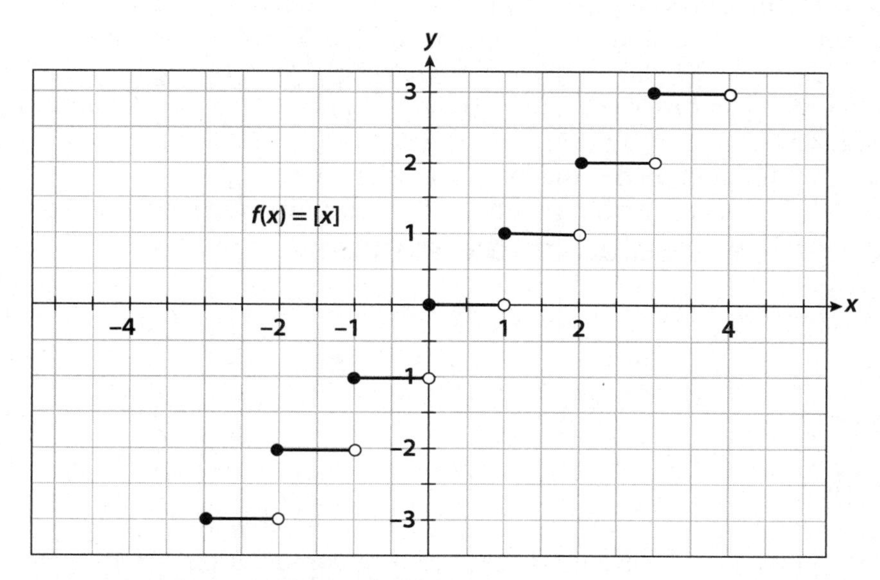

Figure 12.4 Greatest integer function

The greatest integer function is a piecewise function because you can define it as follows:

$$f(x)=[x]=\begin{cases} \vdots \\ -2 & \text{if } -2 \leq x < -1 \\ -1 & \text{if } -1 \leq x < 0 \\ 0 & \text{if } \ 0 \leq x < 1 \\ 1 & \text{if } \ 1 \leq x < 2 \\ 2 & \text{if } \ 2 \leq x < 3 \\ \vdots \end{cases}$$

PROBLEM Suppose $f(x)=[x]$. Find each of the following function values:

a. $f(7.8)$

b. $f(-4.99)$

c. $f(\pi)$

SOLUTION a. $f(7.8) = [7.8] = 7$

b. $f(-4.99) = [-4.99] = -5$

c. $f(\pi) = [\pi] = 3$

EXERCISE 12·2

Suppose $f(x) = [x]$. *Find each of the following function values.*

1. $f(2.99)$
2. $f(-2.99)$
3. $f(0.001)$
4. $f(-12)$
5. $f(-e)$

Absolute value function

The absolute value function is defined by $f(x) = |x|$. The domain is R, and the range is $[0,\infty)$. The only zero occurs at $x = 0$. The x- and y-intercepts are both 0. The absolute value function is a piecewise function because you can write $f(x) = |x|$ as $f(x) = \begin{Bmatrix} x \text{ if } x \geq 0 \\ -x \text{ if } x < 0 \end{Bmatrix}$. Figure 12.5 shows the absolute value function.

The absolute value function has the following properties.

Properties of the absolute value function

$|x| \geq 0$

$|x| = |-x|$

$|xy| = |x||y|$

$\left|\frac{x}{y}\right| = \frac{|x|}{|y|}$

$|u + v| \leq |u| + |v|$

Note: Always perform computations inside absolute value bars *before* you evaluate the absolute value.

$|x| = c \ (c > 0)$ if and only if either $x = c$ or $x = -c$

$|x| < c \ (c > 0)$ if and only if $-c < x < c$

$|x| > c \ (c > 0)$ if and only if either $x < -c$ or $x > c$

$\sqrt{x^2} = |x|$

Note: Properties involving $<$ and $>$ hold if you replace $<$ with $\leq$ and $>$ with $\geq$.

PROBLEM Evaluate each of the following:

a. $|17| - |-17|$

b. $|-9.5| \cdot |2|$

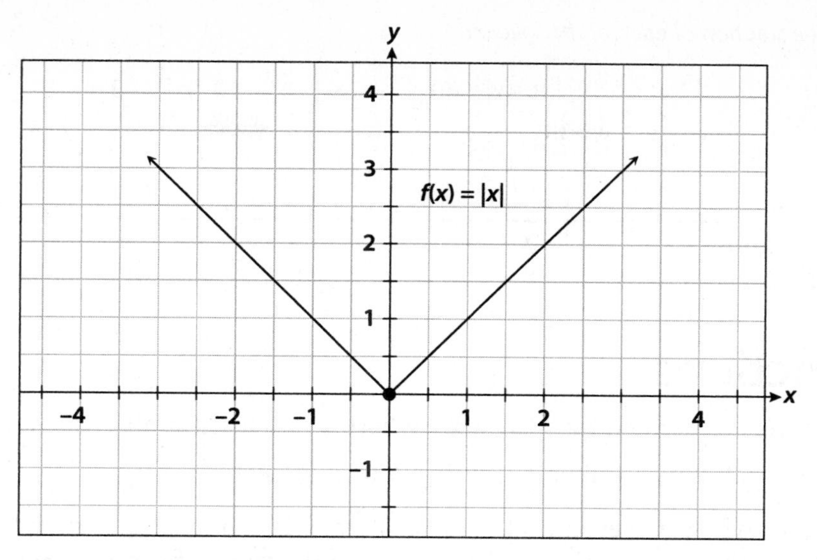

Figure 12.5 Absolute value function

c. $\frac{|-15|+|3|}{|8|-|-5|}$

d. $\sqrt{(-5)^2}$

SOLUTION a. $|17|-|-17|=17-17=0$

b. $|-9.5|\cdot|2|=(9.5)(2)=19$

c. $\frac{|-15|+|3|}{|8|-|-5|}=\frac{15+3}{8-5}=\frac{18}{3}=6$

d. $\sqrt{(-5)^2}=|-5|=5$

PROBLEM Find the solutions of each of the following:

a. $|x|=3.5$

b. $|x|<3.5$

c. $|x|>3.5$

SOLUTION a. $|x|=3.5$ if and only if either $x=3.5$ or $x=-3.5$

b. $|x|<3.5$ if and only if $-3.5<x<3.5$

c. $|x|>3.5$ if and only if either $x<-3.5$ or $x>3.5$

Note: See the section "Absolute value" in Chapter 1, "Real numbers," for a review of absolute value.

EXERCISE

12·3

For 1–5, evaluate each of the following.

1. $|-2.3|-|-4.7|$
2. $|6|\cdot\left|-\frac{2}{3}\right|$
3. $\frac{|-20|+|5|}{|45|-|-5|}$
4. $\frac{|-4|\cdot|-2|}{|24-40|}$
5. $\sqrt{(-10)^2}$

For 6–10, find the solution of each of the following.

6. $|x| = \frac{1}{2}$
7. $|x| < 9.57$
8. $|x| > 40$
9. $|x| > -2$
10. $|x| < -3$

Power functions

Power functions are defined by $f(x) = x^a$, where a is a real number (provided that 0^0 does not occur). If a is a rational number $\frac{p}{q}$ [with p and q integers $(q \neq 0)$ and $\frac{p}{q}$ simplified], then the domain is $[0, \infty)$ when q is even and is $[-\infty, \infty)$ when q is odd. If a is an irrational number, the domain is $[0, \infty)$. The range will vary depending on the value of a, and so will the zeros.

PROBLEM Find the function value for the given power function. Round to two places, when needed.

a. $f(x) = x^{-5}; f(2)$

b. $g(x) = x^{\frac{3}{4}}; g(256)$

c. $h(x) = x^{-\frac{4}{3}}; h(27)$

d. $r(x) = x^{\pi}; r(5)$

e. $s(x) = x^{\sqrt{2}}; s(5)$

f. $t(x) = x^{\sqrt{2}}; t(-5)$

SOLUTION a. $f(2) = (2)^{-5} = \frac{1}{2^5} = \frac{1}{32}$

b. $g(256) = (256)^{\frac{3}{4}} = \left(\sqrt[4]{256}\right)^3 = 4^3 = 64$

c. $h(27) = (27)^{-\frac{4}{3}} = \frac{1}{(27)^{\frac{4}{3}}} = \frac{1}{\left(\sqrt[3]{27}\right)^4} = \frac{1}{3^4} = \frac{1}{81}$

d. $r(5) = 5^{\pi} \approx 156.99$

e. $s(5) = 5^{\sqrt{2}} \approx 9.74$

f. $t(-5) = (-5)^{\sqrt{2}}$ = undefined because -5 is not in the domain of t.

Note: See the section "Exponentiation" in Chapter 1, "Real numbers," for a review of exponents.

The x- and y-intercepts of the graph of a power function defined by $f(x) = x^a$ will vary depending on the value of a. However, the graphs of all power functions pass through (1, 1). Graphs of selected power functions are shown in Figures 12.6–12.8.

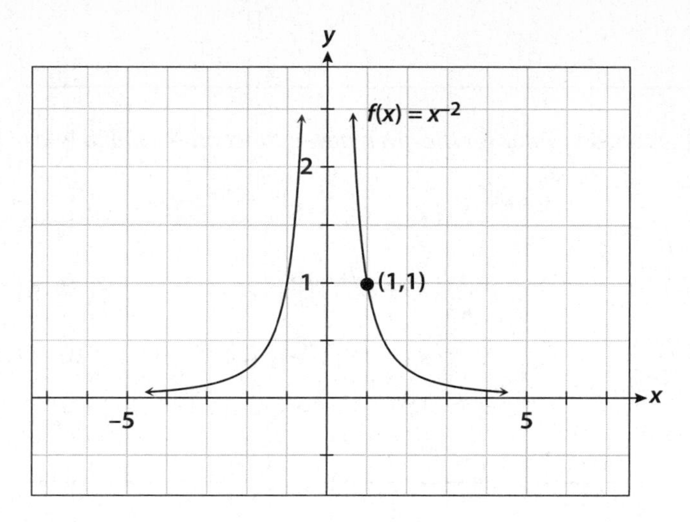

Figure 12.6 Graph of $f(x)=x^{-2}$

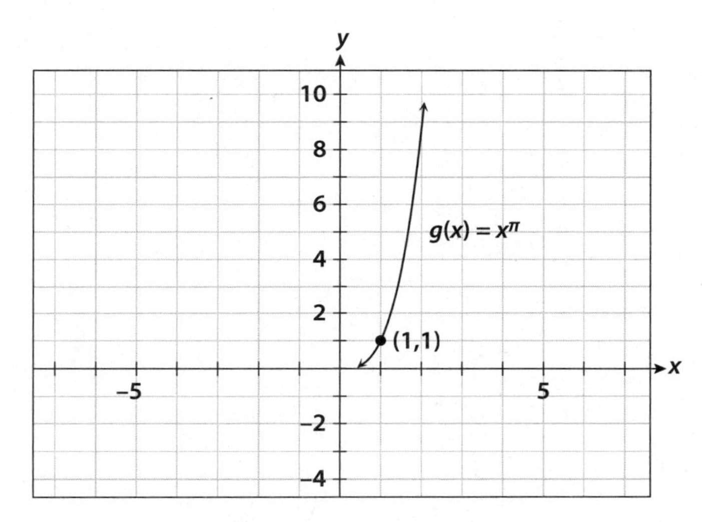

Figure 12.7 Graph of $g(x)=x^{\pi}$

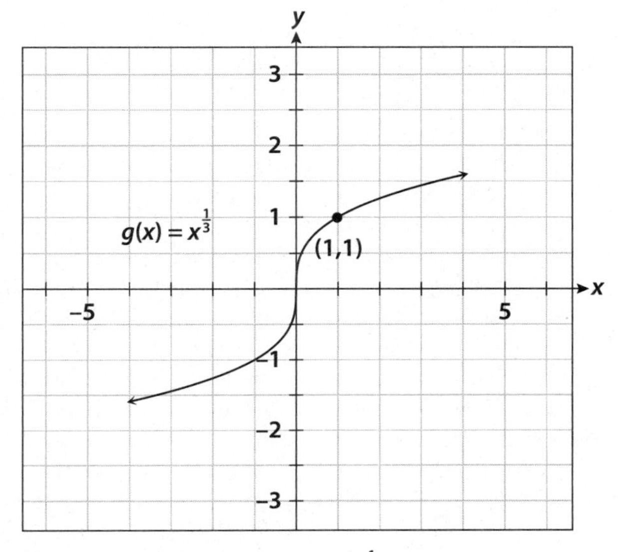

Figure 12.8 Graph of $g(x)=x^{\frac{1}{3}}$

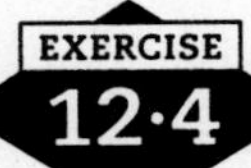

Find the function value for the given power function. Round to two places, when needed.

1. $f(x) = x^{-5}; f\left(\frac{1}{2}\right)$
2. $g(x) = x^{\frac{3}{4}}; g(625)$
3. $h(x) = x^{-\frac{2}{3}}; h(343)$
4. $f(x) = x^{1}; f(4)$
5. $r(x) = x^{\pi}; r(15)$
6. $s(x) = x^{\sqrt{2}}; s(8)$
7. $t(x) = x^{\sqrt{2}}; t(-1)$
8. $f(x) = x^{0}; f(0)$
9. $g(x) = x^{0}; g(\sqrt{3})$
10. $h(x) = x^{-e}; h(2)$

Function models and applications

·13·

Linear function models and applications

You use linear functions to model problem situations that involve change at a constant rate. Here is an example:

PROBLEM A tank contains 500 gallons of water. Suppose water is added to the tank at a constant rate of 150 gallons per hour. Let $f(t)$ be the amount of water in the tank after t hours. (a) Write a formula for $f(t)$. (b) How much water is in the tank at $t = 0$ hours? (c) How much water is in the tank at $t = 8$ hours?

SOLUTION a. $f(t) = \underset{\substack{\uparrow \\ \text{Constant rate} \\ \text{of change}}}{150t} + \underset{\substack{\uparrow \\ \text{Initial amount}}}{500}$

b. $f(0) = 150(0) + 500 = 500$ gallons

c. $f(8) = 150(8) + 500 = 1{,}200 + 500 = 1{,}700$ gallons

EXERCISE 13·1

Solve using an appropriate function model.

1. Suppose $f(t)$ is the distance traveled at time t by a car moving at a constant rate of speed of 70 miles per hour. (a) Write a formula for $f(t)$. (b) What is the distance traveled at time $t = 3\frac{1}{2}$ hours?

2. A tank contains 1,000 gallons of water. Suppose water is drained from the tank at a constant rate of 120 gallons per hour. Let $f(t)$ be the amount of water in the tank after t hours. (a) Write a formula for $f(t)$. (b) How much water is in the tank at $t = 3$ hours? *Hint:* The amount of water is *decreasing*, so the constant rate of change is negative.

3. A train is initially 200 miles from a city and moving toward the city at 50 miles per hour. Let $f(t)$ be the distance from the city at time t. (a) Write a formula for $f(t)$. (b) How far is the train from the city at $t = 2\frac{3}{4}$ hours?
   Hint: The distance is *decreasing*, so the constant rate of change is negative.

4. Hooke's law states that $F = -kx$, where F is the force applied to a spring, x is the distance that the spring stretches from its natural state, and k is the constant of proportionality for the specific spring. Suppose for a certain spring the constant of proportionality is 6. Let $F(x)$ be the force (in pounds) applied when the spring stretches a distance of x inches. (a) Write a formula for $F(x)$. (b) Find the force when the spring stretches 3 inches.

5. The monthly cost of driving a car consists of both fixed and variable costs. Suppose that the fixed monthly cost for driving a certain car is \$250 and the variable cost is \$0.55 per mile driven. Let $f(x)$ be the monthly cost of driving this car for x miles. (a) Write a formula for $f(x)$. (b) Find $f(x)$ when $x = 536$ miles.

Quadratic function models and applications

Quadratic function models are useful in a number of problem situations including determining the maximum or minimum values for the dependent variable and investigating the motion of projectiles. Here are examples:

PROBLEM A farmer has 100 feet of fencing to enclose a rectangular region for a horse. The farmer will use a portion of the side of a large barn as one side of the rectangle, as shown in Figure 13.1. Let $f(W)$ be the area of the rectangular region expressed in terms of its width, W.

a. Write a formula for $f(W)$.

b. Find the dimensions of the rectangle that give the maximum area for the horse.

SOLUTION a. The fence does not go along the barn, so you have:

$$100 = 2W + L;\ \text{thus, } L = 100 - 2W$$

The area of the rectangular region equals length times width, so:

$$f(W) = (100 - 2W)W = 100W - 2W^2 = -2W^2 + 100W$$

b. The graph of $f(W)$ is a parabola opening downward. The vertex formula is $\left(-\frac{b}{2a}, f\left(-\frac{b}{2a}\right)\right)$, so the maximum value for $f(W)$ occurs when $W = -\frac{b}{2a}$ $= -\frac{100}{2(-2)} = 25$ feet. The corresponding length is $L = 100 - 2W = 100 - 2.25$ $= 100 - 50 = 50$ feet. The dimensions that maximize the area are 50 feet by 25 feet.

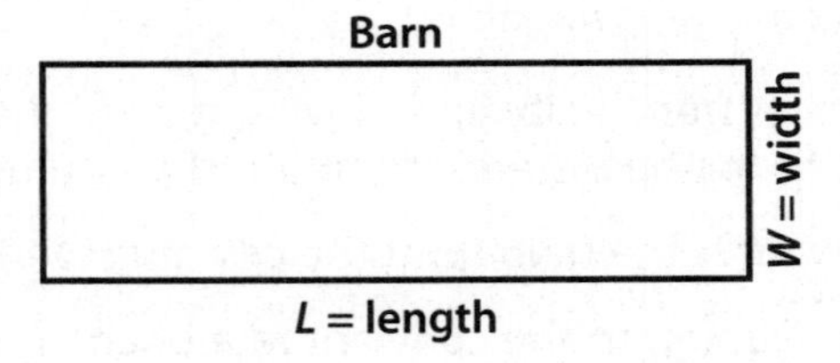

Figure 13.1 Fenced area

PROBLEM A ball is thrown vertically upward with an initial velocity of 56 feet per second (s) from a scaffold that is 30 feet above the ground level. The quadratic function model for the height (in feet) above ground of the ball at time t (in seconds) is given by $h(t) = -16t^2 + 56t + 30$. Find the maximum height above the ground for the ball.

SOLUTION The graph of $h(t)$ is a parabola opening downward. The vertex formula is $\left(-\frac{b}{2a}, f\left(-\frac{b}{2a}\right)\right)$, so the maximum value for $h(t)$ occurs when $t = -\frac{b}{2a} = -\frac{56}{2(-16)} = 1.75$ seconds. Therefore, the maximum height is reached when $t = 1.75$ seconds:

$$h(1.75) = -16(1.75)^2 + 56(1.75) + 30 = 79 \text{ feet}$$

Solve using an appropriate function model.

1. A farmer has 200 feet of fencing to enclose a rectangular area for goats. What dimensions for the rectangle will maximize the area for the goats?
2. The height (in feet) above the ground for an object that is projected vertically upward with a velocity of 80 feet per second from an initial height of 50 feet is given by $h(t) = -16t^2 + 80t + 50$. Find the maximum height of the object above the ground.
3. Suppose the monthly revenue R in thousands of dollars that a company receives from producing x thousand boxes of paper clips is given by $R(x) = -5x^2 + 600x$. How many boxes of paper clips should the company produce each month to maximize $R(x)$?
4. Suppose a yearly cost function C in thousands of dollars is given by $C(x) = x^2 - 48x + 5{,}000$. Find the minimum yearly cost.
5. The height (in feet) above the ground for a ball that is tossed straight upward with a velocity of 80 feet per second from an initial height of 6 feet is given by $h(t) = -16t^2 + 80t + 6$. At what time will the ball reach its maximum height?

Additional function models and applications

You can use rational functions, exponential functions, logarithmic functions, and power functions to model many real-world problem situations. Here are examples:

PROBLEM (rational function) The surface area, denoted $S.A.$, in square units of a right circular cylinder of specified volume, V, is given by $S.A. = f(r) = \frac{2V + 2\pi r^3}{r}$, where r is the radius of the base of the cylinder. Find the surface area for a right circular cylinder that has a volume of 785 centimeters^3 and radius of 5 centimeters.

SOLUTION $S.A. = f(5) = \frac{2 \cdot 785 + 2\pi \cdot 5^3}{5} \approx 471 \text{ cm}^2$

PROBLEM (exponential function) Radioactive elements undergo radioactive decay. The amount remaining, A, of a radioactive substance after t years is given by $A(t) = A_0\left(\frac{1}{2}\right)^{\frac{t}{k}}$, where A_0 is the initial amount of radioactive substance and k is its half-life. How much of a 3-gram sample of radium-228, which has a half-life of 5.75 years, remains after 2 years?

SOLUTION For 3 grams of radium-228, the formula is $A(t) = 3\left(\frac{1}{2}\right)^{\frac{t}{5.75}}$. Therefore, the amount remaining after 2 years is $A(2) = 3\left(\frac{1}{2}\right)^{\frac{2}{5.75}} \approx 2.36$ grams.

PROBLEM (logarithmic function) The formula for the pH of an aqueous solution is given by $f(x) = -\log_{10} x$, where x is the hydronium ion concentration (in moles per liter) of the solution. Find the pH of lemon juice, which has a hydronium concentration of 0.005.

SOLUTION pH of lemon juice $= f(0.005) = -\log_{10}(0.005) \approx 2.3$

PROBLEM (power function) The number of Earth-years that it takes for a planet to orbit the Sun is given by $f(x) = x^{1.5}$, where x is the ratio of the average distance the planet is from the Sun to 93 million miles, which is the average distance that Earth is from the Sun. Approximately how many Earth-years does it take for the planet Jupiter to orbit the Sun if its average distance from the Sun is 483.6 million miles?

SOLUTION For Jupiter, $x = \frac{483.6 \text{ million miles}}{93 \text{ million miles}} = 5.2$; then

$f(5.2) = (5.2)^{1.5} \approx 11.86$ Earth-years.

EXERCISE 13·3

Solve using an appropriate function model.

1. The surface area, *S.A.* (in square units), of a right circular cylinder of specified volume, V, is given by $S.A. = f(r) = \frac{2V + 2\pi r^3}{r}$, where r is the radius of the base of the cylinder. Find the surface area for a right circular cylinder that has volume of 188 centimeters^3 and radius of 2 centimeters.
2. Radioactive elements undergo radioactive decay. The amount remaining, A, of a radioactive substance after t years is given by $A(t) = A_0\left(\frac{1}{2}\right)^{\frac{t}{k}}$, where A_0 is the initial amount of radioactive substance and k is its half-life. How much of a 10-gram sample of carbon-14, which has a half-life of 5,730 years, remains after 9,000 years?
3. The formula for the pH of an aqueous solution is given by $f(x) = -\log_{10} x$, where x is the hydronium ion concentration (in moles per liter) of the solution. Find the pH of vinegar, which has a hydronium concentration of 0.0016.

4. The number of Earth-years that it takes for a planet to orbit the Sun is given by $f(x) = x^{1.5}$, where x is the ratio of the average distance the planet is from the Sun to 93 million miles, which is the average distance that Earth is from the Sun. Approximately how many Earth-years does it take for the planet Neptune to orbit the Sun if its average distance from the Sun is 2,794.4 million miles?
5. The intensity, I, of light from a light source is given by $I(d) = \frac{k}{d^2}$, where d is the distance from the light source and k is a constant specific to the light source. Determine the intensity of light at a distance of 4 meters from a 100-watt light bulb for which k is 7.92.

Matrices and systems of linear equations

Basic concepts of matrices

A matrix is a rectangular array of elements. Matrices are used as convenient ways of organizing certain sets of data. A matrix consists of rows and columns of elements within brackets, and its size (dimension) is determined by the number of rows and columns. The notation $n \times k$ (read "n by k") is used to indicate its size, where n is the number of rows and k is the number of columns. For example, the following is a 2×3 matrix. The subscript is a convenient way of indicating the size of a matrix.

$$\begin{bmatrix} 3 & 0 & 4 \\ 6 & -2 & 7 \end{bmatrix}_{2\times3}$$

A matrix is a **square matrix** if the number of rows is equal to the number of columns. If the matrix has only one row or one column, it is a **row matrix** or a **column matrix**. These one-column matrices are also called **row** or **column vectors**.

The **main diagonal** of a square matrix is the diagonal from the upper left element to the lower right element. An **identity matrix**, denoted ***I***, is a square matrix all of whose elements on the main diagonal are 1s and whose remaining elements are 0s. For example:

$$I_{2\times2} = \begin{bmatrix} 1 & 0 \\ 0 & 1 \end{bmatrix}$$

A square matrix all of whose elements are zero is a **zero matrix**, $0_{n\times n}$.

Identity matrices have the property that $AI = IA = A$, where A and I are square matrices of the same size. Also $0A = A0 = 0$, where again the matrices must be of the same size.

1. Exhibit the 3×1 column vector whose elements are 1, 5, and 7.
2. Exhibit the $I_{3\times3}$ matrix.
3. What is the dimension of the matrix $\begin{bmatrix} 1 & 2 \\ 0 & 0 \\ 4 & 5 \end{bmatrix}$?

Operations with matrices

Equal matrices

Two matrices are **equal** if and only if they are of the same size and their corresponding elements are equal. For example:

$$\begin{bmatrix} 2 & 8 \\ -3 & 1 \end{bmatrix} = \begin{bmatrix} \frac{4}{2} & \sqrt{64} \\ \frac{-21}{7} & 1 \end{bmatrix}, \begin{bmatrix} 0 & 0 \\ 0 & 0 \end{bmatrix} \neq \begin{bmatrix} 0 & 0 & 0 \\ 0 & 0 & 0 \\ 0 & 0 & 0 \end{bmatrix}, \text{and} \begin{bmatrix} 1 & 4 & 6 \end{bmatrix} \neq \begin{bmatrix} 1 \\ 4 \\ 6 \end{bmatrix}$$

Addition (subtraction) of matrices

Matrices can only be added (subtracted) if they are of the same size. The sum (difference) of two matrices is the matrix whose elements are the sums (differences) of the corresponding elements of the two matrices. For example

$$\begin{bmatrix} 1 & 8 \\ 4 & -3 \end{bmatrix} + \begin{bmatrix} -2 & 3 \\ 0 & 7 \end{bmatrix} = \begin{bmatrix} -1 & 11 \\ 4 & 4 \end{bmatrix}$$

$$\begin{bmatrix} 2 & 3 \\ 5 & 1 \end{bmatrix} + \begin{bmatrix} a & b \\ -c & d \end{bmatrix} = \begin{bmatrix} 2+a & 3+b \\ 5-c & 1+d \end{bmatrix}$$

$$\begin{bmatrix} 1 & 0 & 4 \\ -5 & 6 & 2 \end{bmatrix} - \begin{bmatrix} 0 & 4 & 5 \\ -3 & 4 & 11 \end{bmatrix} = \begin{bmatrix} 1 & -4 & -1 \\ -2 & 2 & -9 \end{bmatrix}$$

PROBLEM If $A = \begin{bmatrix} 4 & 1 \\ -3 & 6 \end{bmatrix}$, $B = \begin{bmatrix} 2 & -5 \\ 0 & 3 \end{bmatrix}$, and $C = \begin{bmatrix} 4 & 2 & 9 \end{bmatrix}$, then

a. $A + B = ?$

b. $A + C = ?$

SOLUTION a. $A + B = \begin{bmatrix} 6 & -4 \\ -3 & 9 \end{bmatrix}$

b. Matrices A and C are not the same size and cannot be added.

There are two types of multiplication associated with matrices: *scalar* multiplication and *matrix* multiplication.

Scalar multiplication

The multiplication of a matrix by a constant is called scalar multiplication. The product of a scalar and a matrix is a matrix whose elements are each multiplied by the scalar. Any size matrix can be multiplied by a scalar. For example:

$$3\begin{bmatrix} 4 & 1 \\ 0 & -2 \end{bmatrix} = \begin{bmatrix} 12 & 3 \\ 0 & -6 \end{bmatrix}; \quad c\begin{bmatrix} 7 & 3 \\ 1 & -2 \\ 0 & 5 \end{bmatrix} = \begin{bmatrix} 7c & 3c \\ c & -2c \\ 0 & 5c \end{bmatrix}$$

Matrix multiplication

Matrices must be of certain sizes in order to multiply them. If A and B are matrices and A is of size $n \times k$, then matrix B must have the same number of rows as A has columns and thus must be of size $k \times m$. In this case, the matrices are said to be **compatible** for multiplication. The product matrix AB will be of size $n \times n$. The element in the ath row and bth column of the product matrix is the sum of the products of the elements of the ath row of A by the corresponding elements in the bth column of B (this probably takes longer to explain than to do the actual multiplication). For example:

$$\text{If } A = \begin{bmatrix} 3 & 2 \\ 1 & 5 \\ -3 & 0 \end{bmatrix}_{3\times 2} \text{ and } B = \begin{bmatrix} 4 \\ 3 \end{bmatrix}_{2\times 1}, \text{ then } AB = \begin{bmatrix} 3\cdot 4 + 2\cdot 3 \\ 1\cdot 4 + 5\cdot 3 \\ -3\cdot 4 + 0\cdot 3 \end{bmatrix} = \begin{bmatrix} 18 \\ 19 \\ -12 \end{bmatrix}_{3\times 1}$$

Note here that the commutative law for multiplication is not valid for matrix multiplication. That is, in general, $AB \neq BA$.

PROBLEM If $A = \begin{bmatrix} 1 & 5 \\ 3 & -5 \end{bmatrix}$ and $B = \begin{bmatrix} 0 & 4 & 2 \\ -5 & 3 & 1 \end{bmatrix}$, compute AB and BA.

SOLUTION

$$AB = \begin{bmatrix} 1 & 5 \\ 3 & -5 \end{bmatrix}\begin{bmatrix} 0 & 4 & 2 \\ -5 & 3 & 1 \end{bmatrix} = \begin{bmatrix} 1\cdot 0 + 5\cdot(-5) & 1\cdot 4 + 5\cdot 3 & 1\cdot 2 + 5\cdot 1 \\ 3\cdot 0 + (-5)\cdot(-5) & 3\cdot 4 + (-5)\cdot 3 & 3\cdot 2 + (-5)\cdot 1 \end{bmatrix}$$

$$= \begin{bmatrix} -25 & 19 & 7 \\ 25 & -3 & 1 \end{bmatrix}$$

$BA =$ undefined. The matrices are not compatible for multiplication in this order.

PROBLEM If $A = \begin{bmatrix} 2 & 0 \\ 1 & 3 \end{bmatrix}$ and $B = \begin{bmatrix} 1 & 5 \\ 3 & 2 \end{bmatrix}$, compute AB and BA.

SOLUTION

$$AB = \begin{bmatrix} 2 & 0 \\ 1 & 3 \end{bmatrix}\begin{bmatrix} 1 & 5 \\ 3 & 2 \end{bmatrix} = \begin{bmatrix} 2 & 10 \\ 10 & 11 \end{bmatrix}$$

$$BA = \begin{bmatrix} 1 & 5 \\ 3 & 2 \end{bmatrix}\begin{bmatrix} 2 & 0 \\ 1 & 3 \end{bmatrix} = \begin{bmatrix} 7 & 15 \\ 8 & 6 \end{bmatrix}$$

Both of these examples reinforce the fact that $AB \neq BA$.

Subtraction of matrices can now be defined by the equation $A - B = A + (-B) = A + (-1)B$.

Complete the following calculations.

1. $\begin{bmatrix} 2 & 8 \\ -3 & 0 \end{bmatrix} + \begin{bmatrix} 5 & 4 \\ -2 & 3 \end{bmatrix}$

2. $\begin{bmatrix} 3 & 6 & 0 \\ 4 & -2 & -5 \\ 6 & 0 & -1 \end{bmatrix} - \begin{bmatrix} -1 & 6 & 4 \\ -4 & 3 & -1 \\ 5 & 4 & 2 \end{bmatrix}$

3. $\begin{bmatrix} 1 & 6 & 3 \\ 0 & 5 & 4 \end{bmatrix}\begin{bmatrix} 2 & 4 \\ -3 & 1 \\ 5 & 0 \end{bmatrix}$

4. $\begin{bmatrix} 1 & 0 \\ 0 & 1 \end{bmatrix}\begin{bmatrix} 3 & 6 \\ -4 & 5 \end{bmatrix}$

5. $\begin{bmatrix} 0 & 0 \\ 0 & 0 \end{bmatrix}\begin{bmatrix} 5 & 6 \\ 4 & 9 \end{bmatrix}$

Determinant of a square matrix

The determinant of a matrix is a number associated with the matrix and its elements and is only meaningful if the matrix is square. For 2 × 2 matrices the determinant is simple to calculate and is defined by

$$\det\begin{bmatrix} a & b \\ c & d \end{bmatrix} = \begin{vmatrix} a & b \\ c & d \end{vmatrix} = ad - bc$$

Note that vertical bars are used for determinants.

A third-order (3 × 3) determinant is a little more tedious to calculate, and there are several different methods to perform the calculation. One of these methods uses three 2 × 2 determinants, and this method is the one of choice at this time. This particular method is referred to as **expanding by elements of the first row.** Other methods will be discussed later in this chapter.

$$\det\begin{bmatrix} a_1 & a_2 & a_3 \\ b_1 & b_2 & b_3 \\ c_1 & c_2 & c_3 \end{bmatrix} = \begin{vmatrix} a_1 & a_2 & a_3 \\ b_1 & b_2 & b_3 \\ c_1 & c_2 & c_3 \end{vmatrix} = a_1\begin{vmatrix} b_2 & b_3 \\ c_2 & c_3 \end{vmatrix} - a_2\begin{vmatrix} b_1 & b_3 \\ c_1 & c_3 \end{vmatrix} + a_3\begin{vmatrix} b_1 & b_2 \\ c_1 & c_2 \end{vmatrix}$$

Tip: To find the 2 × 2 determinant associated with a_1, cross out the row and column containing a_1 and the remaining 2 × 2 determinant is the one needed. A similar technique is used to determine the other two determinants. In applying this definition, errors are frequently made by forgetting the negative sign on the second term.

PROBLEM Find the determinants of $A = \begin{bmatrix} 2 & -3 \\ 4 & 5 \end{bmatrix}$ and $B = \begin{bmatrix} 3 & 0 & 4 \\ -1 & 6 & 2 \\ 5 & -3 & 6 \end{bmatrix}$.

SOLUTION $|A| = \begin{vmatrix} 2 & -3 \\ 4 & 5 \end{vmatrix} = 2\cdot 5 - 4\cdot(-3) = 22$

$$|B| = \begin{vmatrix} 3 & 0 & 4 \\ -1 & 6 & 2 \\ 5 & -3 & 6 \end{vmatrix} = 3\begin{vmatrix} 6 & 2 \\ -3 & 6 \end{vmatrix} - 0\begin{vmatrix} -1 & 2 \\ 5 & 6 \end{vmatrix} + 4\begin{vmatrix} -1 & 6 \\ 5 & -3 \end{vmatrix}$$

$$= 3(36+6) - 0(-6-10) + 4(3-30)$$

$$= 126 - 108 = 18$$

Determinants of larger matrices can be similarly calculated, but only these two sizes will be used for the examples and problems.

Note: By reducing calculation time and tedium, graphing calculators and computer algebra systems are valuable tools for working with matrices and determinants.

Find the determinant of the given matrix.

1. $\begin{bmatrix} 2 & 5 \\ 1 & -4 \end{bmatrix}$

2. $\begin{bmatrix} 3 & -6 \\ -1 & 2 \end{bmatrix}$

3. $\begin{bmatrix} 3 & 4 & 2 \\ 0 & 7 & -3 \\ -2 & 1 & 5 \end{bmatrix}$

4. $\begin{bmatrix} 1 & 0 \\ 0 & 1 \end{bmatrix}$

5. $\begin{bmatrix} 1 & 0 & 0 \\ 0 & 1 & 0 \\ 0 & 0 & 1 \end{bmatrix}$

6. $\begin{bmatrix} 2 & 4 \\ -3 & 1 \\ 5 & 0 \end{bmatrix}$

Inverse of a square matrix

The **inverse** of a square matrix A is a square matrix of the same size, A^{-1}, such that $AA^{-1} = A^{-1}A = I$. Finding the inverse of a given matrix is very important in the theory and application of matrices. There are several methods of finding the inverse "by hand." Two methods for 2×2 matrices will be shown here. The inverses for larger matrices can be found using technological devices such as graphing calculators.

Method 1

Let $A = \begin{bmatrix} 1 & 1 \\ 4 & 2 \end{bmatrix}$ and let $A^{-1} = \begin{bmatrix} a & c \\ b & d \end{bmatrix}$. It follows then that $AA^{-1} = \begin{bmatrix} 1 & 1 \\ 4 & 2 \end{bmatrix}\begin{bmatrix} a & b \\ c & d \end{bmatrix} = \begin{bmatrix} a+c & b+d \\ 4a+2c & 4b+2d \end{bmatrix}$ $= \begin{bmatrix} 1 & 0 \\ 0 & 1 \end{bmatrix}$. From this you get the two systems of equations:

$$\begin{array}{ll} a+c=1 & b+d=0 \\ 4a+2c=0 & 4b+2d=1 \end{array}$$

Solving these yields $a=-1$, $b=\frac{1}{2}$, $c=2$, and $d=-\frac{1}{2}$. Thus $A^{-1} = \begin{bmatrix} -1 & 1/2 \\ 2 & -1/2 \end{bmatrix}$.

Method 2

$$A^{-1} = \frac{1}{\det(A)} \cdot \begin{bmatrix} 2 & -1 \\ -4 & 1 \end{bmatrix} = \frac{1}{-2} \cdot \begin{bmatrix} 2 & -1 \\ -4 & 1 \end{bmatrix} = \begin{bmatrix} -1 & 1/2 \\ 2 & -1/2 \end{bmatrix}$$

Warning: This method works only for 2×2 matrices. Also note that this will not work if $\det(A) = 0$. In fact, if the determinant of any square matrix is 0, the matrix will not have an inverse. The matrix here is determined from the original matrix by interchanging the elements on the main diagonal and negating the other elements.

In general, if $A = \begin{bmatrix} a & b \\ c & d \end{bmatrix}$, then $A^{-1} = \frac{1}{ad-bc}\begin{bmatrix} d & -b \\ -c & a \end{bmatrix}$ provided $ad - bc \neq 0$.

For 1–7, let $A = \begin{bmatrix} 1 & 2 \\ -1 & 1 \end{bmatrix}$ *and* $B = \begin{bmatrix} 1 & -1 \\ 2 & 1 \end{bmatrix}$. *Compute each of the following.*

1. A^{-1}
2. B^{-1}
3. $(A\,B)^{-1}$
4. $A^{-1}B^{-1}$
5. $B^{-1}A^{-1}$
6. $A(B^{-1})$
7. $-A(B^{-1})$

For 8–10, explain how the two quantities are related.

8. $(AB)^{-1}$ and $B^{-1}A^{-1}$
9. $A(-B^{-1})$ and $-A(B^{-1})$
10. $(AB)^{-1}$ and $A^{-1}B^{-1}$

Systems of linear equations

Determinants can be used to solve systems of equations by using **Cramer's rule**. This method is illustrated in the following examples. Consider the two generic systems:

$$\begin{aligned} a_1x+b_1y &= k_1 \\ a_2x+b_2y &= k_2 \end{aligned} \quad \text{and} \quad \begin{aligned} a_1x+b_1y+c_1z &= k_1 \\ a_2x+b_2y+c_2z &= k_2 \\ a_3x+b_3y+c_3z &= k_3 \end{aligned}$$

The corresponding solutions are

$$\begin{vmatrix} a_1 & b_1 \\ a_2 & b_2 \end{vmatrix} x = \begin{vmatrix} k_1 & b_1 \\ k_2 & b_2 \end{vmatrix} \quad \text{and} \quad \begin{vmatrix} a_1 & b_1 \\ a_2 & b_2 \end{vmatrix} y = \begin{vmatrix} a_1 & k_1 \\ a_2 & k_2 \end{vmatrix}$$

These can be written as $Dx = X$ and $Dy = Y$.

For the 3×3 system:

$$\begin{vmatrix} a_1 & b_1 & c_1 \\ a_2 & b_2 & c_2 \\ a_3 & b_3 & c_3 \end{vmatrix} x = \begin{vmatrix} k_1 & b_1 & c_1 \\ k_2 & b_2 & c_2 \\ k_3 & b_3 & c_3 \end{vmatrix}, \quad \begin{vmatrix} a_1 & b_1 & c_1 \\ a_2 & b_2 & c_2 \\ a_3 & b_3 & c_3 \end{vmatrix} y = \begin{vmatrix} a_1 & k_1 & c_1 \\ a_2 & k_2 & c_2 \\ a_3 & k_3 & c_3 \end{vmatrix}, \text{ and } \begin{vmatrix} a_1 & b_1 & c_1 \\ a_2 & b_2 & c_2 \\ a_3 & b_3 & c_3 \end{vmatrix} z = \begin{vmatrix} a_1 & b_1 & k_1 \\ a_2 & b_2 & k_2 \\ a_3 & b_3 & k_3 \end{vmatrix}$$

Or $Dx = X$, $Dy = Y$, and $Dz = Z$. The matrix D is called the coefficient matrix.

In either case, if $D \neq 0$, there is a unique solution to the system. The solution can then be written $x = \dfrac{X}{D}$, $y = \dfrac{Y}{D}$, and in the 3×3 system, $z = \dfrac{Z}{D}$.

If $D = 0$ and $X \neq 0$ or $Y \neq 0$ or $Z \neq 0$, then there is no solution to the system. If $D = 0$ and $X = Y = Z = 0$, then there are infinitely many solutions to the system.

PROBLEM Solve the system of equations:

$$\begin{aligned} 4x+2y &= 3 \\ -2x+\ y &= 5 \end{aligned}$$

SOLUTION $x = \dfrac{\begin{vmatrix} 3 & 2 \\ 5 & 1 \end{vmatrix}}{\begin{vmatrix} 4 & 2 \\ -2 & 1 \end{vmatrix}} = \dfrac{3\cdot1-5\cdot2}{4\cdot1+2\cdot2} = \dfrac{-7}{8}$, $y = \dfrac{\begin{vmatrix} 4 & 3 \\ -2 & 5 \end{vmatrix}}{8} = \dfrac{26}{8}$. The calculations $4\left(-\dfrac{7}{8}\right)+2\left(\dfrac{26}{8}\right) = -\dfrac{28}{8}+\dfrac{52}{8} = \dfrac{24}{8} = 3$ and $-2\left(-\dfrac{7}{8}\right)+\dfrac{26}{8} = \dfrac{14}{8}+\dfrac{26}{8} = \dfrac{40}{8} = 5$ verify the solution.

PROBLEM Solve the system of equations:

$$\begin{aligned} x+\ y+\ z &= 4 \\ 2x-y-2z &= -1 \\ x-2y-\ z &= 1 \end{aligned}$$

SOLUTION $D = \begin{vmatrix} 1 & 1 & 1 \\ 2 & -1 & -2 \\ 1 & -2 & -1 \end{vmatrix} = 1\begin{vmatrix} -1 & -2 \\ -2 & -1 \end{vmatrix} - 1\begin{vmatrix} 2 & -2 \\ 1 & -1 \end{vmatrix} + 1\begin{vmatrix} 2 & -1 \\ 1 & -2 \end{vmatrix} = -3+0+2-3 = -6,$

$$X = \begin{vmatrix} 4 & 1 & 1 \\ -1 & -1 & -2 \\ 1 & -2 & -1 \end{vmatrix} = -12,\ Y = \begin{vmatrix} 1 & 4 & 1 \\ 2 & -1 & -2 \\ 1 & 1 & -1 \end{vmatrix} = 6,\ Z = \begin{vmatrix} 1 & 1 & 4 \\ 2 & -1 & -1 \\ 1 & -2 & 1 \end{vmatrix} = -18,$$ and finally,

$x = \dfrac{-12}{-6} = 2$, $y = \dfrac{6}{-6} = -1$, and $z = \dfrac{-18}{-6} = 3$. The calculations $2 + (-1) + 3 = 4$, $2(2) - (-1) - 2(3) = -1$, and $2 - 2(-1) - 3 = 1$ verify the solution.

PROBLEM Solve the system of equations:

$$\begin{aligned} 2x - 4y + 2z &= 3 \\ x + y - z &= 2 \\ 3x - 6y + 3z &= 2 \end{aligned}$$

SOLUTION

$$D = \begin{vmatrix} 2 & -4 & 2 \\ 1 & 1 & -1 \\ 3 & -6 & 3 \end{vmatrix} = 2\begin{vmatrix} 1 & -1 \\ -6 & 3 \end{vmatrix} + 4\begin{vmatrix} 1 & -1 \\ 3 & 3 \end{vmatrix} + 2\begin{vmatrix} 1 & 1 \\ 3 & -6 \end{vmatrix} = -6 + 24 - 18 = 0,$$

$$X = \begin{vmatrix} 3 & -4 & 2 \\ 2 & 1 & -1 \\ 2 & -6 & 3 \end{vmatrix} = -5.$$ Consequently, there is no solution.

PROBLEM Solve the system of equations:

$$\begin{aligned} 2x + 2y - 2z &= 4 \\ x + y - z &= 2 \\ 3x - 6y + 3z &= 2 \end{aligned}$$

SOLUTION

$$D = \begin{vmatrix} 2 & 2 & -2 \\ 1 & 1 & -1 \\ 3 & -6 & 3 \end{vmatrix} = 2(-3) - 2(6) - 2(-9) = 0$$

$$X = \begin{vmatrix} 4 & 2 & -2 \\ 2 & 1 & -1 \\ 2 & -6 & 3 \end{vmatrix} = 4(-3) - 2(8) - 2(-14) = 0$$

$$Y = \begin{vmatrix} 2 & 4 & -2 \\ 1 & 2 & -1 \\ 3 & 2 & 3 \end{vmatrix} = 2(8) - 4(6) - 2(-4) = 0$$

$$Z = \begin{vmatrix} 2 & 2 & 4 \\ 1 & 1 & 2 \\ 3 & -6 & 2 \end{vmatrix} = 2(14) - 2(-4) + 4(-9) = 0$$

Now, try to solve two equations such that the determinant of two of the variables is not 0. In this case, use the second and third equations and write them as $\begin{aligned} x + y &= 2 + z \\ 3x - 6y &= 2 - 3z \end{aligned}$. The solution to this 2×2 system is:

$$x=\frac{\begin{vmatrix}2+z & 1\\ 2-3z & -6\end{vmatrix}}{\begin{vmatrix}1 & 1\\ 3 & -6\end{vmatrix}}=\frac{(2+z)(-6)-(2-3z)}{-9}=\frac{-14-3z}{-9}=\frac{14+3z}{9}$$

$$y=\frac{\begin{vmatrix}1 & 2+z\\ 3 & 2-3z\end{vmatrix}}{-9}=\frac{(2-3z)-3(2+z)}{-9}=\frac{-4-4z}{-9}=\frac{4+4z}{9}$$

Variable z is a "free" variable and can be any real number. Hence, there are infinitely many solutions. One solution is $z=0, x=\frac{14}{9}$, and $y=\frac{4}{9}$.

EXERCISE

14·5

Solve, using Cramer's rule.

1. $2x-3y=16$
   $5x-2y=-4$
2. $x-2y=7$
   $3x+2y=5$
3. $5x-2y=3$
   $x+6y=1$
4. $x-2y-3z=-20$
   $2x+4y-5z=11$
   $3x+7y-4z=33$
5. $x+2y-z=7$
   $4x+3y+2z=1$
   $9x+8y+3z=4$
6. $2x-4y+7z=5$
   $3x+2y-z=2$
   $x-10y+15z=8$

Matrix solutions of systems of linear equations

You also can use matrices to solve systems of linear equations. Technological advances have made this approach more or less the technique of choice since the calculations involved, many of which are tedious, can be done rapidly using appropriate hardware such as computers or sophisticated calculators. To make optimum use of this approach, however, you must completely understand the theory and use of the matrix algebra involved. The basic concepts are introduced here and then exemplified in solving simple equations. Finally, an example is given using a graphing calculator to solve a system of equations.

By using the definitions of matrix multiplication and matrix equality, it follows that the linear system $\begin{matrix}a_1x+b_1y=c_1\\ a_2x+b_2y=c_2\end{matrix}$ is equivalent to the matrix equation $\begin{bmatrix}a_1 & b_1\\ a_2 & b_2\end{bmatrix}\begin{bmatrix}x\\ y\end{bmatrix}=\begin{bmatrix}c_1\\ c_2\end{bmatrix}$. In matrix notation this is of the form $AX=C$. If A^{-1} exists, then $A^{-1}AX=A^{-1}C$; and, thus, $IX=A^{-1}C$ or $X=A^{-1}C$. Then you can easily determine the solution.

PROBLEM Solve:

$3x-2y=4$
$2x-y=3$

SOLUTION Since $A=\begin{bmatrix}3 & -2\\ 2 & -1\end{bmatrix}, A^{-1}=\begin{bmatrix}-1 & 2\\ -2 & 3\end{bmatrix}$. The solution is $\begin{bmatrix}x\\ y\end{bmatrix}=\begin{bmatrix}-1 & 2\\ -2 & 3\end{bmatrix}\begin{bmatrix}4\\ 3\end{bmatrix}=\begin{bmatrix}2\\ 1\end{bmatrix}$ or $x=2$ and $y=1$. The calculations $3(2)-2(1)=4$ and $2(2)-1=3$ verify the solution.

PROBLEM Solve:

$$\begin{aligned} x+y+z&=4\\ 2x-y-2z&=-1\\ x-2y-z&=1 \end{aligned}$$

SOLUTION Since $A=\begin{bmatrix}1&1&1\\2&-1&-2\\1&-2&-1\end{bmatrix}$ and $C=\begin{bmatrix}4\\-1\\1\end{bmatrix}$, you need a technique for finding A^{-1}, and to this point there is no such technique. However, if you enter A and C into a graphing calculator and do the calculation $A^{-1}C$, the solution is $x=2$, $y=-1$, and $z=3$, which can be verified by substitution into the original system. This example illustrates the use and power of the technology.

There are techniques for finding inverses of matrices larger than 2×2, but they are put on hold to discuss other concepts referred to as row transformations and equivalence of matrices.

Two matrices are **equivalent** if and only if one is obtained from the other by performing one or more of the following **row transformations**:

1. interchanging any two rows;
2. multiplying each element of a row by the same nonzero constant; or
3. adding a nonzero constant multiple of the elements of one row to the corresponding elements of another row.

Note: Equivalence is not to be confused with equality. The symbol ~ is used to indicate equivalence.

Column transformations are similar, but row transformations will suffice for the examples.

These concepts are brought up at this time because judicious use of these ideas will enable you to "reduce" a given system of equations to an equivalent system in which the solutions are obvious or at least simpler to solve. A little more vocabulary is introduced here as well.

For the system of the form $\begin{bmatrix}a_1&b_1\\a_2&b_2\end{bmatrix}\begin{bmatrix}x\\y\end{bmatrix}=\begin{bmatrix}c_1\\c_2\end{bmatrix}$, the matrix $\begin{bmatrix}a_1&b_1&c_1\\a_2&b_2&c_2\end{bmatrix}$ is the **augmented matrix** of the system. The augmentations are similar for larger systems of equations. The following examples illustrate the "reduction" referred to earlier.

PROBLEM Solve:

$$\begin{aligned} 3x-2y&=4\\ 2x-y&=3 \end{aligned}$$

SOLUTION

1. Form the augmented matrix. $\begin{bmatrix}3&-2&4\\2&-1&3\end{bmatrix}$
2. Multiply the second row by –2 and add to the first row. $\begin{bmatrix}-1&0&-2\\2&-1&3\end{bmatrix}$
3. Multiply the first row by 2 and add to the second row. $\begin{bmatrix}-1&0&-2\\0&-1&-1\end{bmatrix}$
4. Multiply the first and second rows by –1. $\begin{bmatrix}1&0&2\\0&1&1\end{bmatrix}$

The "reduced" equivalent system is now $x = 2$ and $y = 1$, for which the solution is obvious. The equivalent systems all have the same solution set, so this solution is also the solution of the original system.

PROBLEM Solve the system:

$$\begin{aligned} x-2y+3z&=1\\ x+3y-z&=4\\ 2x+y-2z&=13 \end{aligned}$$

SOLUTION

$$\begin{bmatrix} 1 & -2 & 3 & 1 \\ 1 & 3 & -1 & 4 \\ 2 & 1 & -2 & 13 \end{bmatrix} \sim \begin{bmatrix} 1 & -2 & 3 & 1 \\ 0 & 5 & -4 & 3 \\ 2 & 1 & -2 & 13 \end{bmatrix}$$ Multiply row 1 by –1 and add to row 2.

$$\sim \begin{bmatrix} 1 & -2 & 3 & 1 \\ 0 & 5 & -4 & 3 \\ 0 & 5 & -8 & 11 \end{bmatrix}$$ Multiply row 1 by –2 and add to row 3.

$$\sim \begin{bmatrix} 1 & -2 & 3 & 1 \\ 0 & 5 & -4 & 3 \\ 0 & 0 & -4 & 8 \end{bmatrix}$$ Multiply row 2 by –1 and add to row 3.

$$\sim \begin{bmatrix} 1 & -2 & 3 & 1 \\ 0 & 5 & 0 & -5 \\ 0 & 0 & -4 & 8 \end{bmatrix}$$ Multiply row 3 by –1 and add to row 2.

$$\sim \begin{bmatrix} 1 & -2 & 3 & 1 \\ 0 & 1 & 0 & -1 \\ 0 & 0 & 1 & -2 \end{bmatrix}$$ Multiply row 2 by $\frac{1}{5}$ and row 3 by $-\frac{1}{4}$.

The solution is $z = -2, y = -1$, and $x - 2y + 3z = x - 2(-1) + 3(-2) = 1$ so that $x = 5$. The calculations $5 + 3(-1) - (-2) = 4$ and $2(5) + (-1) - 2(-2) = 13$ verify the solution.

EXERCISE 14·6

For 1–2, solve using inverse matrices.

1. $\begin{aligned} 3x+2y&=4\\ x-2y&=3 \end{aligned}$

2. $\begin{aligned} 5x-y&=3\\ 2x+2y&=2 \end{aligned}$

For 3–5, solve using row transformations.

3. $\begin{aligned} 3x+2y&=4\\ x-2y&=3 \end{aligned}$

4. $\begin{aligned} x-3y+z&=2\\ 2x-y-2z&=1\\ 3x+2y-z&=5 \end{aligned}$

5. $\begin{aligned} 2x-y&=0\\ 2y-z&=0\\ x+2y-z&=3 \end{aligned}$

Sequences

Sequences and series

A sequence is a function whose domain is a subset of the integers, usually the natural numbers $N = \{1, 2, 3, 4, \ldots\}$ or the whole numbers $W = \{0,1,2,3,4, \ldots\}$. A sequence can be thought of as an ordered set of numbers in the sense that there is a first term, a second term, and so on. A listing of the whole numbers $0,1,2,3,\ldots$ is an example of a sequence. If there is a pattern in the listing, then it is very important that the pattern be characterized in some way. In practice, there are essentially two ways of characterizing sequences: **closed-form** characterizations and **recursive form** characterizations. Examples of each will be given after some initial notation and vocabulary is presented.

In the listing $a_1, a_2, a_3, a_4, \ldots$ the subscripts denote the term number, and the a values are called the *terms* of the sequence. For example, the third term of the sequence is a_3, and the nth term is a_n. The nth term is also called the *general term* of the sequence.

The characterization of a sequence is a connection between the term number and the term itself, usually by a formula relating the two. For example, the sequence $0,1,2,3,\ldots$ can be characterized by the formula $a_n = n - 1$, where $n = 1, 2, 3, 4, \ldots$. Notice that any term of the sequence can be determined by this formula. This is an example of a **closed-form** characterization.

A famous sequence, called the *Fibonacci sequence*, $1,1,2,3,5,8,13,\ldots$ can be characterized by $a_1 = 1$, $a_2 = 1$, $a_n = a_{n-1} + a_{n-2}$, where $n = 3, 4, 5, \ldots$ This is an example of a **recursive form** characterization. In general, a recursive form requires a beginning term(s), sometimes called a seed term(s), and some previous terms to generate the terms that follow.

A recursive form has the disadvantage that to generate a term, you need to know some of the previous terms. However, the recursive form is usually preferred when you are writing computer programs.

If a formula for a sequence is known, its terms can be generated with relative ease. However, one of the challenges of working with sequences lies in divining the formula when the listing is known but the formula is missing. Success in this endeavor requires some experience and being comfortable with notation. The problems and examples in this chapter are designed to help with both of these requirements. In particular, two special types of sequences are discussed to help hone your abilities and because they are found in many different disciplines with numerous applications.

PROBLEM Write the closed formula for the sequences.

a. $1, \frac{1}{2}, \frac{1}{3}, \frac{1}{4}, \ldots$

b. $3, 5, 7, 9, \ldots$

SOLUTION a. $a_n = \frac{1}{n}$

b. $a_n = 2n + 1$

PROBLEM Write the recursive formula for the sequences.

a. $2, 5, 8, 11, \ldots$

b. $1, 4, 5, 9, 14, 23, \ldots$

SOLUTION a. $a_1 = 2,\ a_{n+1} = a_n + 3,\ n = 1, 2, 3, \ldots$

b. $a_1 = 1,\ a_2 = 4,\ a_n = a_{n-1} + a_{n-2},\ n = 3, 4, 5, \ldots$

For 1–3, write the closed formula for the sequence.

1. 5,6,7,8,...
2. 3,6,9,12,...
3. 4,16,64,256,...

For 4–6, write the recursive formula for the sequence.

4. 1,2,3,5,8,13,...
5. 6,8,10,12,...
6. 5,10,15,20,...

Arithmetic and geometric sequences

An **arithmetic sequence** begins with a first (seed) term, and then you find each successive term by adding a fixed constant d, the **common difference**, to the preceding term. For example, $2, 5, 8, 11, 14, \ldots$ is an arithmetic sequence with first term 2 and common difference 3.

All arithmetic sequences have the same form: $s, s+d, s+2d, s+3d, s+4d, \ldots, s+(n-1)d, \ldots$, where s is the first term and d is the common difference. If the sequence is identified by using the subscript notation, then the form is $a_1, a_1+d, a_1+2d, a_1+3d, a_1+4d, \ldots, a_1+(n-1)d, \ldots$. This form leads to a **closed formula** for an arithmetic sequence: $a_n = a_1 + (n-1)d,\ n = 1, 2, 3, 4, \ldots$. The **recursive formula** for an arithmetic sequence is $a_1 = s$ and $a_{n+1} = a_n + d$, where s is the first term and $n = 1, 2, 3, 4, \ldots$. The common difference, d, is so named because $d = a_{n+1} - a_n$.

A **geometric sequence** begins with a first (seed) term, and then you find each successive term by multiplying the preceding term by a fixed constant r, the **common ratio**. For example, $3, 6, 12, 24, 48, \ldots$ is a geometric sequence with seed term 3 and the common ratio 2.

All geometric sequences have the same form: $s, sr, sr^2, sr^3, sr^4, \ldots, sr^{n-1}, \ldots$, where s is the seed term and r is the common ratio. If the sequence is identified using the subscript notation, then the form is $a_1, a_1r, a_1r^2, a_1r^3, \ldots, a_1r^{n-1}, \ldots$. This form leads to a **closed formula** for a geometric sequence: $a_n = a_1 r^{n-1}$, $n = 1, 2, 3, 4, \ldots$. The **recursive formula** for a geometric sequence is $a_1 = s$ and $a_{n+1} = ra_n$, where s is the seed term and $n = 1, 2, 3, 4, \ldots$. The common ratio, r, is so named because $r = \frac{a_{n+1}}{a_n}$.

PROBLEM Write the closed and recursive formulas for the following sequences:

a. $3, 7, 11, 15, 19, \ldots$

b. $4, 2, 1, \frac{1}{2}, \frac{1}{4}, \ldots$

SOLUTION a. This is an arithmetic sequence with $s = 3$ and $d = 4$. The closed formula is $a_n = 3 + (n-1)4 = 4n - 1$. The recursive formula is $a_1 = 3$ and $a_{n+1} = a_n + 4$.

b. This is a geometric sequence with $s = 4$ and $r = \frac{1}{2}$. The closed formula is $a_n = 4\left(\frac{1}{2}\right)^{n-1}$. The recursive formula is $a_1 = 4$ and $a_{n+1} = \frac{1}{2}a_n$.

PROBLEM Determine whether the following sequences are arithmetic, geometric, both, or neither:

a. $1, 0.5, 0.25, 0.125, \ldots$

b. $\frac{1}{4}, \frac{3}{4}, \frac{7}{8}, 1, \ldots$

c. $5, 4, 3, 2, \ldots$

d. $2, 2, 2, 2, \ldots$

SOLUTION Use the common difference $s = a_{n+1} - a_n$ and common ratio $r = \frac{a_{n+1}}{a_n}$ formulas to check.

a. $0.5 - 1 = -0.5$, $0.25 - 0.5 = -0.25$. These differences are not the same, so the sequence is not arithmetic.

$\frac{0.5}{1} = 0.5$, $\frac{0.25}{0.5} = 0.5$, $\frac{0.125}{0.25} = 0.5$. The ratios are the same, so the sequence is geometric.

b. $\frac{3}{4} - \frac{1}{4} = \frac{2}{4} = \frac{1}{2}$, $\frac{7}{8} - \frac{3}{4} = \frac{1}{8}$. These differences are not the same, so the sequence is not arithmetic.

$\frac{\frac{3}{4}}{\frac{1}{4}} = 3$, $\frac{\frac{7}{8}}{\frac{3}{4}} = \frac{7}{8} \cdot \frac{4}{3} = \frac{7}{6}$. These ratios are not the same, so the sequence is not geometric.

c. $4 - 5 = -1$, $3 - 4 = -1$, $2 - 3 = -1$. These differences are the same, so the sequence is arithmetic.

$\frac{4}{5} \neq \frac{3}{4}$. These ratios are not the same, so the sequence is not geometric.

d. The sequence $2, 2, 2, 2, \ldots$ is arithmetic with $d = 0$ and geometric with $r = 1$.

PROBLEM

a. What is the 20th term of the sequence 1, 4, 7, 10, 13,...?

b. What is the 12th term of the sequence 1, 2, 4, 8, ...?

c. How many terms are indicated in the listing 2, 7, 12, 17, ... ,147?

d. How many terms are indicated in the listing 2, 4, 8, 16, ... ,4,096?

SOLUTION

a. This is an arithmetic sequence whose formula is $a_n = 1 + (n-1)3 = 3n - 2$. The 20th term is $a_{20} = 3(20) - 2 = 58$.

b. This is a geometric sequence whose formula is $a_n = 1(2)^{n-1} = 2^{n-1}$. The 12th term is $a_{12} = 2^{12-1} = 2^{11} = 2{,}048$.

c. These are terms of an arithmetic sequence whose formula is $a_n = 2 + (n-1)5 = 5n - 3$. The number 147 must satisfy this formula. Thus, $147 = 5n - 3$. Solving this equation for n yields $n = \frac{147+3}{5} = \frac{150}{5} = 30$. There are 30 terms.

d. These are terms of a geometric sequence whose formula is $a_n = 2(2)^{n-1} = 2^n$. The number 4,096 must satisfy this formula. Thus, $4{,}096 = 2^n$. Trial-and-error guessing leads to the solution $n = 12$. (This problem can also be solved by using logarithms; see Chapter 11, "Exponential and logarithmic functions.")

EXERCISE 15·2

For 1–5, identify the seed term and the common difference, and write the closed formula for the sequence. Use the formula to find the 10th and 12th terms of the sequence.

1. $2, 7, 12, 17, \ldots$
2. $\frac{3}{2}, \frac{9}{4}, 3, \frac{15}{4}, \ldots$
3. $3.10, 3.25, 3.40, \ldots$
4. $8, 6, 4, 2, \ldots$
5. $\frac{5}{7}, \frac{3}{14}, \frac{-2}{7}, \frac{-11}{14}, \ldots$

For 6–10, identify the seed term and the common ratio, and write the closed formula for the sequence. Use the formula to find the 10th term of the sequence.

6. $243, 81, 27, 9, \ldots$
7. $\frac{1}{3}, -1, 3, -9, \ldots$
8. $\frac{-7}{8}, \frac{7}{4}, \frac{-7}{2}, 7, \ldots$
9. $2, 4, 8, 16, \ldots$
10. $\frac{1}{2}, \frac{1}{3}, \frac{2}{9}, \frac{4}{27}, \ldots$

Series and summation notation

If a finite number of terms of a sequence are given by $a_1, a_2, a_3, a_4, \ldots, a_n$, the sum of the terms $s_n = a_1 + a_2 + a_3 + a_4 + \ldots + a_n$ is called a **finite series** or a **partial sum**. This sum can also be written using the summation (sigma) notation $s_n = \sum_{k=1}^{n} a_k$. The k subscript is the summing index. This notation is a convenient shorthand and is used extensively when working with series. Some useful properties of the summation (sigma) notation are as follows:

1. $\sum_{k=1}^{n} c = nc$. A constant, c, added n times is nc.
2. $\sum_{k=1}^{n}(a_k \pm b_k) = \sum_{k=1}^{n} a_k \pm \sum_{k=1}^{n} b_k$. The terms can be summed or subtracted individually.
3. $\sum_{k=1}^{n} ca_k = c\sum_{k=1}^{n} a_k$. A constant can be factored out of a summation.
4. $\sum_{k=1}^{n} a_k = \sum_{k=1}^{m} a_k + \sum_{k=m+1}^{n} a_k$; $1 \le m < n$. Parts can be summed individually.
5. $\sum_{k=1}^{n} a_k = \sum_{j=1}^{n} a_j = \sum_{m=0}^{n-1} a_{m+1}$. The subscripting can be adjusted to fit special purposes.
6. $\sum_{k=1}^{n} a_k = a_1 + a_2 + \sum_{k=3}^{n-1} a_k + a_n$. Terms can be "taken out" of the sum if needed.

PROBLEM Given the sequence whose formula is $a_n = 5n - 3$:

a. Write the sum of the first 15 terms, using the sigma notation where the first two terms have been taken out of the sum.

b. Write the sum of the first n terms, using the sigma notation such that the summing index begins with 2.

c. Write the sum of the first 18 terms and apply properties 1, 2, and 3 above to rewrite the sum.

d. Write the extended form of the sum $\sum_{k=1}^{n}(5k-3)$.

SOLUTION a. $2 + 7 + \sum_{k=3}^{15}(5k-3)$

b. $\sum_{k=1}^{n}(5k-3) = \sum_{j=2}^{n+1}(5j-8)$. A symbol substitution was made: $k = j - 1$. When $k = 1, j = 2$ and when $k = n, j = n + 1$.

c. $\sum_{k=1}^{18}(5k-3) = 5\sum_{k=1}^{18} k - \sum_{k=1}^{18} 3 = 5\sum_{k=1}^{18} k - 18(3) = 5\sum_{k=1}^{18} k - 54$

d. $\sum_{k=1}^{n}(5k-3) = 2 + 7 + 12 + 17 + \ldots + (5n-3)$

PROBLEM Write the sum $3 + 5 + 7 + \ldots + 873$, using the sigma notation.

SOLUTION This is an arithmetic sequence whose formula is $a_n = 2n + 1$. The term 873 must satisfy the formula so $873 = 2n + 1$. Solve to get $n = 436$. Then $3 + 5 + 7 + \ldots + 873 = \sum_{k=1}^{436}(2k+1)$.

EXERCISE

15·3

For 1–5, use the sigma notation to write the sum of the first n terms of the sequence, given the following information.

1. $a_1 = 4$ and $a_{n+1} = a_n + 5$
2. $a_n = 4\left(\frac{1}{3}\right)^{n-1}$
3. $a_n = 4 + (n-1)6$
4. $5, 3, 1, -1, \ldots$
5. $\frac{1}{6}, \frac{1}{15}, \frac{4}{150}, \frac{8}{750}, \ldots$
6. Change the sum $\sum_{k=1}^{n}(3k+6)$ to an equivalent sum whose index begins with 0.
7. Change the sum $\sum_{k=1}^{n}(3k+6)$ to an equivalent sum with the last two terms taken out.
8. Change the sum $\sum_{k=1}^{n}(3k+6)$ to an equivalent sum, using properties 1, 2, and 3 above.
9. Change the sum $\sum_{k=1}^{n}4\left(\frac{1}{5}\right)^{k}$ to two sums with the first being the sum from 1 to 18.
10. Change the sum $\sum_{k=1}^{n}4\left(\frac{1}{5}\right)^{k}$ to an equivalent sum whose index begins with 3.

Arithmetic and geometric series

The partial sums of arithmetic and geometric sequences can be expressed in extended form, by sigma notation, and by closed formulas. The extended form and the sigma notation are not closed formulas.

The three forms for the partial sum of an arithmetic sequence, the latter being the closed form, are $s_n = a_1 + a_2 + \ldots + a_n = \sum_{k=1}^{n} a_k = \frac{n(a_1 + a_n)}{2}$.

The three forms for the partial sum of a geometric sequence, the latter being the closed form, are $s_n = a_1 + a_1 r + \ldots + a_1 r^{n-1} = \sum_{k=1}^{n} a_1 r^{k-1} = \frac{a_1 - a_1 r^n}{1-r} = \frac{a_1(1-r^n)}{1-r}$ provided $r \neq 1$. If $r = 1$, then $s_n = na_1$.

If you were to add the first 70 positive odd numbers by "brute force," it would take a considerable amount of time even with the use of a calculator. However, the odd numbers are an arithmetic sequence whose formula is $a_n = 2n - 1$. Thus, $s_{70} = 1 + 3 + 5 + \ldots + 139 = \frac{70(1+139)}{2} = 4{,}900$.

One celebrated sum that you encounter frequently in math is the sum of the first n natural numbers. This sum can be characterized by using the closed form for arithmetic sequences: $1+2+3+\ldots+n=\sum_{k=1}^{n} k=\frac{n(n+1)}{2}$. Many other summation formulas can be generated by using the closed formula for arithmetic series or for geometric series.

PROBLEM

a. Calculate the indicated sum $2+4+6+8+\ldots+888$.

b. Calculate the indicated sum $3+7+11+15+\ldots+239$.

c. Calculate the indicated sum $\frac{1}{10}+\frac{1}{100}+\frac{1}{1000}+\ldots+\frac{1}{10{,}000{,}000}$.

d. Calculate the indicated sum $1+3+5+7+\ldots+2n-1$.

e. Write the first four terms of the geometric sequence given $r=3$ and $s_6=1{,}456$.

SOLUTION

a. $s_n=\frac{444(2+888)}{2}=197{,}580$

b. $s_n=\frac{60(3+239)}{2}$. The value $n=60$ was determined by solving $239=4n-1$.

c. $s_n=\frac{1}{10}+\frac{1}{10}\left(\frac{1}{10}\right)+\ \ldots+\frac{1}{10}\left(\frac{1}{10}\right)^6=\frac{\frac{1}{10}\left(1-\left(\frac{1}{10}\right)^7\right)}{1-\frac{1}{10}}=\frac{\frac{1}{10}}{\frac{9}{10}}\left(\frac{10^7-1}{10^7}\right)$

$=\frac{1}{9}\left(\frac{999999}{10^7}\right)=\left(\frac{111111}{10^7}\right)$

d. $s_n=1+3+5+\ldots+(2n-1)=\frac{n(2n-1+1)}{2}=\frac{n(2n)}{2}=n^2$. The sum of the first n odd numbers is n^2. Pretty neat.

e. Since $s_n=\frac{a_1(1-r^n)}{1-r}$, it follows that $1456=\frac{a_1(1-3^6)}{1-3}=\frac{a_1(728)}{2}=364a_1$. Hence, $a_1=4$. The first four terms are $4,12,36,108$.

EXERCISE 15·4

Use the summation formulas to evaluate the indicated sums.

1. $\sum_{k=1}^{30}(3k-4)$
2. $\sum_{j=1}^{32}\left(\frac{1}{2}j-5\right)$
3. $\sum_{n=2}^{30}(8-3n)$
4. $\sum_{k=2}^{10}3\left(\frac{1}{3}\right)^k$
5. $\sum_{k=0}^{10}2^k$
6. $-12+(-9.5)+(-7)+(-4.5)+\ldots+23$

7. $\sqrt{2}+2\sqrt{2}+3\sqrt{2}+4\sqrt{2}+\ldots+20\sqrt{2}$

8. Compute s_{20} for $\frac{9}{2}+\frac{7}{2}+\frac{5}{2}+\ldots$

9. Compute s_8 for $\frac{1}{5}+\frac{1}{25}+\frac{1}{125}+\ldots$

10. Write the first four terms of the arithmetic sequence given $a_1=5$ and $s_{10}=230$.

Mathematical induction

The principle of mathematical induction is a technique for proving that a statement involving natural numbers is true. In many instances, you observe a pattern involving natural numbers, and then you use induction to verify that the pattern you observed is indeed true for all natural numbers.

The principle of mathematical induction

Let S_n be a statement involving natural numbers, where $n=1, 2, 3, \ldots, k, k+1, \ldots$. If:

1. S_1 is true, and
2. if S_k being true implies S_{k+1} is true,

then S_n is true for all natural numbers n.

It is important to realize that a proof by mathematical induction is a two-part proof. The statement "S_k is true" is termed the *induction hypothesis* (IH). In the examples, IH will be indicated to point out where the induction hypothesis is used in the proof.

An example of a statement S_n is "$1+2+3+4+\ldots+n=\frac{n(n+1)}{2}$."

PROBLEM Prove S_n: $1+2+3+4+\ldots+n=\frac{n(n+1)}{2}$ is true for all natural numbers n.

SOLUTION Proof by induction:

1. $1=\frac{1(1+1)}{2}$. Thus, S_1 is true.
2. If S_k is true, then $1+2+3+\ldots+k=\frac{k(k+1)}{2}$. Thus:

$$\begin{aligned}
S_{k+1} &= 1+2+3+\ldots+(k+1)=1+2+3+\ldots+k+(k+1)\\
&= \frac{k(k+1)}{2}+(k+1) \text{ IH}\\
&= (k+1)\left(\frac{k}{2}+1\right)=(k+1)\left(\frac{k+2)}{2}\right)\\
&= \frac{(k+1)((k+1)+1)}{2}
\end{aligned}$$

Thus, S_{k+1} is true; consequently, S_n is true for all natural numbers n.

PROBLEM Prove: The sum of the first n terms of an arithmetic sequence $a_1, a_1+d, a_1+2d, \ldots$ is $s_n = \frac{n(a_1+a_n)}{2}$.

SOLUTION Proof by induction:

1. $a_1 = \frac{1(a_1+a_1)}{2}$. Thus, S_1 is true.
2. If S_k is true, then $S_k = a_1 + a_2 + \ldots + a_k = \frac{k(a_1+a_k)}{2}$. Then:

$$S_{k+1} = a_1 + a_2 + \ldots + a_{k+1} = a_1 + a_2 + \ldots + a_k + a_{k+1}$$
$$= \frac{k(a_1+a_k)}{2} + a_{k+1} \text{ IH}$$
$$= \frac{k(a_1+a_k)}{2} + (a_1 + kd)$$
$$= \frac{ka_1+ka_k}{2} + \frac{2a_1+2kd}{2} = \frac{ka_1+a_1+a_1+ka_k+2kd}{2}$$
$$= \frac{ka_1+a_1+a_1+ka_k+kd+kd}{2}$$
$$= \frac{ka_1+a_1+a_1+kd+ka_k+kd}{2}$$
$$= \frac{ka_1+a_1+(a_1+kd)+k(a_k+d)}{2}$$
$$= \frac{ka_1+a_1+a_{k+1}+ka_{k+1}}{2} \text{ using closed and recursive forms of } a_k$$
$$= \frac{(k+1)a_1+(k+1)a_{k+1}}{2}$$
$$= \frac{(k+1)(a_1+a_{k+1})}{2}$$

Hence, $S_{k+1} = \frac{(k+1)(a_1+a_{k+1})}{2}$. Consequently, S_n is true for all natural numbers n.

PROBLEM Prove: $1+4+9+16+\ldots+n^2 = \frac{n(n+1)(2n+1)}{6}$ is true for all natural numbers n.

SOLUTION Proof by induction:

1. $\frac{1(1+1)(2(1)+1)}{6} = \frac{2(3)}{6} = 1 = 1^2$. Thus, the sum is true for $n = 1$.
2. If the sum is true for $n = k$, then $1+4+9+16+\ldots+k^2 = \frac{k(k+1)(2k+1)}{6}$.

$$S_{k+1} = 1+4+9+16+\ldots+(k+1)^2 = 1+4+9+16+\ldots+k^2+(k+1)^2$$
$$= \frac{k(k+1)(2k+1)}{6} + (k+1)^2 \text{ IH}$$
$$= \frac{k(k+1)(2k+1)}{6} + \frac{6(k+1)^2}{6}$$

$$= \frac{(k+1)(2k^2+k+6k+6)}{6}$$

$$= \frac{(k+1)(2k^2+7k+6)}{6}$$

$$= \frac{(k+1)(k+2)(2k+3)}{6}$$

$$= \frac{(k+1)((k+1)+1)(2(k+1)+1)}{6}$$

The sum is true for S_{k+1}; and, consequently, S_n is true for all natural numbers n.

PROBLEM Prove: $3^n \geq 2n+1$ for all natural numbers n.

SOLUTION Proof by induction:

1. $3^1 = 2(1)+1$. The statement is true for $n=1$. $3^2 = 9 > 5 > 2(2)+1$ and the statement is also true for $n=2$.
2. If the statement is true for $n=k$, then $3^k \geq 2k+1$.

$$3^{k+1} = 3(3^k)$$

$$> 3(2k+1) \text{ IH}$$

$$\geq 6k+3 \geq 2k+3 = 2(k+1)+1$$

Thus, $3^{k+1} \geq 2(k+1)+1$; thus, the statement is true for all natural numbers n.

EXERCISE

15·5

Use mathematical induction to prove the statement is true for all natural numbers n.

1. $2+4+6+\ldots+2n = n(n+1)$
2. $5+9+13+\ldots+(4n+1) = n(2n+3)$
3. $2+4+8+\ldots+2^n = 2^{n+1}-2$
4. $\frac{1}{3}+\frac{1}{15}+\frac{1}{35}+\ldots+\frac{1}{(2n-1)(2n+1)} = \frac{n}{2n+1}$
5. $3+9+27+\ldots+3^n = \frac{3(3^n-1)}{2}$
6.
7. $\frac{1}{2}+\frac{1}{6}+\frac{1}{12}+\ldots+\frac{1}{n(n+1)} = \frac{n}{n+1}$
8. $s+sr+sr^2+sr^3+\ldots+sr^{n-1} = \frac{s(1-r^n)}{1-r}$, provided $r \neq 1$
9. $1+8+27+64+\ldots+n^3 = \frac{n^2(n+1)^2}{4}$
10. $3(4^{n-1}) \leq 4^n - 1$

PRECALCULUS TRIGONOMETRY

Trigonometry is the outcome of concerted attempts to understand certain ratios, periodic occurrences, surveying measurements, astronomical measurements, and ideas associated with angle measurements. One of the earliest records of some of the trigonometric ideas was found on a Babylonian tablet referred to as Plimpton 322 (c. 1900–1600 B.C.).

Trigonometry provides a very practical use of mathematics in any type of problem that involves angles. It does not require the sophistication of calculus for many of its applications. It is, however, a powerful tool that helps explain many of the concepts encountered in the study of calculus. To be comfortable using trigonometry as a tool, it is essential that you have a thorough understanding of its basic ideas. The basic ideas are essentially the degree and radian measure of angles, six special angles, three special ratios, two special triangles, the unit circle, and three special graphs. If you can master these ideas, your study of trigonometry will be greatly simplified.

Trigonometric functions

Angles and their measure

What is an angle? Although it may seem obvious, the answer to that question has been altered throughout history. The description of an angle given by Euclid (325–265 B.C.), as interpreted by some authors, is the following: "an angle is the inclination of one to another of two lines that meet in a plane." One definition widely used today employs the concept of a ray. A **ray** is a line extending from a point. When two rays meet at a common point, they form an **angle**. Euclid's description is essentially the same. The point where the rays meet is called the **vertex** of the angle. See Figure 16.1.

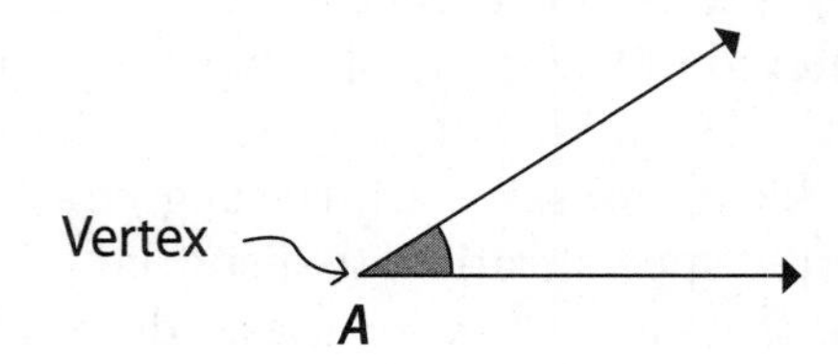

Figure 16.1 Angle *A*

The unit for the measurement of angles varies with the user, but in mathematics the two common units studied are the degree and the radian. Both of these units are associated with a circle. The symbol ° is used to denote degrees.

If the circumference of a circle is divided into 360 equal parts, then one **degree** is the measure of the angle that subtends an arc on the circle of length $\frac{1}{360}$th of the circumference. A **radian** is the measure of an angle that subtends an arc of length equal to the radius of a circle. Since both of these measures are widely used, it is important to understand the connection between them.

The number π is defined as the ratio of the circumference of a circle to its diameter and is constant regardless of the size of the circle. This ratio is written $\pi = \frac{C}{d}$, with C being the measure of the circumference and d the measure of the diameter. Since the diameter is twice the radius, $d = 2r$. Thus $C = 2\pi r$ and this in turn implies that there are 2π radians in a circle. Since there is 360° in a circle, it follows that $360° = 2\pi$ radians. This connection can be shortened to $180° = \pi$ radians. You use this equality to quickly convert between the two measurements. In fact, you have:

$$x° = \frac{x\pi}{180} \text{ radians and } x \text{ radians} = \left(\frac{180x}{\pi}\right)^{\circ}$$

An angle of measure 90° is a **right angle**. An **acute angle** is an angle of measure less than 90° and greater than 0°. An **obtuse angle** is an angle of measure greater than 90° and less than 180°. A **straight angle** is an angle of measure 180°. Two angles that sum to 90° are **complementary angles**, and two that sum to 180° are **supplementary angles**.

Table 16.1 shows the degree and radian measure of the **six special angles of trigonometry**. It is helpful to know these angles in both types of measurements.

Table 16.1 Degree and radian measure of special angles

Degrees	0°	30°	45°	60°	90°	180°
Radians	0	$\frac{\pi}{6}$	$\frac{\pi}{4}$	$\frac{\pi}{3}$	$\frac{\pi}{2}$	π

PROBLEM

a. Convert 27° to radians.

b. Convert $\frac{\pi}{8}$ radian to degrees.

SOLUTION

a. $27^\circ = \left(\frac{27\pi}{180}\right) \approx 0.47124$ radian

b. $\frac{\pi}{8}$ radian $= \left(\frac{180 \cdot \frac{\pi}{8}}{\pi}\right)^\circ = \left(\frac{22.5\pi}{\pi}\right)^\circ = 22.5^\circ$

When you need finer measurements of angles, each degree is divided into 60 equal **minutes** and each minute is divided into 60 equal **seconds**. The angle whose measure is 42 degrees, 21 minutes, and 13 seconds is written 42°21′ 13″. Most users today use decimal measures for angles, but conversions from one form to the other are needed at times so an example of each conversion is illustrated.

PROBLEM

a. Convert 42° 21′13″ to decimal form.

b. Convert 36.7426° to degree, minute, second form.

SOLUTION

a. Since $1' = \frac{1}{60}$ of a degree and $1'' = \frac{1}{3{,}600}$ of a degree,

$$42^\circ 21' 13'' = \left[42 + 21\left(\frac{1}{60}\right) + 13\left(\frac{1}{3{,}600}\right)\right]^\circ \approx 42.3536^\circ.$$

b.
$$\begin{aligned} 36.7426^\circ &= 36^\circ + 0.7426^\circ \\ &= 36^\circ + 0.7426^\circ(60)' \\ &= 36^\circ + 44.556' \\ &= 36^\circ + 44' + 0.556' \\ &= 36^\circ + 44' + 0.556'\,(60)'' \\ &\approx 36^\circ + 44' + 33'' \end{aligned}$$

EXERCISE

16·1

For 1–8, convert the angle as indicated.

1. Convert 1° to radians.
2. Convert 1 radian to degrees.
3. Convert 30° to radians.

4. Convert $\frac{\pi}{6}$ radians to degrees.
5. Convert 147° to radians.
6. Convert 22 radians to degrees.
7. Convert 27° 53′ 25′ to decimal form.
8. Convert 57.5692° to degree, minute, second form.

For 9–10, answer true or false.

9. True or false? The angle 39° is an obtuse angle.
10. True or false? The angle 457° is an obtuse angle.

Right angle trigonometry

To define the six trigonometric ratios, you begin with a right triangle ABC (see Figure 16.2) and label its parts as follows:

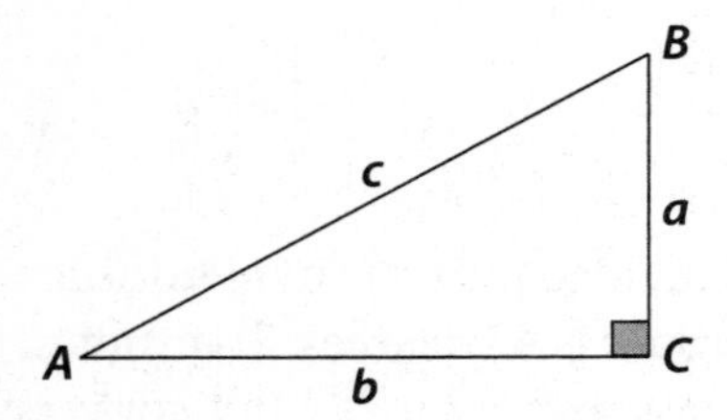

Figure 16.2 Right triangle *ABC*

A = measure of $\angle A$

B = measure of $\angle B = 90° - A$

C = measure of $\angle C = 90°$

a = side opposite $\angle A$

b = side adjacent to $\angle A$

c = side opposite the right angle hypotenuse

The ratios relative to angle A in the right triangle ABC are as follows:

$$\text{sine of } \angle A = \sin A = \frac{\text{side opposite}}{\text{hypotenuse}} = \frac{a}{c} \qquad \text{cosecant of } \angle A = \csc A = \frac{\text{hypotenuse}}{\text{side opposite}} = \frac{c}{a}$$

$$\text{cosine of } \angle A = \cos A = \frac{\text{side adjacent}}{\text{hypotenuse}} = \frac{b}{c} \qquad \text{secant of } \angle A = \sec A = \frac{\text{hypotenuse}}{\text{side adjacent}} = \frac{c}{b}$$

$$\text{tangent of } \angle A = \tan A = \frac{\text{side opposite}}{\text{side adjacent}} = \frac{a}{b} \qquad \text{cotangent of } \angle A = \cot A = \frac{\text{side adjacent}}{\text{side opposite}} = \frac{b}{a}$$

The cosecant is the reciprocal of the sine, the secant is the reciprocal of the cosine, and the cotangent is the reciprocal of the tangent. Therefore, it is necessary to remember only the **three basic ratios: sine, cosine,** and **tangent.**

The **two special right triangles of trigonometry** referred to time after time are shown in Figure 16.3.

The trigonometric ratios associated with these triangles are given in Table 16.2.

Rather than trying to memorize this table, it is better to remember the triangles and how the ratios are defined. All triangles similar to these two have the same ratio values. Several examples should enforce your familiarity with these special triangles.

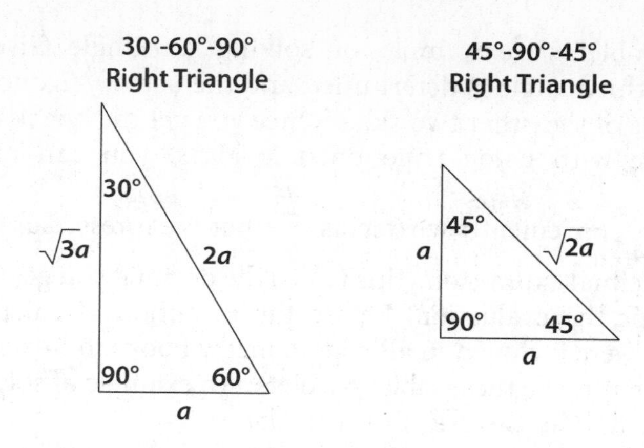

Figure 16.3 Special triangles

Table 16.2 Special ratios

Angle	30°	45°	60°
Sine	$\frac{1}{2}$	$\frac{1}{\sqrt{2}}=\frac{\sqrt{2}}{2}$	$\frac{\sqrt{3}}{2}$
Cosine	$\frac{\sqrt{3}}{2}$	$\frac{\sqrt{2}}{2}$	$\frac{1}{2}$
Tangent	$\frac{1}{\sqrt{3}}=\frac{\sqrt{3}}{3}$	1	$\sqrt{3}$

PROBLEM Determine the unknown values given $\angle C = 90°$ and the following information:

a. The length of sides a and b given $\angle A = 30°$ and $c = 48$
b. The length of sides a and c given $\angle A = 45°$ and $b = 25$
c. The length of sides a and c given $\angle A = 60°$ and $b = 48$

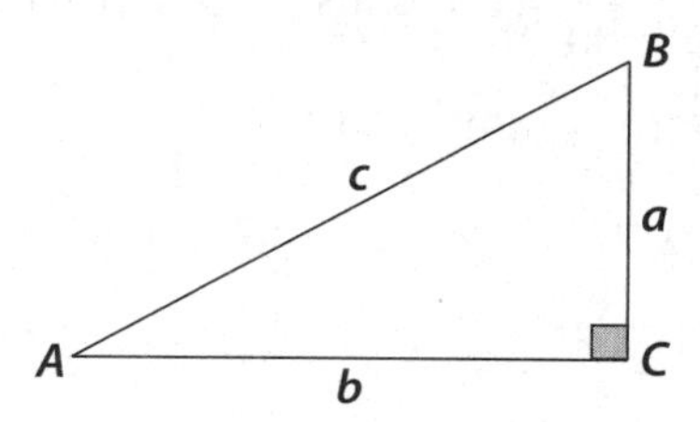

SOLUTION a. Since $\angle A = 30°$, $\angle B = 60°$. $\sin\angle A = \frac{a}{c} = \frac{a}{48} = \frac{1}{2}$. Thus, $a = 24$ and since the triangle is similar to the basic 30°–60°–90° right triangle, $b = 24\sqrt{3}$.

b. Since $\angle A = 45°$, $\angle B = 45°$. The triangle is similar to the basic 45°–45°–90° right triangle; thus, $a = 25$ and $c = 25\sqrt{2}$.

c. Since $\angle A = 60°$, $\angle B = 30°$. The triangle is similar to the basic 30°–60°–90° right triangle, thus $\cos\angle A = \frac{b}{c} = \frac{5}{c} = \frac{1}{2}$. Hence, $c = 10$. Also, $\sin\angle A = \frac{a}{c} = \frac{a}{10} = \frac{\sqrt{3}}{2}$. Hence, $a = 5\sqrt{3}$.

The previous problems are examples of "solving" a triangle. Given an angle and a side of a right triangle, the third angle is determined and the use of trigonometry allows you to determine the measures of the other two sides. Once you get the knack of solving right triangles and are comfortable with using trigonometric ideas, you can move on to solving any triangle.

The sine of 60° is conveniently written as $\frac{\sqrt{3}}{2}$ but is expressed as 0.866025 when converted to an approximate decimal expression. This is also the decimal you get by using the sine function of your calculator for the evaluation. Before the invention of calculators, lengthy tables of trigonometric values were included in all trigonometry books to be used for numerical calculations. The calculator has made these tables obsolete. An example of solving a right triangle using nonspecial angles exemplifies the use of the calculator:

PROBLEM Determine the lengths of sides b and c in a right triangle, given $\angle A = 13^\circ$, $\angle C = 90^\circ$, and side $a = 12$.

SOLUTION $\frac{a}{c} = \frac{12}{c} = \sin \angle A = \sin(13^\circ) \approx 0.225$. Hence, $c = \frac{12}{\sin 13^\circ} \cong 53.345$. Now $\angle B = 77^\circ$ and, using the Pythagorean theorem, $b = \sqrt{(53{,}345)^2 - (12)^2} \approx 51.978$. You also could have determined b by using the trig ratios $\frac{b}{c} = \sin(77^\circ)$ or $\frac{b}{c} = \cos(13^\circ)$ or $\frac{a}{b} = \tan(13^\circ)$.

Given any angle of a right triangle, you can determine the sine, cosine, or tangent of that angle by using your calculator. Additionally, if you have the reverse problem—determining the angle given a trig ratio of an angle of a right triangle—you can find the angle by using the **inverse trigonometric functions** denoted $\theta = \sin^{-1} x$, $\theta = \cos^{-1} x$, and $\theta = \tan^{-1} x$. Using these functions enables you to solve a right triangle, given the measure of at least one side and any two other parts. An example follows:

PROBLEM Solve the right triangle given $a = 30$, $b = 45$, and $\angle C = 90^\circ$.

SOLUTION $\tan \angle A = \frac{b}{a} = \frac{45}{30} = 1.5$. Thus $\angle A = \tan^{-1}(1.5) \approx 56^\circ$. It follows, then, that $\angle B \approx 34^\circ$ and $c = \sqrt{(30)^2 + (45)^2} \approx 54.1$.

EXERCISE

16·2

Solve the right triangles given $\angle C = 90^\circ$ and the following information.

1. Side $a = 35$ and $\angle A = 60^\circ$
2. Side $c = 25$ and $\angle B = 35^\circ$
3. Side $b = 126$ and side a = 200
4. Side $c = 25$ and $\angle A = 30^\circ$
5. Side $a = 30$ and $\angle B = 45^\circ$
6. Side $c = 10$ and side a = 3
7. Side $c = 2$ and side $a = 1$
8. Side $a = 1$ and side $b = 1$
9. Side $a = 60$ and side $c = 120$
10. Side $a = 38$ and side $b = 38$

The unit circle and trigonometric functions

The **unit circle** is a circle centered at (0, 0), with radius equal to 1. This is a convenient circle with which to study trigonometric (trig) ratios and some of their associated ramifications. If the circle is plotted on the Cartesian coordinate system, then the trig functions can be associated with the circle in the manner shown in Figure 16.4.

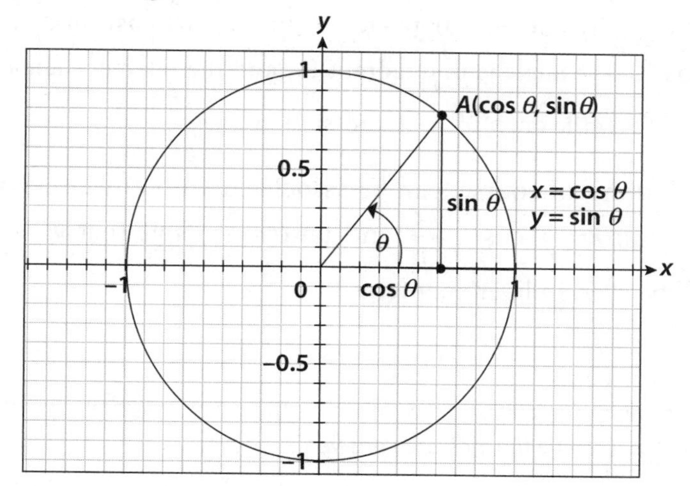

Figure 16.4 The unit circle

Since the radius is 1, cos(θ) becomes the x-coordinate and sin(θ) becomes the y-coordinate of the point on the circle intercepted by the ray determined by the central angle θ. If the point on the circle is rotated counterclockwise to the y-axis, then sin(θ) = 1 and cos(θ) = 0. Thus sin(90°) = 1 and cos(90°) = 0. By the same token, sin(0) = 0 and cos(0) = 1. These are just some of the relationships you can deduce from the unit circle. Many more trig ideas will be associated with the unit circle as we progress.

Note: If an angle is placed on the Cartesian coordinate plane such that the vertex is at the origin and one side is along the x-axis, then the angle is said to be in **standard position**. The side along the x-axis is the initial side, and the other side is the terminal side. An angle can be thought of as being formed by a rotation of the x-axis. If the rotation is counterclockwise, the angle is positive; and if the rotation is clockwise, the angle is negative.

The **reference angle** for an angle in standard position is the positive acute angle between the x-axis and the terminal side of the angle. For example, the reference angle for 120° is 60°. Figure 16.5 illustrates angle $\angle A$ and its associated reference angle $\angle A_r$.

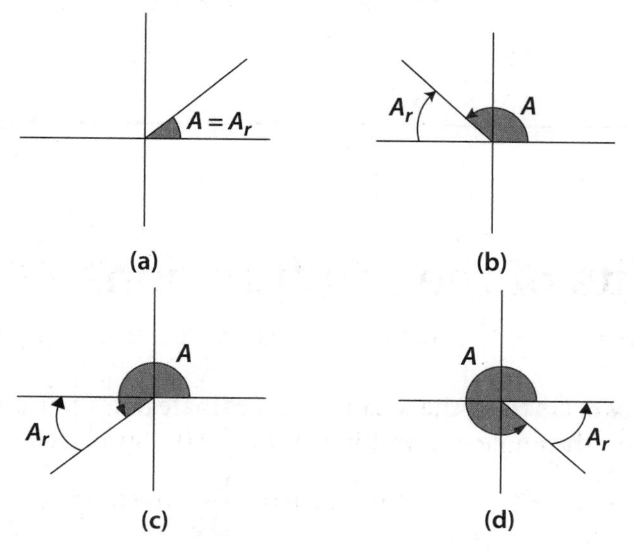

Figure 16.5 Reference angles

Referring to the unit circle, $y = \sin\theta$ and $x = \cos\theta$. Realizing that the trig ratios can be positive or negative, the algebraic sign of the three basic trig functions in each quadrant is easily remembered by using this mnemonic: All Students Take Calculus. All three are positive in quadrant I, the Sine is positive in quadrant II, the Tangent is positive in quadrant III, and the Cosine is positive in quadrant IV.

PROBLEM Use reference angles to determine $\sin\theta$, $\cos\theta$, and $\tan\theta$ for $\theta = 315°$.

SOLUTION The angle is in quadrant IV, and the reference angle is 45°. Thus, $\cos 315° = \frac{\sqrt{2}}{2}$ while $\tan 315° = -1$ and $\sin 315° = -\frac{\sqrt{2}}{2}$.

The trig functions, of course, can be associated with any circle of radius r. The ratios then take on the forms $\sin\theta = \frac{y}{r}$, $\cos\theta = \frac{x}{r}$, and $\tan\theta = \frac{y}{x}$. Note that $y = r\sin\theta$ and $x = r\cos\theta$. These connections will be used in later chapters.

For 1–6, provide the indicated answer.

1. Name the reference angle for 138°.
2. Name the reference angle for 500°.
3. If angle A is in standard position and $(4\sqrt{3}, 4)$ is on the terminal side of A, then find the values of the six trig functions.
4. Find the exact value of $\sin\theta$, $\cos\theta$, and $\tan\theta$ for $\theta = \frac{4\pi}{3}$.
5. Find the exact value of $\sin\theta$, $\cos\theta$, and $\tan\theta$ for $\theta = 390°$.
6. Find the values of x, y, and r for the point on a circle of radius r if θ is the angle that determines the point such that $\sin\theta < 0$ and $\cos\theta = \frac{4}{5}$.

For 7–10, state the quadrant of the terminal side and the sign of the function in that quadrant, and use a calculator to evaluate the expression. Round to three decimal places.

7. $\sin 723°$
8. $\tan 995°$
9. $\cos 528°$
10. $\csc 687°$

Zeros and units of the trig functions

The zeros of the sine and cosine functions can easily be observed by referring to the unit circle (see Figure 16.6).

Thinking of $(\cos\theta, \sin\theta)$ as the x- and y-coordinates of a point on the unit circle, you see that $y = \sin\theta = 0$ when the angle is a multiple of π or 180°. Similarly, $x = \cos\theta = 0$ when the angle is an odd multiple of $\frac{\pi}{2}$ or 90°. Also, since $\tan\theta = \frac{\sin\theta}{\cos\theta}$, the tangent function has the same zeros as the sine function.

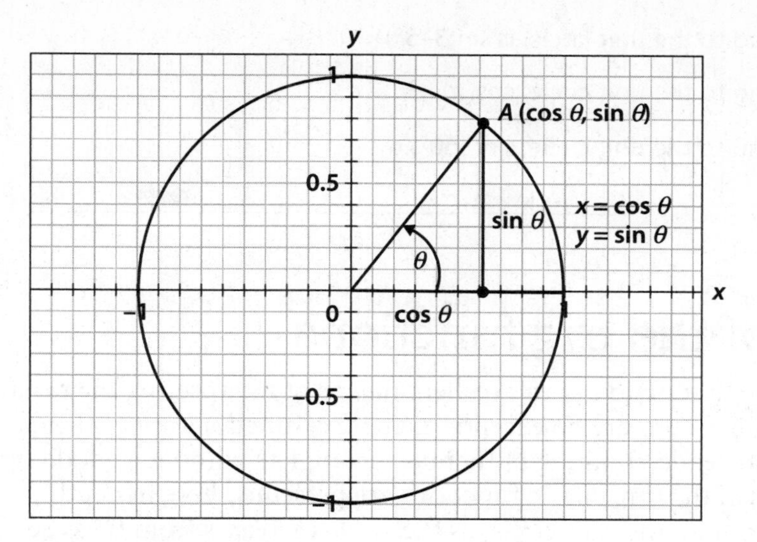

Figure 16.6 Unit circle

The sine function $y = \sin\theta = \pm 1$ when the angle is an odd multiple of $\frac{\pi}{2}$, where the sign depends on the quadrant of the angle. Similarly $y = \cos\theta = \pm 1$ when the angle is a multiple of π, where again the sign depends on the quadrant of the angle. Finally, $y = \tan\theta = \pm 1$ when the angle is an odd multiple of $\frac{\pi}{4}$ or 45°, and the sign depends on the quadrant of the angle. Table 16.3 shows these special values for angles between 0 and π.

Table 16.3 Special trigonometric values

θ	0	$\frac{\pi}{4}$	$\frac{\pi}{2}$	$\frac{3\pi}{4}$	π
$\sin\theta$	0	$\frac{\sqrt{2}}{2}$	1	$\frac{\sqrt{2}}{2}$	0
$\cos\theta$	1	$\frac{\sqrt{2}}{2}$	0	$-\frac{\sqrt{2}}{2}$	−1
$\tan\theta$	0	1	undefined	−1	0

EXERCISE

16·4

For 1–10, provide the indicated answer.

1. What are the radian equivalents of the angles 0°,30°,45°,60°,90°, 120°,135°,150°,180°,270°, and 360°?
2. At what angle(s) are the sine and cosine equal?
3. At what angle(s) do the sine and cosine differ only by the algebraic sign?
4. What is the reference angle for 225°?
5. What is the reference angle for $\frac{3\pi}{4}$?
6. What is the algebraic sign of tan θ if θ is in quadrant IV?
7. Referring to the unit circle, why is $(\sin\theta)^2 + (\cos\theta)^2 = 1$?

8. Referring to the unit circle, is $\sin(315°) < 0$?
9. Referring to the unit circle, $\cos(\pi) = ?$
10. Referring to the unit circle, $\sin(360°) = ?$

Graphs of the trig functions

The graph of a function is an excellent means of interpreting the behavior of a function, and the trig function graphs are no exception. The trig functions are applicable to many real-world problems, and an understanding of them is incomplete without a good familiarity with their graphs. The graphs of the three basic functions are presented here in depth.

Referring to the unit circle, as the angle increases from 0° to 360°, the sine function begins at 0 , then to 1, back to 0, then to −1, then back to 0 and begins to repeat itself. The cosine starts at 1, then to 0, then to −1, then to 0, and back to 1. The periodic behavior of both functions is reflected by this repetition. Since the tangent function is a quotient, a vertical asymptote is encountered at angles where the cosine is 0. The tangent function is also periodic.

In graphing the trig functions, the radian measure for the angles is used almost exclusively in the study of calculus.

The graphs of $y = \sin x$, $y = \cos x$, and $y = \tan x$—the three main trigonometric functions—are shown in Figures 16.7 and 16.8.

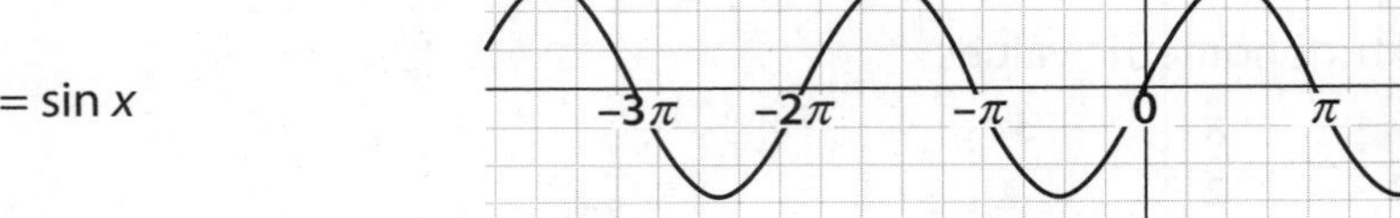

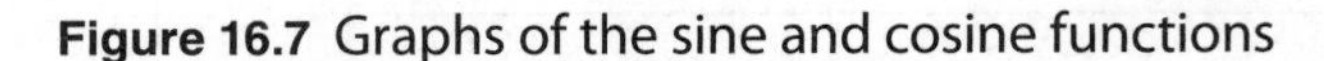

Figure 16.7 Graphs of the sine and cosine functions

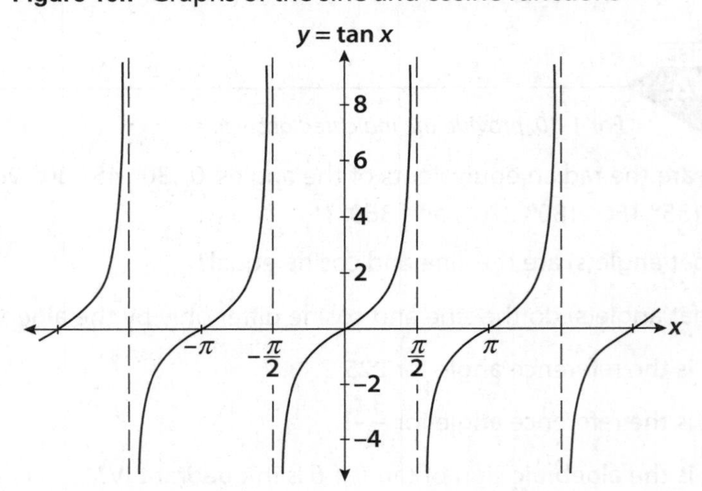

Figure 16.8 Graph of the tangent function

The graphs of the other three functions, shown in Figures 16.9–16.11, are used much less than the three basic graphs and are included here for completeness.

Knowing the shapes of the graphs of the basic trig functions will enable you to graph functions that are similar but are variations of the basic graphs.

A function f is **periodic** if there is a positive number P such that $f(x+P)=f(x)$ for all x in the domain. The least number P for which this is true is the **period** of f.

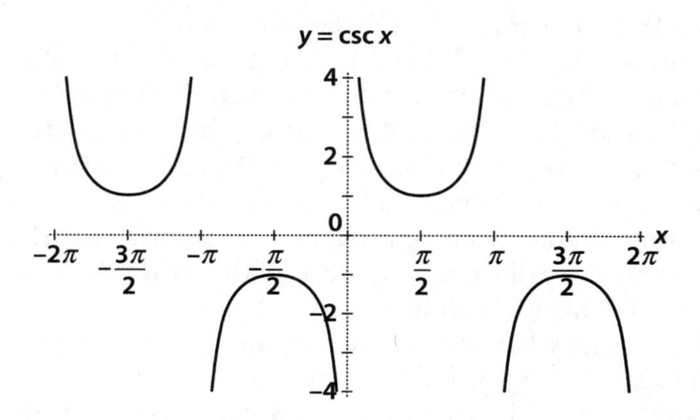

Figure 16.9 Graph of the cosecant function

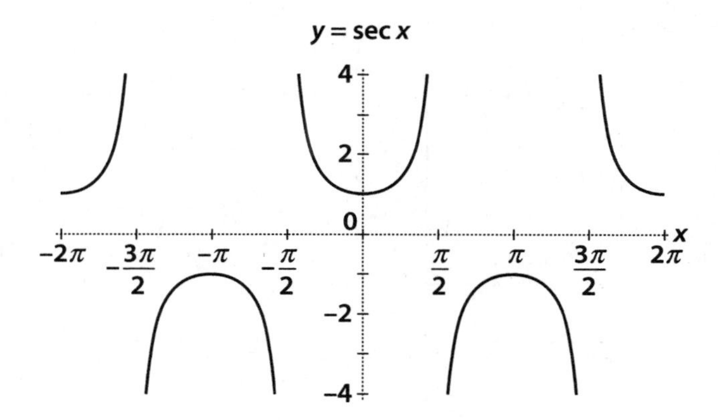

Figure 16.10 Graph of the secant function

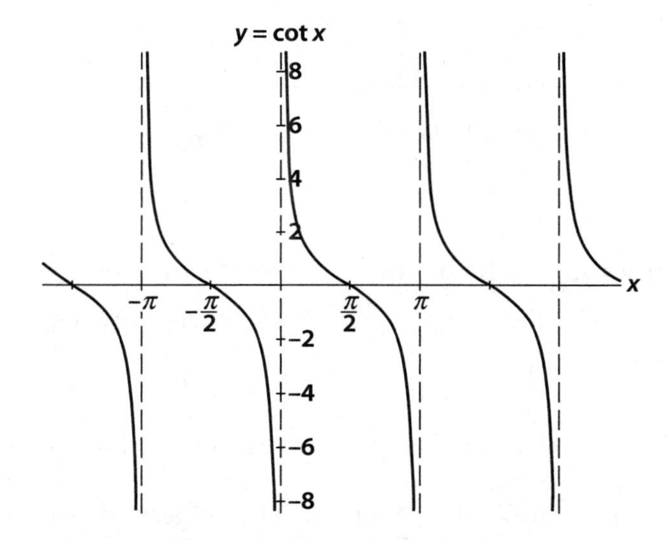

Figure 16.11 Graph of the cotangent function

The sine and cosine functions are periodic with period 2π. The tangent function is periodic with period π.

Transformations of trigonometric functions

Transformations of functions are covered in Chapter 6, "Function transformations and symmetry," and this section is a study of how some of those transformations apply to trig functions. A quick reference to that chapter may be beneficial.

The functions $y = A\sin(Bx + C)$ and $y = A\cos(Bx + C)$ are transformations of the basic sine and cosine functions. The constant A induces a vertical compression or stretch; and if $A < 0$, a reflection about the x-axis. The constant B induces a horizontal stretch or compression, and the constant C induces a horizontal shift. The number $|A|$ is the **amplitude** of both the sine and cosine functions. The constant B, by inducing a horizontal transformation, automatically alters the period. Since the cosine function can be thought of as a horizontal slide of the sine function, the sine function will be used to illustrate the effects of these three types of transformations, since the effects are exactly the same for both functions.

The function $y = \sin x$ has a maximum of 1 and a minimum of –1, whereas the function $y = A\sin x$ has a maximum of $|A|$ and minimum of $-|A|$. The zeros and periods of $y = \sin x$ and $y = A\sin x$ remain the same. The graph of the new function is easily drawn with this information. One period of the function $y = 3\sin x$ is shown in Figure 16.12.

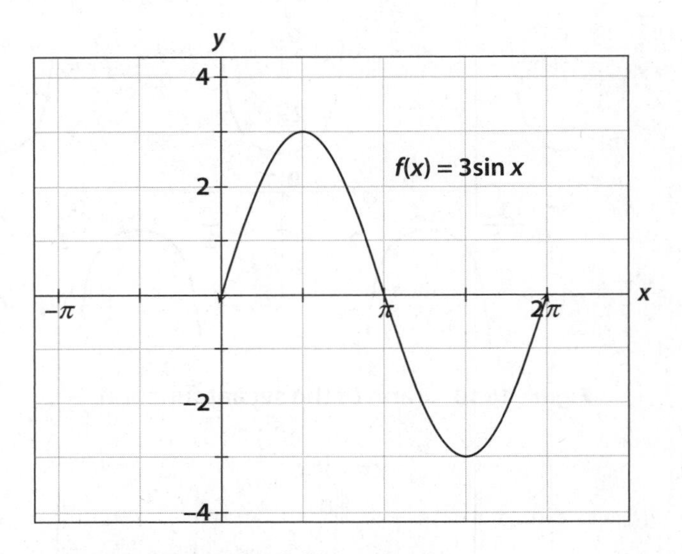

Figure 16.12 Graph of $f(x) = 3\sin x$

The period of $y = \sin x$ is 2π, but the period of the function $y = \sin(Bx)$ is $P = \frac{2\pi}{|B|}$. This is the way the constant B alters the original period. The partial graphs of $y = \sin 4x$ and $y = \sin x$ are shown in Figure 16.13.

The graph of $y = \sin\left(\frac{1}{2}x\right)$ is shown in Figure 16.14. You can see the horizontal compression (Figure 16.13) and stretch (Figure 16.14) due to the facts that $4 > 1$ and $\frac{1}{2} < 1$, respectively. Note also the period changes in each graph.

Ocean waves are sinusoidal and tend to have extended periods whereas electromagnetic waves have short periods. Trigonometric ideas are used in these studies.

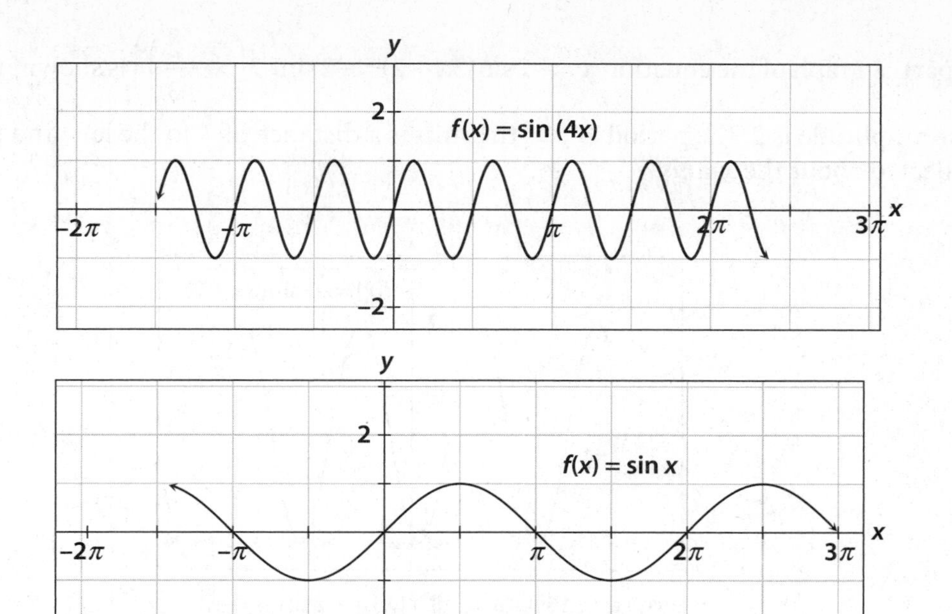

Figure 16.13 Graphs of $\sin(4x)$ and $\sin x$

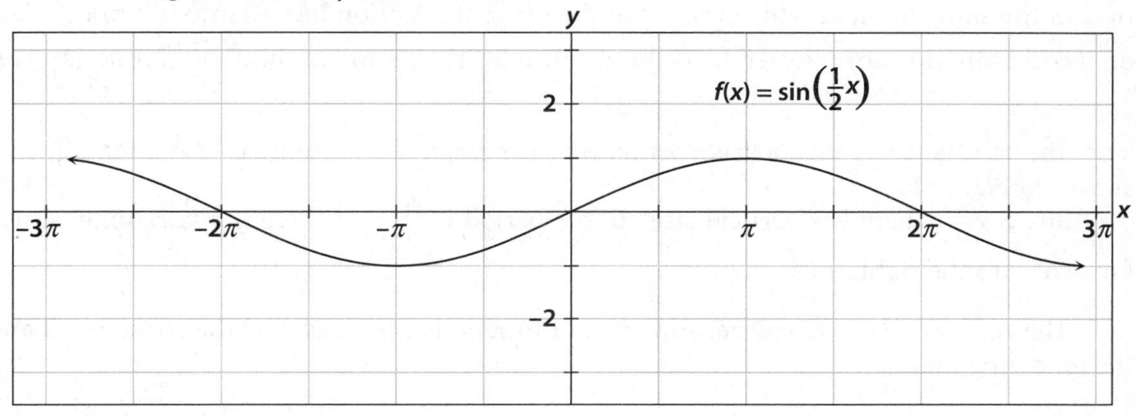

Figure 16.14 Graph of $f(x) = \sin\left(\frac{1}{2}x\right)$

The constant C induces a horizontal shift as illustrated by the graph of $y = \sin(x - \pi)$ shown in Figure 16.15. The original graph has been shifted to the right by a distance of π.

When all these constants are incorporated into one equation, it is relatively simple to sketch a graph by knowing the effect of each constant. However, there is one little hitch you have to remember. If the equation is written in **standard form** as $y = A\sin(Bx + C)$ you need to rewrite it in **shift form** as $y = A\sin\left(B\left(x + \frac{C}{B}\right)\right)$ because the horizontal shift is actually a distance of $\left|\frac{C}{B}\right|$.

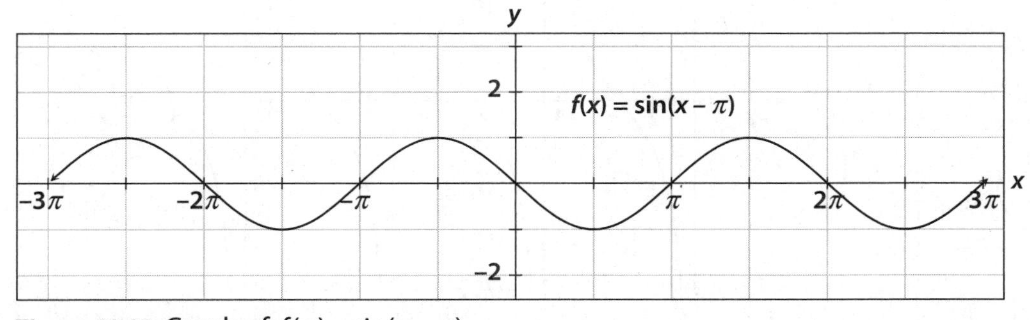

Figure 16.15 Graph of $f(x) = \sin(x - \pi)$

A partial graph of the equation $y = -2\sin(3x+2) = -2\sin\left(3\left(x+\frac{2}{3}\right)\right)$ is shown in Figure 16.16.

The amplitude is 2, the period is $\frac{2\pi}{3}$, the shift is a distance of $\frac{2}{3}$ to the left, and the graph has been reflected about the x-axis.

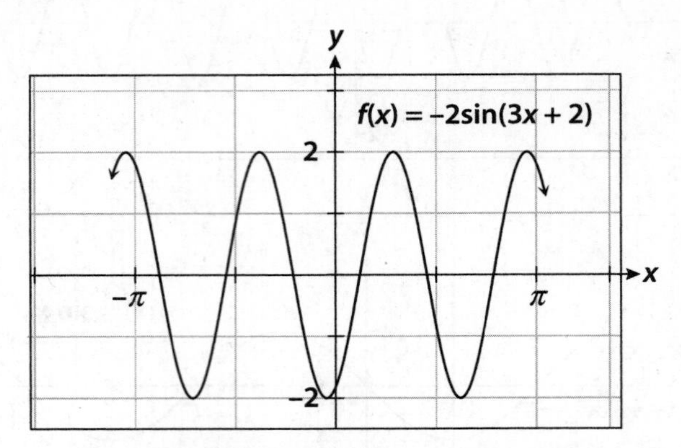

Figure 16.16 Graph of $f(x) = -2\sin(3x+2)$

The constants A, B, and C in the equation $y = A\tan(Bx + C)$ alter the basic tangent function in the same manner. However, since the tangent function has no maximum or minimum on its domain, A is not referred to as the amplitude. The period of the altered function is $P = \frac{\pi}{|B|}$ since the basic tangent function has a period of π. The graph of $y = 6\tan(2x-3) = 6\tan\left(2\left(x-\frac{3}{2}\right)\right)$ is shown in Figure 16.17.

Since $6 > 1$, there is a vertical stretch, the period is $\frac{\pi}{2}$, and there is a horizontal shift a distance of $\frac{3}{2}$ to the right.

The zeros of the sine, cosine, and tangent functions are altered by the constants B and C in the following manner:

$$y = A\sin(Bx+C) = 0 \text{ when } Bx + C = n\pi \text{ or when } x = \frac{n\pi - C}{B}$$

Similarly:

$$y = A\cos(Bx+C) = 0 \text{ when } Bx + C = \frac{(2n+1)\pi}{2} \text{ or when } x = \frac{(2n+1)\pi}{2B} - \frac{C}{B}$$

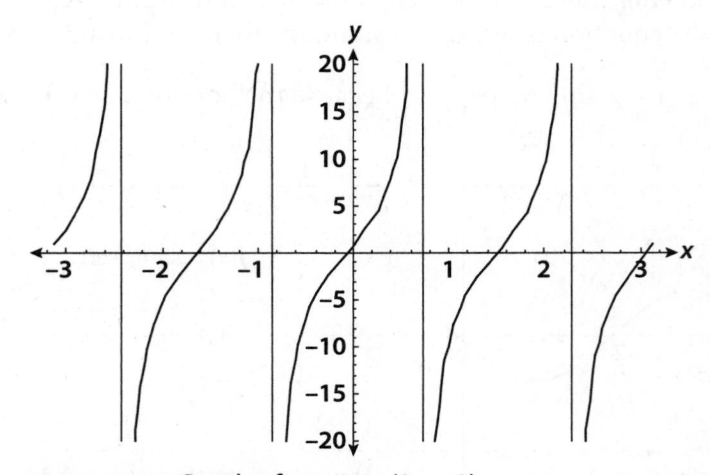

Figure 16.17 Graph of $y = 6\tan(2x-3)$

Recall that the tangent function has the same zeros as the sine function. Rather than memorizing these values, you will find it simpler just to solve the $Bx+C$ equations when needed.

The angles at which the generalized sine and cosine functions attain their maxima and minima are the angles at which the basic sine and cosine functions are ± 1. These angles for the sine and cosine are exactly reversed from the angles that give the zeros.

Finally, the asymptotes for the tangent function occur at the angles where the cosine is 0.

Putting all this together enables you to analyze any trig function and in fact sketch a reasonable quick graph of the function. Any study where trigonometry applies invariably involves solving for the zeros, finding the angles that give maxima and minima, and determining the periods of the functions. If you are able to consistently do these three things correctly, you will have made a huge dent in mastering trigonometry. An example will round out the discussion:

PROBLEM For the function $y=-6\cos(3x-9)$, find (a) the amplitude, (b) the period, (c) the zeros, (d) the shift, and (e) the angles at which the function attains its maximum and minimum. Graph two periods of the function.

SOLUTION

a. Amplitude $=|-6|=6$

b. Period $=\dfrac{2\pi}{3}$

c. Zeros are found by solving $3x-9=\dfrac{(2n+1)\pi}{2}$ so that $x=\dfrac{(2n+1)\pi}{6}+3$.

Four zeros are $3-\dfrac{3\pi}{6}\approx 1.4$, $3-\dfrac{\pi}{6}\approx 2.5$, $3+\dfrac{\pi}{6}\approx 3.5$, and $3+\dfrac{3\pi}{6}\approx 4.6$.

d. The shift is a distance of $\dfrac{9}{3}=3$ to the right.

e. Maximum and minimum angles are found by solving $3x-9=n\pi$ or $x=\dfrac{n\pi}{3}+3$.

Four such angles are $3-\dfrac{2\pi}{3}\approx 0.9$, $3-\dfrac{\pi}{3}\approx 1.95$, 3, and $3+\dfrac{\pi}{3}\approx 4.1$.

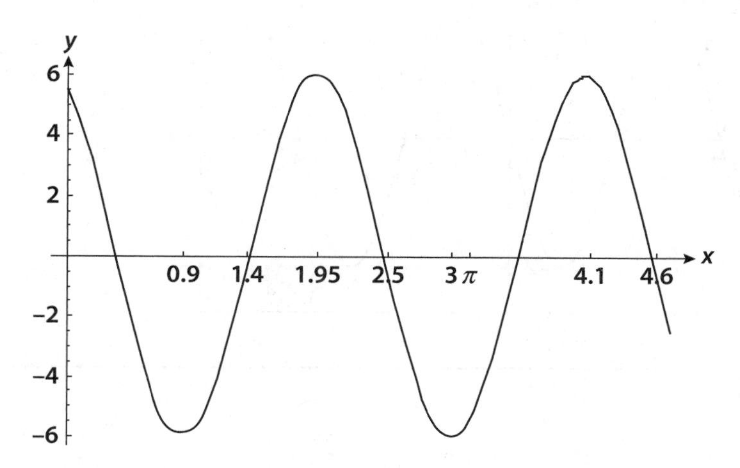

The final transformation is the vertical shift discussed in Chapter 6. This simply raises or lowers the graph and is of the form $y=A\sin(Bx+C)+D$.

The following exercise is designed to bring together all the transformations discussed in this section. You should do your best to try call up the transformations from memory as you work each problem. Practice on problems of this type should "cement" your graphing abilities of trig functions.

For 1–3, find (a) the amplitude (if applicable), (b) the period, (c) the zeros, (d) the shift, and (e) the angles at which the function attains its maximum and minimum (if applicable).

1. $y = 3\cos(2x - \pi)$
2. $y = \sin\left(\frac{\pi}{6}x - \frac{\pi}{3}\right)$
3. $y = 3\tan(4x + 2\pi)$

For 4–5, write the equations for the graphs.

4. 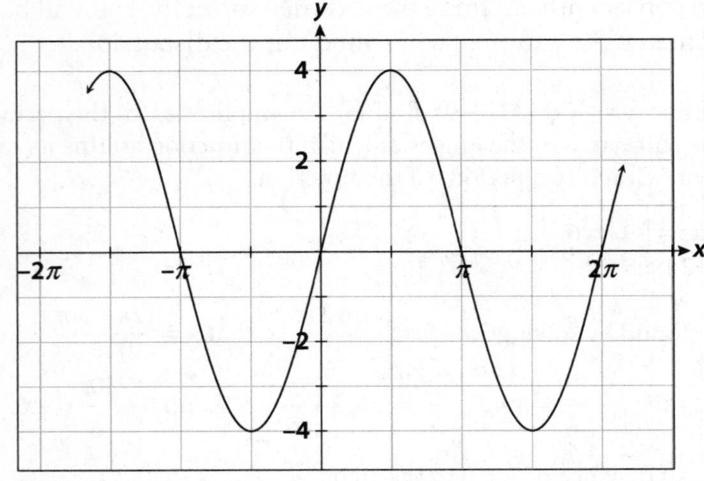

5. 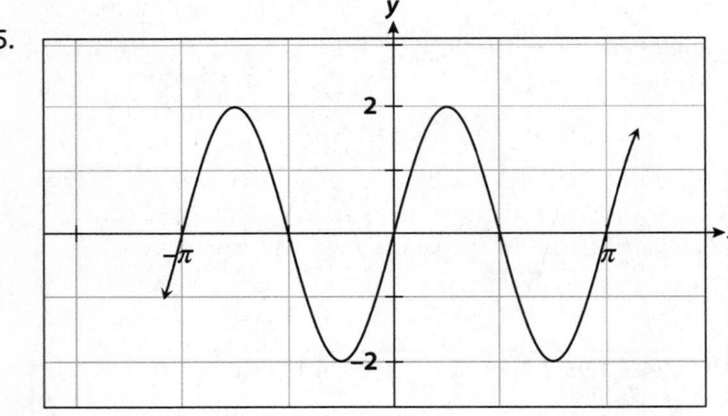

Analytic trigonometry

Identities

An identity of functions is an equation that is true for all values of the elements in the common domains. This concept is encountered in trigonometry, probably more than in any other area of mathematics. It is also an area in which many students falter in their study of trig. Working with trig identities requires good algebra skills, tenacity of purpose, and a mastery of the fundamental trig identities. One of the most useful formulas from algebra is the addition-subtraction formula for fractions: $\frac{A}{B} \pm \frac{C}{D} = \frac{AD \pm BC}{BD}$. This formula is used extensively in the work with identities.

A convenient notation for $(\sin\theta)^2$ is $\sin^2\theta$. This notation will also be used on the other trig functions.

The fundamental trigonometric identities are those that are easily determined by referring to the unit circle and the original definitions of the trig functions. See Table 17.1.

The reciprocal and ratio identities result from the definitions of the trig functions, the Pythagorean identities result from the use of the Pythagorean theorem for right triangles, and the even-odd identities are due to symmetry. The sine and tangent functions are odd functions, and the cosine function is even.

Not only should you memorize these identities, but also you should *own* them. That is, they should be an instinctive part of your mathematical knowledge. Moreover, you should be able to recognize and use these in their various forms such as $\sin\theta\,\csc\theta = 1$ or $\cot\theta = \frac{1}{\tan\theta}$.

Table 17.1 Fundamental trigonometric identities

Reciprocal identities	Ratio identities	Pythagorean identities	Even-odd function identities
$\sin\theta = \frac{1}{\csc\theta}$	$\tan\theta = \frac{\sin\theta}{\cos\theta}$	$\sin^2\theta + \cos^2\theta = 1$	$\sin(-\theta) = -\sin\theta$
$\cos\theta = \frac{1}{\sec\theta}$	$\cot\theta = \frac{\cos\theta}{\sin\theta}$	$1 + \tan^2\theta = \sec^2\theta$	$\cos(-\theta) = \cos\theta$
$\tan\theta = \frac{1}{\cot\theta}$	$\tan\theta = \frac{\sec\theta}{\csc\theta}$	$1 + \cot^2\theta = \csc^2\theta$	$\tan(-\theta) = -\tan\theta$

In using these fundamental identities, other useful identities are developed, many of which are categorized so that they are easily referenced. Textbook problems are designed to strengthen your abilities to use and work with the fundamental identities.

Proving or verifying an identity *cannot* be accomplished by repeatedly substituting in domain values to get a true equation. Proving identities *can* be accomplished by the use of good logic, referencing the fundamental identities, and using algebra skills and substitution principles. Here are examples to illustrate some of these ideas:

PROBLEM Verify (prove) that $\cos\theta(\sec\theta-\cos\theta)=\sin^2\theta$ is an identity.

SOLUTION The approach is to logically transform one side of the equation to the other side. Justifications are provided for clarity, and you should make a habit of doing the same on problems you work on.

$$\cos\theta(\sec\theta-\cos\theta)=\cos\theta\sec\theta-\cos^2\theta \quad \text{distribute}$$
$$=1-\cos^2\theta \quad \text{substitute 1 for } \cos\theta\sec\theta$$
$$=\sin^2\theta \quad \text{Pythagorean identity}$$

PROBLEM Verify that $(\cos x-\sin x)^2=1-2\sin x\cos x$ is an identity.

SOLUTION
$$(\cos x-\sin x)^2=\cos^2 x-2\cos x\sin x+\sin^2 x \quad \text{squaring a binomial}$$
$$=\cos^2 x+\sin^2 x-2\cos x\sin x \quad \text{rearrange terms}$$
$$=1-2\cos x\sin x \quad \text{Pythagorean identity}$$
$$=1-2\sin x\cos x \quad \text{commute terms}$$

PROBLEM Write $\cos x$ in terms of $\tan x$.

SOLUTION
$$\cos x=\frac{1}{\sec x} \quad \text{reciprocal}$$
$$\sec^2 x=1+\tan^2 x \quad \text{Pythagorean identity}$$
$$\sec x=\pm\sqrt{1+\tan^2 x} \quad \text{square roots}$$
$$\cos x=\frac{1}{\pm\sqrt{1+\tan^2 x}} \quad \text{substitution}$$

PROBLEM Verify that $\sin(\theta+\beta)=\sin\theta+\sin\beta$ is *not* an identity.

SOLUTION It takes only one counterexample to establish that the equation is not an identity. Let $\theta=\beta=\frac{\pi}{3}$. Then $\sin(\theta+\beta)=\sin\left(\frac{2\pi}{3}\right)=\frac{\sqrt{3}}{2}$; but $\sin\theta+\sin\beta=\sin\frac{\pi}{3}+\sin\frac{\pi}{3}$ $=\frac{\sqrt{3}}{2}+\frac{\sqrt{3}}{2}=\sqrt{3}$. Hence, the equation is not an identity.

PROBLEM Verify that $\dfrac{\sin x\cos x+\sin x}{\cos x+\cos^2 x}=\tan x$ is an identity.

SOLUTION

$$\frac{\sin x\cos x+\sin x}{\cos x+\cos^2 x}=\frac{\sin x(\cos x+1)}{\cos x(1+\cos x)} \quad \text{factoring}$$

$$=\frac{\sin x}{\cos x} \quad \text{cancellation}$$

$$=\tan x \quad \text{ratio identity}$$

PROBLEM Verify that $\dfrac{\cot x}{\sec x}-\dfrac{\cos x}{\sin x}=\dfrac{\cos x-1}{\tan x}$ is an identity.

SOLUTION

$$\frac{\cot x}{\sec x}-\frac{\cos x}{\sin x}=\frac{\cot x\sin x-\sec x\cos x}{\sec x\sin x} \quad \text{subtraction of fractions}$$

$$=\frac{\frac{\cos x}{\sin x}\sin x-1}{\frac{1}{\cos x}\sin x} \quad \text{reciprocal identities (three used)}$$

$$=\frac{\cos x-1}{\frac{\sin x}{\cos x}} \quad \text{multiplication and cancellation}$$

$$=\frac{\cos x-1}{\tan x} \quad \text{ratio identity}$$

When you are verifying identities, it does not matter which side of the identity you start with; but as a rule, you generally start with the side that appears to be the most complicated. Some guidelines to keep in mind are as follows:

1. In general, work with only one side of the equation. You cannot assume the equation is true, so equality properties cannot be applied.
2. Converting all functions to sines and cosines may be helpful.
3. Know and use your algebra skills. Review them, daily if necessary.
4. ***Own*** **the fundamental identities.**

EXERCISE

17·1

For 1–9, verify the equation is an identity.

1. $\sin x\cot x=\cos x$
2. $\sin x(\csc x-\sin x)=\cos^2 x$
3. $\tan x(\csc x+\cot x)=\sec x+1$
4. $\tan^2 x\csc^2 x-\tan^2 x=1$
5. $\dfrac{\sin x\cos x+\cos x}{\sin x+\sin^2 x}=\cot x$
6. $\dfrac{(\sin x+\cos x)^2}{\cos x}=\sec x+2\sin x$
7. $(1+\sin x)(1+\sin(-x))=\cos^2 x$
8. $\dfrac{\cos^2 x}{\sin x}+\dfrac{\sin x}{1}=\csc x$
9. $\dfrac{\tan x}{\csc x}-\dfrac{\sin x}{\cos x}=\dfrac{\sin x-1}{\cot x}$
10. Verify that $\cos(2\theta)=2\cos\theta$ is not an identity.

Sum and difference identities

Many textbook identities are in fact contrived by various authors to be used as problems on which you can improve your skills in proof techniques and solidify your grasp of the fundamental identities. However, there is a set of about forty identities (formulas) that are used for special purposes, and these are found in scientific tables, in appendices, on textbook covers, and so forth. In this text a number of these identities are collected and listed in Appendix B.

The complete development of one of these identities is a good example illustrating the importance of the unit circle.

Referring to Figure 17.1, if A and B are points on the unit circle on the terminal sides of angles α and β, respectively, then since $r = 1$, the coordinates of A are $(\cos\alpha, \sin\alpha)$ and those of B are $(\cos\beta, \sin\beta)$.

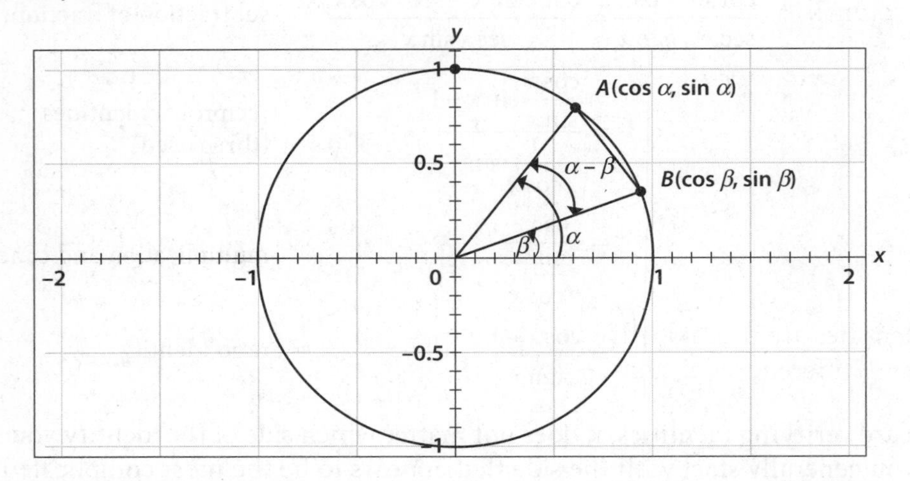

Figure 17.1 The unit circle

Using the distance formula, the distance, $\overline{AB}$, between A and B is:

$$\overline{AB} = \sqrt{(\cos\alpha - \cos\beta)^2 + (\sin\alpha - \sin\beta)^2} \quad \text{distance formula}$$

$$= \sqrt{\cos^2\alpha - 2\cos\alpha\cos\beta + \cos^2\beta + \sin^2\alpha - 2\sin\alpha\sin\beta + \sin^2\beta} \quad \text{squaring binomials}$$

$$= \sqrt{2 - 2\cos\alpha\cos\beta - 2\sin\alpha\sin\beta} \quad \text{Pythagorean identity and grouping terms}$$

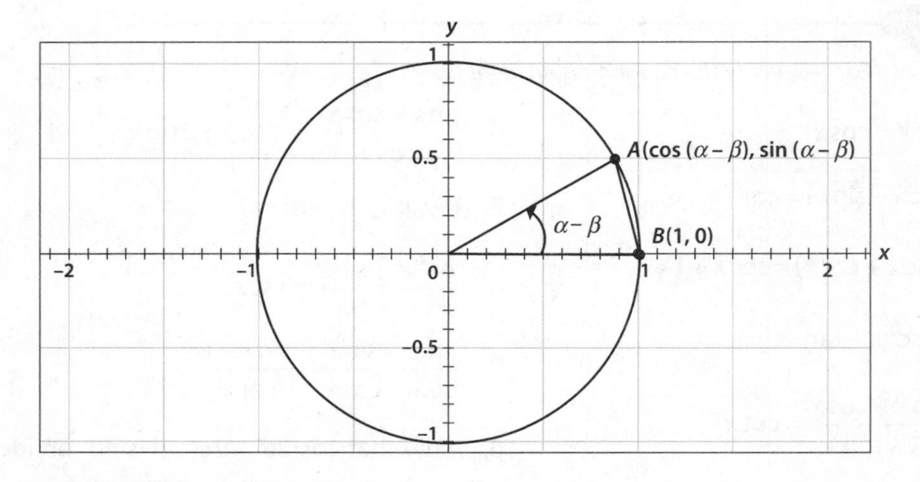

Figure 17.2 Sector rotation

Now if you rotate the sector down to the x-axis (see Figure 17.2), the coordinates of B are (1,0) but $\overline{AB}$ is unchanged.

Thus,

$$\overline{AB} = \sqrt{(\cos(\alpha-\beta)-1)^2 + (\sin(\alpha-\beta)-0)^2} \quad \text{distance formula}$$

$$= \sqrt{\cos^2(\alpha-\beta) - 2\cos(\alpha-\beta) + 1 + \sin^2(\alpha-\beta)} \quad \text{squaring binomials}$$

$$= \sqrt{2 - 2\cos(\alpha-\beta)} \quad \text{Pythagorean identity and collecting terms}$$

Then equating the two expressions for $\overline{AB}$ yields $\sqrt{2-2\cos(\alpha-\beta)} = \sqrt{2-2\cos\alpha\cos\beta - 2\sin\alpha\sin\beta}$. Squaring each side and simplifying yields the difference identity:

$$\cos(\alpha-\beta) = \cos\alpha\,\cos\beta + \sin\alpha\,\sin\beta$$

Also,

$$\cos(\alpha+\beta) = \cos(\alpha-(-\beta)) \quad \text{equivalent expressions}$$

$$= \cos\alpha\cos(-\beta) + \sin\alpha\sin(-\beta) \quad \text{difference identity}$$

$$= \cos\alpha\cos\beta - \sin\alpha\sin\beta \quad \cos(-\beta) = \cos\beta \text{ and } \sin(-\beta) = -\sin\beta$$

These two identities are normally displayed as $\cos(\alpha \pm \beta) = \cos\alpha\cos\beta \mp \sin\alpha\sin\beta$. They are the cosine sum and difference identities.

Similar identities for the sine and tangent functions are:

$$\sin(\alpha \pm \beta) = \sin\alpha\cos\beta \pm \cos\alpha\sin\beta$$

and

$$\tan(\alpha \pm \beta) = \frac{\tan\alpha \pm \tan\beta}{1 \mp \tan\alpha\tan\beta}$$

Using the sum and difference identities gives rise to the cofunction identities:

$$\sin\left(\frac{\pi}{2} - x\right) = \cos x \text{ and } \cos\left(\frac{\pi}{2} - x\right) = \sin x$$

PROBLEM Prove the cofunction identities.

SOLUTION

$$\sin\left(\frac{\pi}{2} - x\right) = \sin\left(\frac{\pi}{2}\right)\cos x - \cos\left(\frac{\pi}{2}\right)\sin x \quad \text{difference identity}$$

$$= 1\cos x - 0\sin x \quad \text{special angle values}$$

$$= \cos x \quad \text{simplify}$$

$$\cos\left(\frac{\pi}{2} - x\right) = \cos\left(\frac{\pi}{2}\right)\cos x + \sin\left(\frac{\pi}{2}\right)\sin x \quad \text{difference identity}$$

$$= 0\cos x + 1\sin x \quad \text{special angle values}$$

$$= \sin x \quad \text{simplify}$$

EXERCISE 17·2

For 1–8, verify the identities.

1. $\sin(x+2\pi)=\sin x$
2. $\cos(x+2\pi)=\cos x$
3. $\tan(x+\pi)=\tan x$
4. $\sin(\alpha+\beta)\sin(\alpha-\beta)=\sin^2\alpha-\sin^2\beta$
5. $\cos\left(x+\frac{\pi}{4}\right)=\frac{\sqrt{2}}{2}(\cos x-\sin x)$
6. $\sin(2t)=2\sin t\cos t$
7. $\cos(\alpha+\beta)+\cos(\alpha-\beta)=2\cos\alpha\cos\beta$
8. $\tan\left(x+\frac{\pi}{4}\right)=\frac{1+\tan x}{1-\tan x}$
9. Verify that $\tan\left(\frac{2\pi}{3}+\frac{\pi}{4}\right)=\frac{1-\sqrt{3}}{1+\sqrt{3}}$.

Double- and half-angle identities

The sum and difference identities enable you to derive many other useful identities. The double- and half-angle identities are used to convert products of angles to single-angle form, which is needed and useful in many applications. For instance:

$$\sin(2\theta)=\sin(\theta+\theta)=\sin\theta\cos\theta+\cos\theta\sin\theta=2\sin\theta\cos\theta$$

The derivations of $\cos(2\theta)$ and $\tan(2\theta)$ are similarly developed and left as exercises.

The cosine double-angle identity has three forms that are usually included in identity tables. See Table 17.2.

Table 17.2 The double-angle identities

$\cos(2\theta)=\cos^2\theta-\sin^2\theta$	$\sin(2\theta)=2\sin\theta\cos\theta$
$\cos(2\theta)=1-2\sin^2\theta$	
$\cos(2\theta)=2\cos^2\theta-1$	$\tan(2\theta)=\frac{2\tan\theta}{1-\tan^2\theta}$

PROBLEM Verify that $(\sin x+\cos x)^2=1+\sin(2x)$.

SOLUTION

$(\sin x+\cos x)^2=\sin^2x+2\sin x\cos x+\cos^2x$ — squaring a binomial

$=1+2\sin x\cos x$ — Pythagorean identity

$=1+\sin(2x)$ — double-angle identity

PROBLEM Verify the identity $\cos(3\theta)=4\cos^3\theta-3\cos\theta$.

SOLUTION

$\cos(3\theta)=\cos(2\theta+\theta)$ — angle addition

$=\cos(2\theta)\cos\theta-\sin(2\theta)\sin\theta$ — sum identity

$=(2\cos^2\theta-1)\cos\theta-(2\sin\theta\cos\theta)\sin\theta$ — double-angle identities

$$= 2\cos^3\theta - \cos\theta - 2\sin^2\theta\cos\theta \qquad \text{distribution}$$

$$= 2\cos^3\theta - \cos\theta - 2(1-\cos^2\theta)\cos\theta \qquad \text{Pythagorean identity}$$

$$= 4\cos^3\theta - 3\cos\theta \qquad \text{simplification}$$

The half-angle identities, shown in Table 17.3, can be derived from the double-angle identities. The derivation for the half-angle cosine identity follows.

Table 17.3 The half-angle identities

$\sin\left(\frac{x}{2}\right) = \pm\sqrt{\frac{1-\cos x}{2}}$	$\cos\left(\frac{x}{2}\right) = \pm\sqrt{\frac{1+\cos x}{2}}$	$\tan\left(\frac{x}{2}\right) = \pm\sqrt{\frac{1-\cos x}{1+\cos x}}$
		$\tan\left(\frac{x}{2}\right) = \frac{1-\cos x}{\sin x}$
		$\tan\left(\frac{x}{2}\right) = \frac{\sin x}{1+\cos x}$

$$\cos x = \cos\left(2\left(\frac{x}{2}\right)\right) \qquad \text{substitution}$$

$$= 2\cos^2\left(\frac{x}{2}\right) - 1 \qquad \text{double-angle identity}$$

Thus,

$$\cos^2\left(\frac{x}{2}\right) = \frac{1+\cos x}{2} \qquad \text{simplifying}$$

Consequently,

$$\cos\left(\frac{x}{2}\right) = \pm\sqrt{\frac{1+\cos x}{2}} \qquad \text{square roots}$$

PROBLEM Verify the identity $1 - 2\sin^2\left(\frac{x}{4}\right) = \cos\left(\frac{x}{2}\right)$.

SOLUTION $\cos\left(\frac{x}{2}\right) = \cos\left(2\left(\frac{x}{4}\right)\right) = 1 - 2\sin^2\left(\frac{x}{4}\right)$ by the double-angle identity.

PROBLEM Verify the identity $\csc x = \frac{1}{2}\csc\left(\frac{x}{2}\right)\sec\left(\frac{x}{2}\right)$.

SOLUTION

$$\frac{1}{2}\csc\left(\frac{x}{2}\right)\sec\left(\frac{x}{2}\right) = \frac{1}{2\sin\left(\frac{x}{2}\right)\cos\left(\frac{x}{2}\right)} \qquad \text{reciprocal identities}$$

$$= \frac{1}{\sin 2\left(\frac{x}{2}\right)} \qquad \text{double-angle identity}$$

$$= \frac{1}{\sin x} \qquad \text{simplifying}$$

$$= \csc x \qquad \text{reciprocal identity}$$

EXERCISE

17·3

Verify the identity.

1. $\tan(2x)=\dfrac{2\tan x}{1-\tan^2 x}$

2. $\cos(2x)=\cos^2 x-\sin^2 x$

3. $\cos(8x)=\cos^2(4x)-\sin^2(4x)$

4. $\dfrac{\cos(2x)}{\sin^2 x}=\cot^2 x-1$

5. $\sin(4x)=4\sin x\cos x(1-2\sin^2 x)$

6. $\tan(2x)=\dfrac{2}{\cot x-\tan x}$

7. $\tan(2x)=\dfrac{2\sin x\cos x}{\cos^2 x-\sin^2 x}$

8. $(\sin^2 x-1)^2=\sin^4 x+\cos(2x)$

9. $\cos^2\left(\dfrac{x}{2}\right)-\sin^2\left(\dfrac{x}{2}\right)=\cos x$

10. $\dfrac{\cos(2x)}{\sin(2x)}=\dfrac{\cot x-\tan x}{2}$

Trigonometric equations

Solving equations frequently involves using the concept and properties of inverse functions. For trigonometric equations, some restrictions must be made to allow you to use inverses. The restrictions in this case are on the domains. The restricted domains on the three basic functions are shown in Table 17.4.

Table 17.4 Restricted domains

$\sin\theta$	$-\dfrac{\pi}{2}\leq\theta\leq\dfrac{\pi}{2}$
$\cos\theta$	$0\leq\theta\leq\pi$
$\tan\theta$	$-\dfrac{\pi}{2}<\theta<\dfrac{\pi}{2}$

On the restricted domains, each of the functions are one-to-one and have inverses. The graphs of each function and its inverse are shown in Figure 17.3. Three common notations for the inverse functions are $\sin^{-1}\theta$, $\arcsin\theta$, and $\text{Arcsin}\theta$.

The domain of both $\sin^{-1} x$ and $\cos^{-1} x$ is the interval $[1,1]$, and the domain of $\tan^{-1} x$ is $(-\infty,\infty)$.

The equation relating the sine function and its inverse takes on the form $y=\sin\theta \Leftrightarrow \theta=\sin^{-1} y$. A helpful way to read $\theta=\sin^{-1} y$ is "θ is the angle whose sine is y." Similar equations apply for the other trig functions.

Note: The symbol "$\Leftrightarrow$" is read "if and only if."

PROBLEM Determine the angle for the values given.

a. $\theta=\sin^{-1}\left(-\dfrac{1}{2}\right)$

b. $x=\arcsin\left(\dfrac{\sqrt{3}}{2}\right)$

c. $\theta=\sin^{-1} 3$

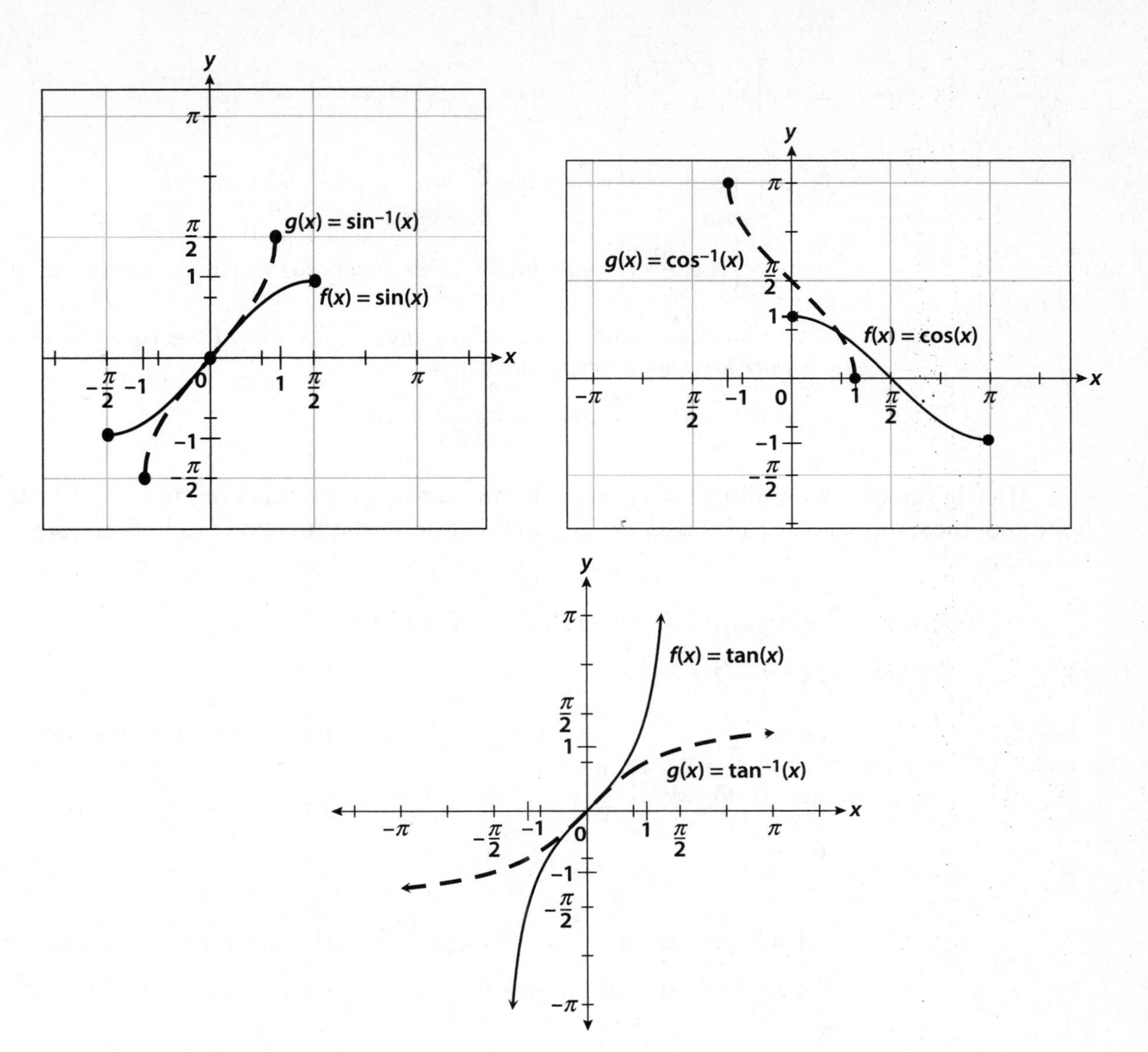

Figure 17.3 Graphs of trig functions and their inverses

SOLUTION

a. Angle θ is the angle whose sine is $-\frac{1}{2}$. Since $\sin\left(-\frac{\pi}{6}\right)=-\frac{1}{2}$, $\theta=-\frac{\pi}{6}$.

b. Angle x is the angle whose sine is $\frac{\sqrt{3}}{2}$. Since $\sin\left(\frac{\pi}{3}\right)=\frac{\sqrt{3}}{2}$, $x=\frac{\pi}{3}$.

c. Angle θ is the angle whose sine is 3. But there is no angle whose sine is 3. There is no solution. Since 3 is not in the domain of the inverse sine function, $\theta=\sin^{-1}3$ is undefined.

PROBLEM

Evaluate each expression.

a. $\sin\left[\sin^{-1}\left(\frac{\sqrt{3}}{2}\right)\right]$

b. $\cos^{-1}\left[\cos\left(\frac{5\pi}{2}\right)\right]$

c. $\tan^{-1}\left(-\sqrt{3}\right)$

SOLUTION a. $\sin\left[\sin^{-1}\left(\frac{\sqrt{3}}{2}\right)\right]=\frac{\sqrt{3}}{2}$ since $\frac{\sqrt{3}}{2}$ is in the interval $[-1,1]$.

b. $\cos^{-1}\left[\cos\left(\frac{5\pi}{2}\right)\right]\neq\frac{5\pi}{2}$ since $\frac{5\pi}{2}$ is not in $\left[-\frac{\pi}{2},\frac{\pi}{2}\right]$. However,

$\cos^{-1}\left[\cos\left(\frac{5\pi}{2}\right)\right]=\cos^{-1}(0)=\frac{\pi}{2}$. *Note*: You need to be cautious about using the inverse relations $f\left(f^{-1}(x)\right)=x$ and $f^{-1}\left(f(x)\right)=x$. These equations only work for x in the proper domain.

c. $\tan^{-1}\left(-\sqrt{3}\right)=-\frac{\pi}{3}$ since $\tan\left(-\frac{\pi}{3}\right)=-\sqrt{3}$.

If a trig equation has multiple solutions, you can use the inverse functions to find a *principal* solution. Then you use identities and the periodic nature of the functions to find all required solutions.

PROBLEM Solve $2\sin\theta-\sqrt{2}=0$ for all solutions in $[0,3\pi]$.

SOLUTION

$$2\sin\theta-\sqrt{2}=0$$

$$\sin\theta=\frac{\sqrt{2}}{2}$$

$$\sin^{-1}(\sin\theta)=\sin^{-1}\left(\frac{\sqrt{2}}{2}\right)$$

$$\theta=\frac{\pi}{4}$$

All solutions are $\frac{\pi}{4}$, $\frac{3\pi}{4}$, $\frac{9\pi}{4}$, and $\frac{11\pi}{4}$. Remembering the unit circle and the algebraic sign of the sine function in the various quadrants makes the solutions easy to visualize.

PROBLEM Solve $3\sin^2\theta+\sin\theta-2=0$ for all solutions in $(0, 360^\circ)$.

SOLUTION This is a quadratic equation in the symbol $\sin\theta$. Factoring this, you have $(\sin\theta+1)(3\sin\theta-2)=0$. Thus $\sin\theta=-1$ or $\sin\theta=\frac{2}{3}$. Solving these equations separately gives:

$\sin\theta=-1$	$\sin\theta=\frac{2}{3}$
$\sin^{-1}(\sin\theta)=\sin^{-1}(-1)$	$\sin^{-1}(\sin\theta)=\sin^{-1}\left(\frac{2}{3}\right)$
$\theta=-90^\circ$	$\theta\approx 41.8^\circ$

In this problem, $\theta\approx 41.8^\circ$ was found with a calculator. Since -90° is not in $[0,360^\circ)$, the solution in $[0,360^\circ)$ is 270°. The other approximate solutions are

41.8° and 138.2°. This latter angle was found by using the identity $\sin\theta = \sin(\pi - \theta)$.

Your knowledge of right triangle trig will help you evalute some expressions that are neither fundamental angles nor fundamental trig values.

PROBLEM Evaluate $\sin\left[\cos^{-1}\left(\frac{7}{25}\right)\right]$.

SOLUTION Sketch a right triangle to fit the angle $\theta = \cos^{-1}\left(\frac{7}{25}\right)$.

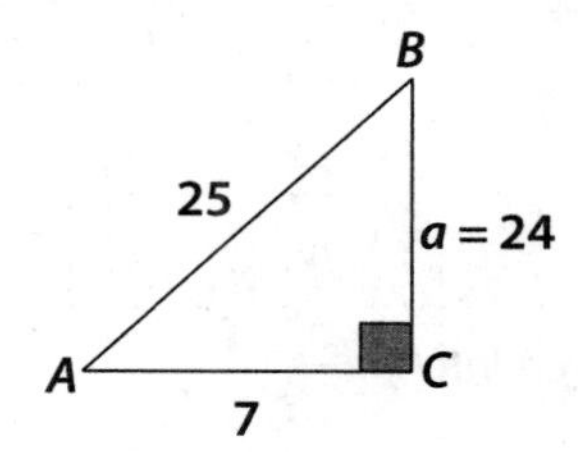

The Pythagorean theorem is used to find $a = 24$. The answer then is $\sin\left[\cos^{-1}\left(\frac{7}{25}\right)\right] = \frac{24}{25}$.

PROBLEM Evaluate $\tan\left[\sec^{-1}\left(\frac{\sqrt{9+x^2}}{x}\right)\right]$.

SOLUTION Sketch the triangle and use the Pythagorean theorem.

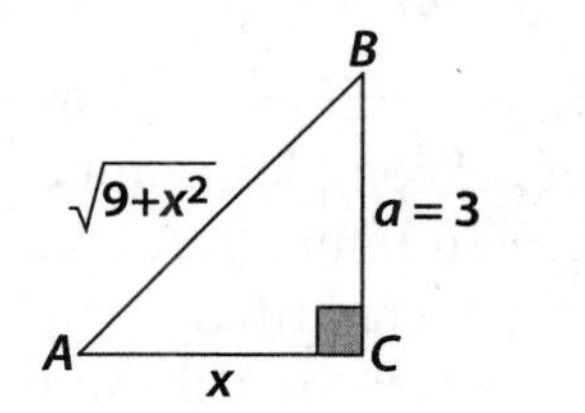

$$\tan\left[\sec^{-1}\left(\frac{\sqrt{9+x^2}}{x}\right)\right] = \tan\left[\cos^{-1}\left(\frac{x}{\sqrt{9+x^2}}\right)\right] \quad \text{reciprocal identity}$$

$$= \tan\left[\sin^{-1}\left(\frac{3}{\sqrt{9+x^2}}\right)\right] \quad \text{from the right triangle and the Pythagorean theorem}$$

$$= \frac{3}{x} \quad \text{from the right triangle}$$

EXERCISE
17·4

For 1–5, evaluate the given expression.

1. $\cos^{-1}\left[\cos\left(\frac{\pi}{4}\right)\right]$

2. $\cos\left[\arccos\left(-\frac{\sqrt{2}}{2}\right)\right]$

3. $\sin\left[\arcsin\left(\frac{3}{5}\right)\right]$

4. $\sin\left[\tan^{-1}\left(\frac{\sqrt{5}}{2}\right)\right]$

5. $\tan\left[\operatorname{arcsec}\left(\frac{5}{2x}\right)\right]$

For 6–10, solve for all the angles in the given interval.

6. $4\cos^2\theta - 3 = 0$ in $[0, 2\pi]$

7. $2\cos^3\theta + \cos^2\theta = 0$ in $[-\pi, \pi]$

8. $3\cos\theta = 1$ in $[0, \pi]$

9. $\frac{1}{2}\sin(2\theta) = \frac{1}{3}$ in $[-90°, 90°]$

10. $\sqrt{2}\sec\theta + 3 = 7$ in $[0, 180°]$

Trigonometric form of a complex number

The use of trigonometry in the study of complex numbers often simplifies the process of computing products, quotients, and powers of complex numbers (see Chapter 2, "Complex numbers," for an additional discussion of complex numbers). A complex number $z = a + bi$ is in rectangular form since it can be graphed on the rectangular complex plane (Figure 17.4). The trigonometric form of z can be realized from the associated right triangle.

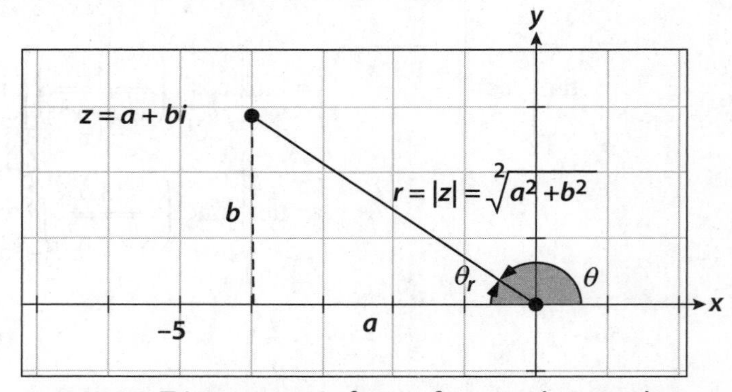

Figure 17.4 Trigonometric form of a complex number

From Figure 17.4, $a = r\cos\theta$, $b = r\sin\theta$, $\tan\theta = \frac{b}{a}$, and $r = \sqrt{a^2 + b^2}$. Hence, the **trigonometric form of z** is $z = r\cos\theta + r\sin\theta i = r(\cos\theta + i\sin\theta)$. The latter form is the most common form used. A shorthand notation for $r(\cos\theta + i\sin\theta)$ is $\boldsymbol{r\,cis}$.

If $z_1 = r_1(\cos\alpha + i\sin\alpha)$ and $z_2 = r_2(\cos\beta + i\sin\beta)$, then the product:

$$\begin{aligned} z_1z_2 &= r_1(\cos\alpha + i\sin\alpha)r_2(\cos\beta + i\sin\beta) \\ &= r_1r_2(\cos\alpha + i\sin\alpha)(\cos\beta + i\sin\beta) \\ &= r_1r_2(\cos\alpha\cos\beta + i\sin\beta\cos\alpha + i\sin\alpha\cos\beta + i^2\sin\alpha\sin\beta) \\ &= r_1r_2(\cos\alpha\cos\beta + i\sin\beta\cos\alpha + i\sin\alpha\cos\beta - \sin\alpha\sin\beta) \\ &= r_1r_2\left[(\cos\alpha\cos\beta - \sin\alpha\sin\beta) + i(\sin\alpha\cos\beta + \cos\alpha\sin\beta)\right] \\ &= r_1r_2\left[\cos(\alpha+\beta) + i\sin(\alpha+\beta)\right] \end{aligned}$$

Also, a similar derivation will give the trigonometric form of the quotient. To summarize, see Table 17.5.

Table 17.5 Products and quotients of complex numbers in trigonometric form

$z_1 = r_1(\cos\alpha + i\sin\alpha)$ and $z_2 = r_2(\cos\beta + i\sin\beta)$
$z_1z_2 = r_1r_2[\cos(\alpha+\beta) + i\sin(\alpha+\beta)]$
$\dfrac{z_1}{z_2} = \dfrac{r_1}{r_2}[\cos(\alpha-\beta) + i\sin(\alpha-\beta)], z_2 \neq 0$

In words, if two complex numbers are in trigonometric form, then to multiply the two, you multiply the moduli and add the angles; and to divide the two, you divide the moduli and subtract the angles. Continued multiplication of a complex number by itself leads to DeMoivre's theorem.

DeMoivre's theorem: If $z = r(\cos\theta + i\sin\theta)$ and n is a positive integer, then $z^n = r^n\left[\cos(n\theta) + i\sin(n\theta)\right]$.

This is a very elegant and useful theorem. One note of practical information here is that if n is large, recall that $\theta \pm 2\pi$ is coterminal with θ, so use the coterminal angle θ for which $0 \leq \theta < 2\pi$ or $0 \leq \theta < 360^\circ$ if θ is measured in degrees.

PROBLEM Compute the product and quotient, using the trigonometric form for $z_1 = 7(\cos 120^\circ + i\sin 120^\circ)$ and $z_2 = 2(\cos 300^\circ + i\sin 300^\circ)$.

SOLUTION $z_1z_2 = 14(\cos 420^\circ + i\sin 420^\circ) = 14(\cos 60^\circ + i\sin 60^\circ) = 14\left(\dfrac{1}{2} + i\dfrac{\sqrt{3}}{2}\right) = 7 + 7\sqrt{3}i$

PROBLEM Use DeMoivre's theorem to compute $(-2+2i)^6$.

SOLUTION First write in trigonometric form: $r = \sqrt{4+4} = \sqrt{8} = 2\sqrt{2}$. Thus,

$-2+2i = 2\sqrt{2}(\cos 135^\circ + i\sin 135^\circ)$. Hence, $(-2+2i)^6 = \left(2\sqrt{2}\right)^6\left(\cos 810^\circ + i\sin 810^\circ\right)$
$= 512(\cos 90^\circ + i\sin 90^\circ) = 512i.$

EXERCISE

17·5

For 1–5, compute the product and quotient, using the trigonometric form. Write the answer, in exact form if possible. If not, round all values to one decimal place.

1. $z_1 = \sqrt{3} + i$ and $z_2 = 1 + \sqrt{3}i$

2. $z_1 = 10(\cos 60° + i \sin 60°)$
   $z_2 = 4(\cos 30° + i \sin 30°)$

3. $z_1 = 5\sqrt{2}\,\text{cis}\,210°$
   $z_2 = 2\sqrt{2}\,\text{cis}\,30°$

4. $z_1 = 9\left[\cos\left(\frac{\pi}{15}\right) + i\sin\left(\frac{\pi}{15}\right)\right]$
   $z_2 = 1.8\left[\cos\left(\frac{2\pi}{3}\right) + i\sin\left(\frac{2\pi}{3}\right)\right]$

5. $z_1 = 6\,\text{cis}\,82°$
   $z_2 = 1.5\,\text{cis}\,27°$

For 6–10, use DeMoivre's theorem to compute. Write the answer in exact form if possible. If not, round all values to one decimal place.

6. $(3 + 3i)^4$

7. $\left(\frac{1}{2} - \frac{\sqrt{3}}{2}i\right)^5$

8. $(4\cos 300° + 4i \sin 300°)^3$

9. $\left(\frac{\sqrt{2}}{2}\text{cis}\,135°\right)^8$

10. $(\sqrt{3} - i)^3$

Solving triangles

·18·

Law of sines

In Chapter 17 you solved right triangles by using trigonometric ratios and the Pythagorean theorem. Many applications of trigonometry involve triangles that are not right triangles but are acute or obtuse. You also can use trigonometric concepts to solve these latter triangles. One of these concepts is the law of sines.

The **law of sines**: For any triangle ABC, $\frac{\sin A}{a} = \frac{\sin B}{b} = \frac{\sin C}{c}$; that is, the ratio of an angle to the side opposite that angle is constant.

These equalities are proportional statements. To solve proportional statements, three of the four parts must be known. To solve triangles, you must find all six parts—the three angles and the three sides. You can solve the proportions if you know one of the following:

1. Two angles and a side opposite one of them (AAS)
2. Two angles and an included side (ASA). *Note*: If two angles are known, the third angle is determined and case 1 applies.
3. Two sides and an angle opposite one of these sides (SSA)

In cases 1 and 2, unique triangles are formed, and the solutions are thus unique. However, case 3 does not necessarily form unique triangles. This case is the *ambiguous case* and will be handled separately. Finally, when you are solving triangles, it is always helpful to sketch the triangles.

PROBLEM Solve the triangle given $\angle B = 100°$, $\angle C = 38°$, and $a = 40$.

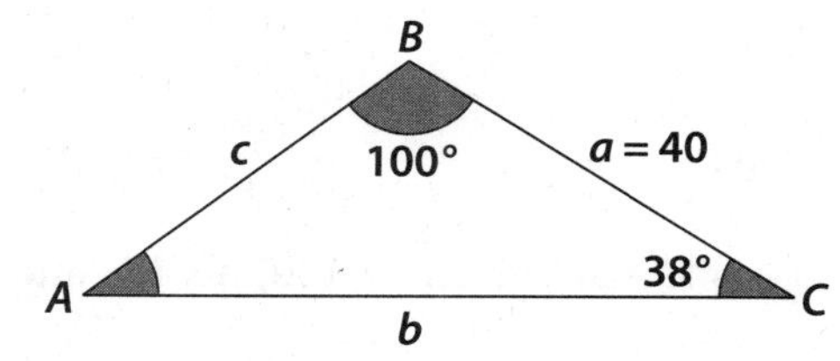

SOLUTION Since $\angle B$ and $\angle C$ are given, $\angle A = 42°$. Thus, $\frac{\sin 42°}{40} = \frac{\sin 100°}{b}$.

Solving this proportion gives $b = \frac{40 \sin 100°}{\sin 42°} \approx 58.9$. Also, $\frac{\sin 42°}{40} = \frac{\sin 38°}{c}$, and solving this proportion gives $c = \frac{40 \sin 38°}{\sin 42°} \approx 36.8$. Therefore, the triangle is solved.

PROBLEM Solve the triangle, given $\angle A = 40°$, $a = 50$, and $b = 30$.

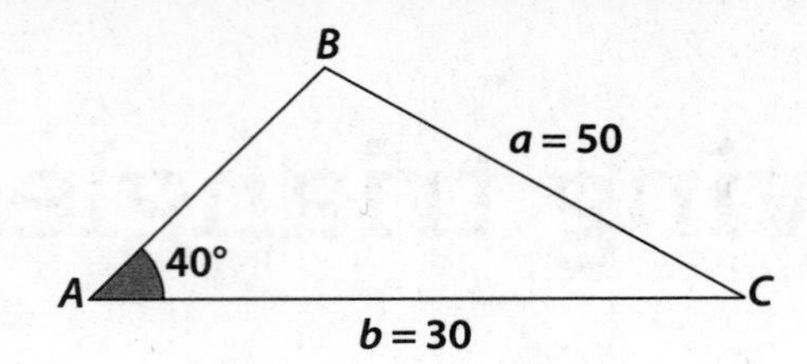

SOLUTION $\frac{\sin 40°}{50} = \frac{\sin B}{30}$. Thus $\sin B = \frac{30 \sin 40°}{50} \approx 0.3857$, and using the inverse function on the calculator, $\angle B_r \approx 22.7°$. Thus $\angle B \approx 22.7°$ or $\angle B \approx 157.3°$. The latter value will not work since the sum of the angles of a triangle is 180°. It follows then that $\angle C \approx 180° - 62.7° = 117.3°$. Using this result, $\frac{\sin 40°}{50} = \frac{\sin 117.3°}{c}$ and $c = \frac{50 \sin 117.3°}{\sin 40°} \approx 69.1$. Therefore, the triangle is solved.

The ambiguous case: In solving SSA triangles, the possibilities are one solution, two solutions, or no solution. Figure 18.1 illustrates the three possibilities. In the triangles, side c, side a, and the angle $\angle A$ opposite side a are known. If $a = 1$, the triangle is the familiar $30° - 60° - 90°$ triangle, and you have a unique solution; if $a > 1$, you have two possible solutions; and if $a < 1$, there is no solution.

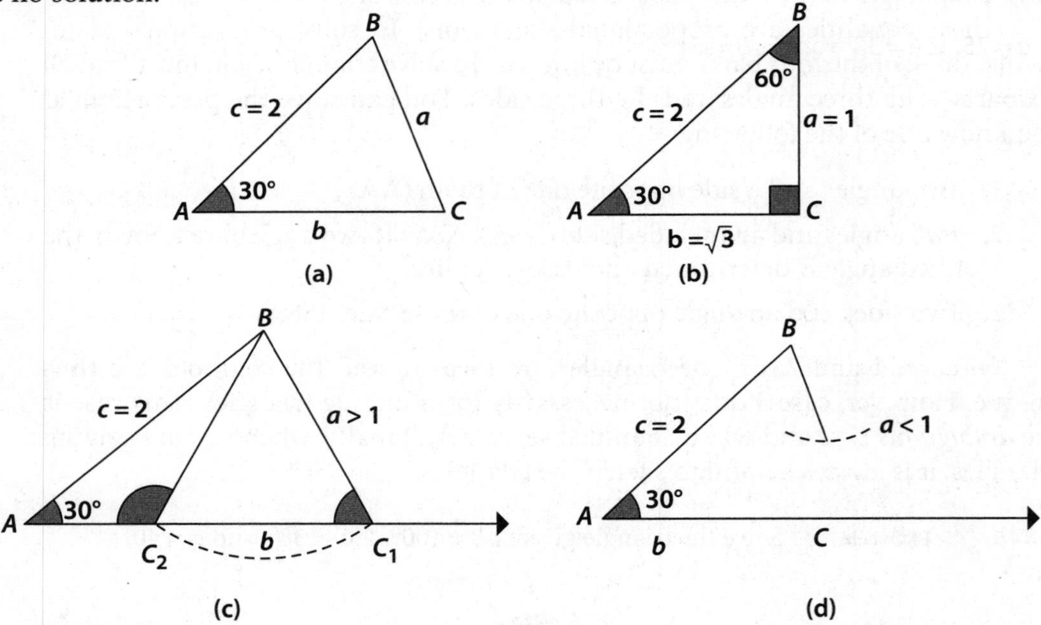

Figure 18.1 Ambiguous case triangles

PROBLEM Solve the triangle ABC, given $c = 100$, $a = 60$, and $\angle A = 28°$.

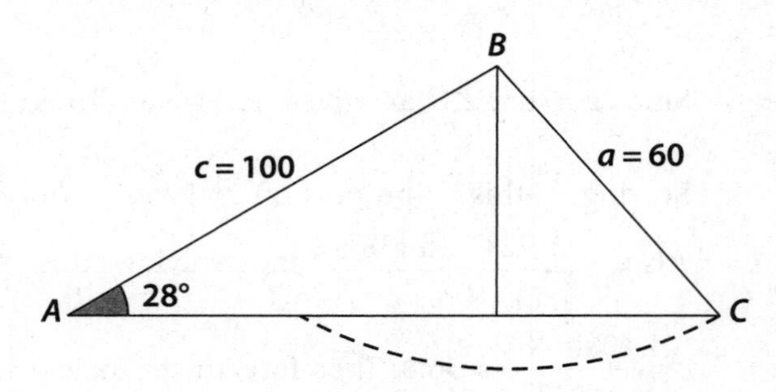

SOLUTION A figure is drawn to indicate there may be two solutions. Using the law of sines, $\frac{\sin 28^\circ}{60} = \frac{\sin C}{100}$ or $\sin C = \frac{100 \sin 28^\circ}{60} \approx 0.7825$. Then $\angle C_1 \approx 51.5^\circ$ or $\angle C_2 \approx (180 - 51.5)^\circ = 128.5^\circ$. These values indicate that $\angle B_1 \approx (180 - 51.5 - 28)^\circ = 100.5^\circ$ or $\angle B_2 \approx (180 - 28 - 128.5)^\circ = 23.5^\circ$. The two possible sides are determined by applying the law of sines twice. $b_1 = \frac{60 \sin 100.5^\circ}{\sin 28^\circ} \approx 125.7$ or $b_2 = \frac{60 \sin 23.5^\circ}{\sin 28^\circ} \approx 51.0$. The solutions are indicated in Table 18.1.

Table 18.1 Ambiguous case solutions

Angles	Sides	Angles	Sides
$\angle A = 28^\circ$	$a = 60$	$\angle A = 28^\circ$	$a = 60$
$\angle B_1 \approx 100.5^\circ$	$b_1 \approx 125.7$	$\angle B_2 \approx 23.5^\circ$	$b_2 \approx 51.0$
$\angle C_1 \approx 51.5^\circ$	$c = 100$	$\angle C_2 \approx 128.5^\circ$	$c = 100$

EXERCISE

18·1

Solve the triangle ABC, given the information, or state why the law of sines cannot be used.

1. $a = 75$, $\angle A = 38^\circ$, and $\angle B = 64^\circ$
2. $b = 10\sqrt{3}$, $\angle A = 30^\circ$, and $\angle B = 60^\circ$
3. $b = 385$, $\angle B = 47^\circ$, and $\angle A = 108^\circ$
4. $a = 7.2$, $\angle A = 27^\circ$, and $\angle B = 98^\circ$
5. $c = 12.9$, $\angle A = 20.4^\circ$, and $\angle B = 63.4^\circ$
6. $c = 126$, $\angle A = 13^\circ$, and $\angle B = 22^\circ$
7. $c = 0.8$, $\angle A = 56^\circ$, and $\angle B = 112^\circ$
8. $\angle A = 45^\circ$, $\angle B = 45^\circ$, and $c = 15\sqrt{2}$
9. $a = 42.7$, $\angle B = 103.4^\circ$, and $\angle C = 19.6^\circ$
10. $b = 67$, $c = 58$, and $\angle C = 59^\circ$

Two other possibilities, SSS and SAS, cannot be solved using the law of sines. Another law that is used in these cases is the law of cosines.

Law of cosines

The **law of cosines**: For any triangle *ABC* with sides *a*, *b*, and *c*:

$$a^2 = b^2 + c^2 - 2bc \cos A$$

$$b^2 = a^2 + c^2 - 2ac \cos B$$

$$c^2 = a^2 + b^2 - 2ab \cos C$$

In each case, the squared term on the left is the side opposite the angle. Notice also if $\angle C = 90^\circ$, the cosine is 0 and the expression is the Pythagorean theorem.

PROBLEM Solve the triangle *ABC*, given $a = 15$, $b = 25$, and $c = 28$.

SOLUTION

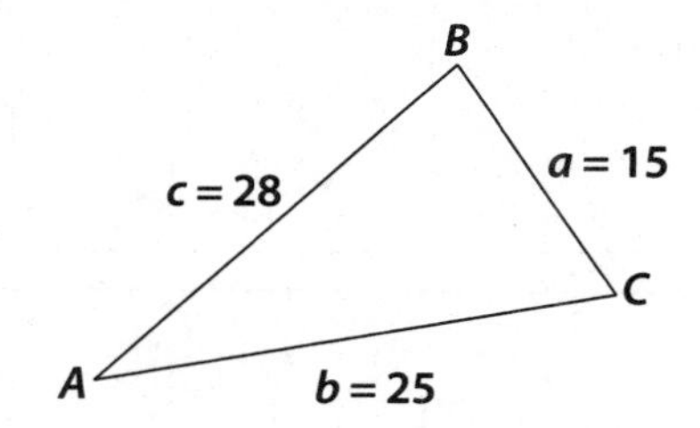

$$15^2 = 25^2 + 28^2 - 2(25)(28)\cos A$$
$$225 = 625 + 784 - 1{,}400\cos A$$
$$225 = 1{,}409 - 1{,}400\cos A$$
$$-1{,}184 = -1{,}400\cos A$$
$$\cos A \approx 0.8457$$
$$\angle A \approx 32.25^\circ$$

Now, you can use the law of sines to solve for $\angle B$. Then $\angle C$ is easily determined.

$$\frac{\sin 32.25^\circ}{15} = \frac{\sin B}{25}$$
$$\sin B = \frac{25\sin 32.25^\circ}{15} \approx 0.8894$$
$$\angle B \approx 62.8^\circ \text{ using calculator } \sin^{-1}$$
$$\angle C \approx 85^\circ$$

PROBLEM Solve the triangle ABC, given $a = 16$, $c = 7$, and $\angle B = 95^\circ$.

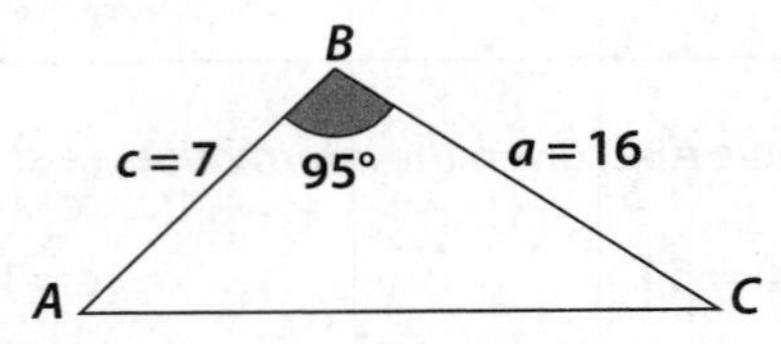

SOLUTION
$$b^2 = 7^2 + 16^2 - 2(7)(16)\cos 95^\circ$$
$$b^2 \approx 49 + 256 - 224(-0.0872) \approx 324.53$$
$$b \approx 18$$
$$\frac{\sin C}{c} = \frac{\sin B}{b}$$
$$\frac{\sin C}{7} = \frac{\sin 95^\circ}{18}$$
$$\sin C = \frac{7\sin 95^\circ}{18} \approx 0.3874$$
$$C \approx 22.8^\circ \text{ using calculator } \sin^{-1}$$

EXERCISE 18·2

For 1–2, solve the triangle ABC.

1. $b = 3.2$, $c = 1.5$, and $\angle A = 95.7^\circ$
2. $a = 75$, $b = 32$, and $\angle C = 38^\circ$

For 3–4, find the other side.

3. $a = 12.9$, $c = 25.8$, and $\angle B = 30^\circ$
4. $a = 6.7$, $c = 10.9$, and $\angle B = 98^\circ$

For 5–6, find $\angle A$.

5. $a = 15\sqrt{3}$, $b = 6\sqrt{3}$, and $c = 10\sqrt{3}$
6. $a = 282$, $b = 129$, and $c = 300$

Vectors in two dimensions

·19·

Basic vector concepts

Concepts such as area, temperature, and volume can be adequately described by a single number, their magnitude. Quantities such as these are called **scalar quantities**. Other concepts such as velocity and force require two quantities, magnitude and direction, to describe their behavior. **Vector quantities** are physical concepts that require more than a single quantity to describe their attributes.

Vector quantities are represented in various ways. To begin, consider concepts that require only two quantities for description. One such concept is velocity, which requires magnitude and direction for complete description. There are several ways to represent these vector quantities. Two of the most common ways are by ordered pairs and by directed line segments or "arrows."

The geometric representation of vectors uses directed line segments (Figure 19.1). The length of the segment represents the magnitude, and the arrowhead indicates the direction of "travel." The notation $\overrightarrow{AB}$ denotes the vector with initial point A and terminal point B. Lowercase, bold letters are used to represent vectors.

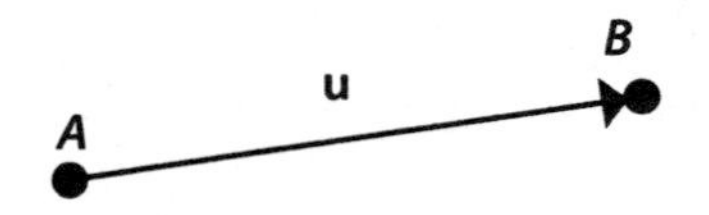

Figure 19.1 The vector $\mathbf{u} = \overrightarrow{AB}$

Vectors are equal if they have the same magnitude and direction. The multiplication of a vector by a constant is called scalar multiplication. Scalar multiplication may alter the magnitude or direction or both. Multiplying by a negative scalar reverses the direction. Multiplying by a scalar with absolute value less than 1 will compress the vector, and multiplying by a scalar with absolute value greater than 1 will stretch the vector. Hence, $\mathbf{w} = 2\mathbf{v}$ is a vector in the same direction as $\mathbf{v}$ with twice the magnitude. Vector representations are very convenient means of illustrating the actions of several vectors acting on a single point, such as forces of different magnitudes acting on a single point from different directions. Graphical representations of vectors in a coordinate plane use some conventions of which you should be cognizant. Figure 19.2, on the next page, shows vectors with different placements in the plane. The vectors are equal even though they have different initial and terminal points. Thus the placement in the plane is irrelevant.

Vectors may also be represented by ordered pairs. The vector $\mathbf{v}$ with (0, 0) as the initial point and (a, b) as the terminal point is called the **position vector** for $\mathbf{v}$. To avoid confusion with ordered pair notation, you represent the position vector for $\mathbf{v}$ as $\mathbf{v} = \langle a, b \rangle$. If $\mathbf{w}$ has initial point (x_1, y_1) and terminal point (x_2, y_2), then the

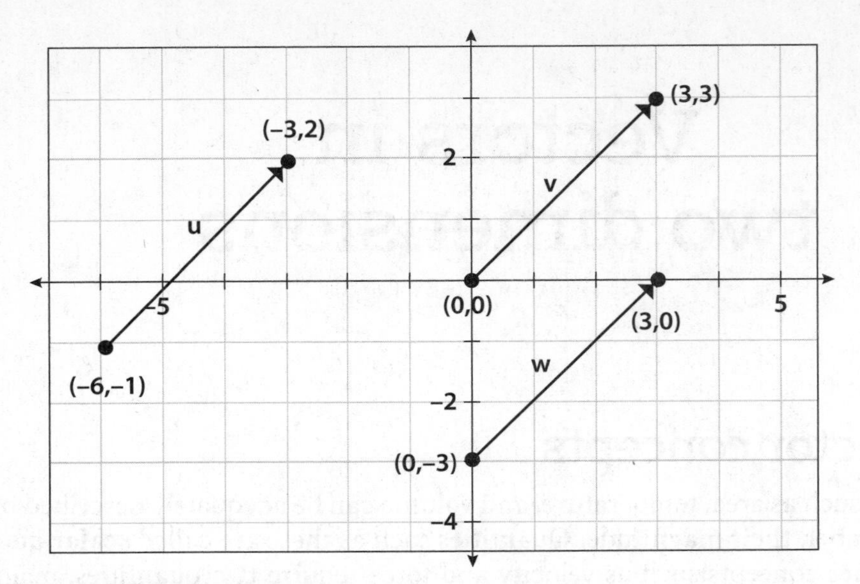

Figure 19.2 Equal vectors

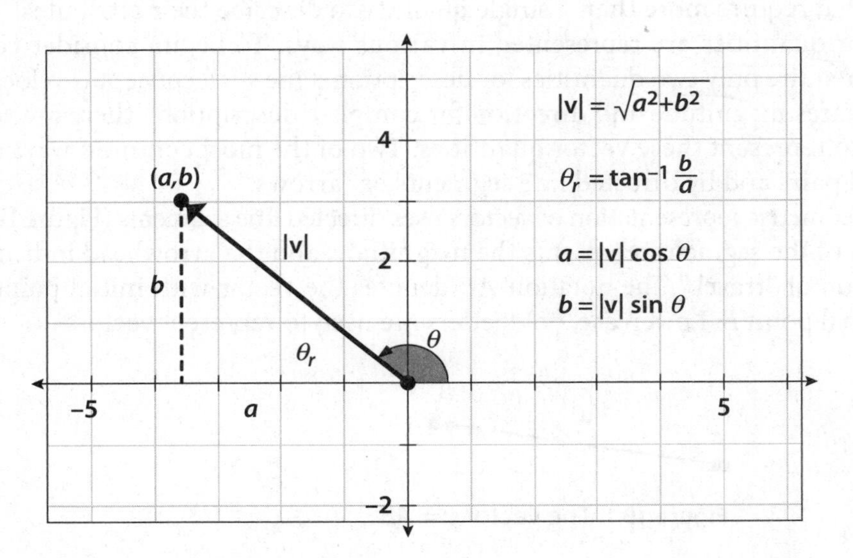

Figure 19.3 Trigonometric components

position vector for **w** is $\mathbf{w} = \langle x_2 - x_1, y_2 - y_1 \rangle$. If $\mathbf{v} = \langle a,b \rangle$, then a is the **horizontal component** of **v**, b is the **vertical component** of **v**, and $|\mathbf{v}|$ is the **magnitude** of **v**. Concepts from trigonometry immediately apply to vectors due to the right triangle associated with the vector **v**. Figure 19.3 relates the components of **v** in trigonometric form.

This form of the components is very useful in solving application problems. Often only partial information is known about a vector quantity, and you must find the remaining information, using mathematics including trigonometry.

PROBLEM Find the magnitude and direction angle of the vector $\mathbf{v} = \langle -2.5, -6 \rangle$.

SOLUTION Sketch the vector and apply the component relations in the following figure.

$|\mathbf{v}| = \sqrt{(-2.5)^2 + (-6)^2} = \sqrt{42.25} = 6.5$. $\theta_r = \tan^{-1}\left(\dfrac{-6}{-2.5}\right) = \tan^{-1}(2.4)$. Hence, $\theta_r \approx 67.4°$ and since θ is in quadrant III, $\theta \approx 247.4°$.

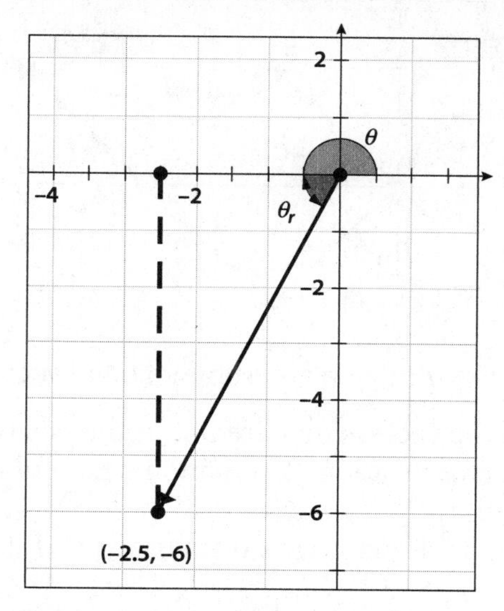

PROBLEM Find the horizontal and vertical components of the vector that has magnitude $|\mathbf{v}| = 21$ and direction angle 205°.

SOLUTION Sketch the vector and solve, using the trig relations. $\theta_r = 205° - 180° = 25°$. Also, $a = |\mathbf{v}|\cos(205°) = 21\cos(205°) \approx -19$ and $b = |\mathbf{v}|\ \sin(205°) = 21\sin(205°) \approx -8.9$. As a check, $|\mathbf{v}| = \sqrt{(-19)^2 + (-8.9)^2} \approx 21$.

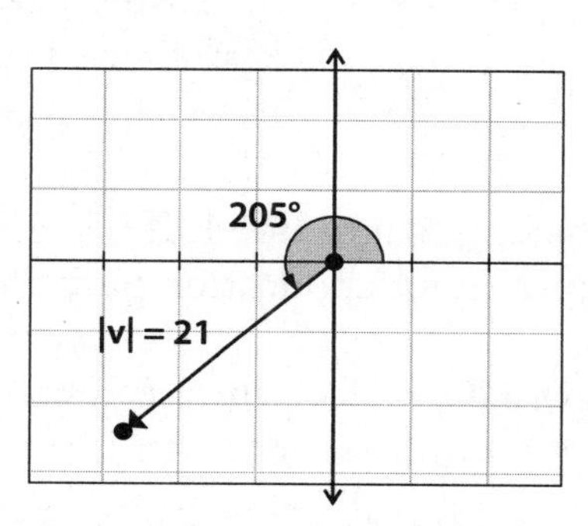

Basic arithmetic for working with vectors is addition, subtraction, and multiplying by scalars. The *sum* or **resultant** of vectors **u** and **v** is a vector **w** formed by placing the initial point of **v** on the terminal point of **u** and then joining the initial point of **u** to the terminal point of **v** (see Figure 19.4). The sum is written **u** + **v**.

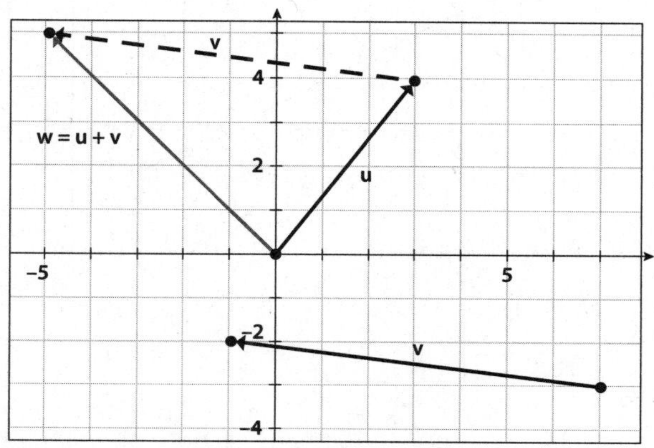

Figure 19.4 Addition of vectors

The difference (subtraction) of vectors is defined as $\mathbf{u} - \mathbf{v} = \mathbf{u} + (-\mathbf{v})$. It is convenient to see the resultant and difference vectors displayed on the same figure (see Figure 19.5).

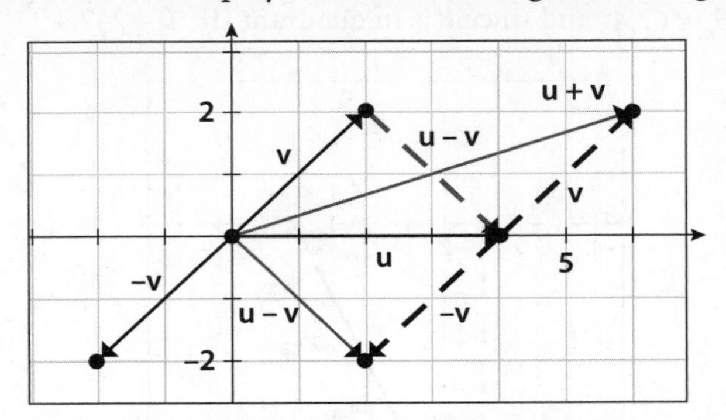

Figure 19.5 Addition and subtraction of vectors

The dashed vectors are the vectors that are not in the position vector location, and the solid vectors are in the position vector location. A mnemonic helpful in sketching the vector $\mathbf{u} - \mathbf{v}$ is to read it as "the vector from $\mathbf{v}$ to $\mathbf{u}$."

The properties of vector arithmetic are summarized in Table 19.1.

Table 19.1 Laws of vector arithmetic

If $\mathbf{u}$, $\mathbf{v}$, and $\mathbf{w}$ are vectors and m and n are scalars, then the following hold:

1. $\mathbf{u} + \mathbf{v} = \mathbf{v} + \mathbf{u}$	Commutative Law for Addition
2. $\mathbf{u} + (\mathbf{v} + \mathbf{w}) = (\mathbf{u} + \mathbf{v}) + \mathbf{w}$	Associative Law for Addition
3. $m\mathbf{u} = \mathbf{u}m$	Commutative Law for Multiplication
4. $m(n\mathbf{u}) = (mn)\mathbf{u}$	Associative Law for Multiplication
5. $(m + n)\mathbf{u} = m\mathbf{u} + n\mathbf{u}$	Distributive Law
6. $m(\mathbf{u} + \mathbf{v}) = m\mathbf{u} + m\mathbf{v}$	Distributive Law

A unit vector is a vector having magnitude 1. If $\mathbf{u} \neq 0$, then $\dfrac{\mathbf{u}}{|\mathbf{u}|}$ is a unit vector in the same direction as $\mathbf{u}$. This is very important for application purposes.

PROBLEM If $\mathbf{u} = \langle 2, -1 \rangle$ and $\mathbf{v} = \langle 2, 4 \rangle$, then represent the following graphically:

a. $-3\mathbf{u}$ c. $\mathbf{u} + \mathbf{v}$

b. $\mathbf{u} - \mathbf{v}$ d. $-3\mathbf{u} + \dfrac{1}{2}\mathbf{v}$

SOLUTION a.

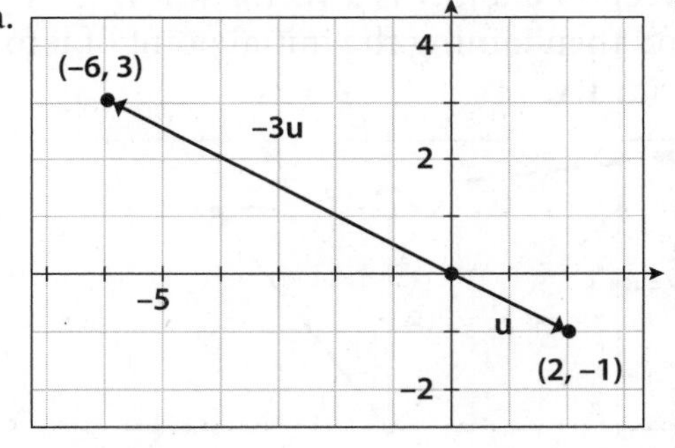

b.

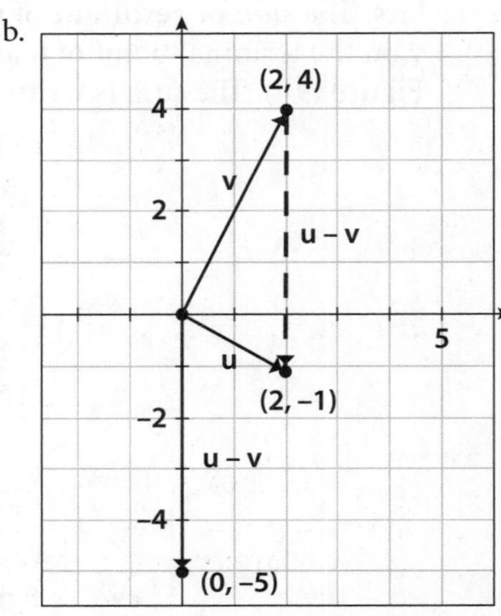

c. 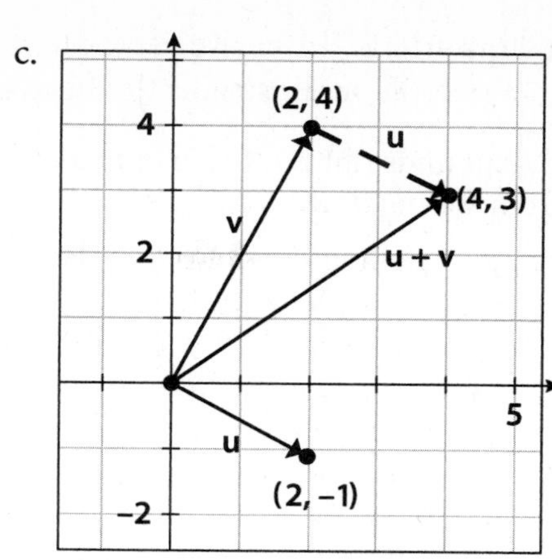

d. 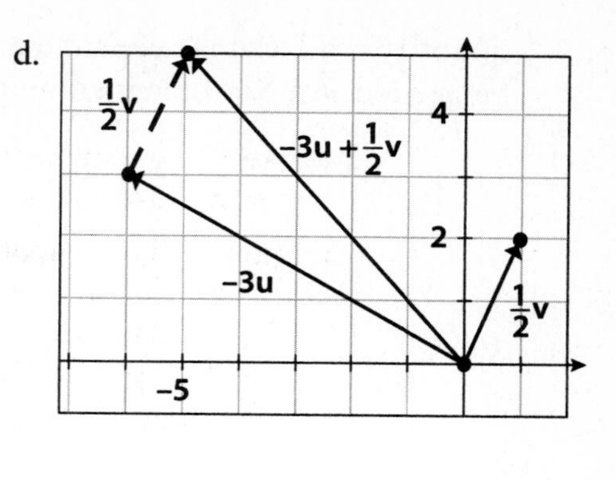

If **u** and **v** are expressed in component form $\mathbf{u} = \langle a, b \rangle$ and $\mathbf{v} = \langle c, d \rangle$ and k is a scalar, then:

1. $\mathbf{u} + \mathbf{v} = \langle a + c, b + d \rangle$
2. $\mathbf{u} - \mathbf{v} = \langle a - c, b - d \rangle$
3. $k\mathbf{u} = \langle ka, kb \rangle$

A form of vectors used extensively in some of the physical sciences and better suited for algebraic manipulations uses special unit vectors associated with the rectangular coordinate system. In two dimensions, these special vectors are $\mathbf{i} = \langle 1, 0 \rangle$ and $\mathbf{j} = \langle 0, 1 \rangle$. You can represent all two-dimensional vectors as a linear combination of these two unit vectors. In particular if $\mathbf{u} = \langle a, b \rangle$, then $\mathbf{u} = a\mathbf{i} + b\mathbf{j}$ and this latter form is the algebraic form preferred by many users. Using this form, if $\mathbf{u} = a\mathbf{i} + b\mathbf{j}$ is a nonzero vector, then $\frac{\mathbf{u}}{|\mathbf{u}|} = \frac{a}{\sqrt{a^2 + b^2}}\mathbf{i} + \frac{b}{\sqrt{a^2 + b^2}}\mathbf{j}$. As mentioned earlier, this is a unit vector in the direction of **u**.

One of the important ideas in applications is that of finding a vector **w** that is coincident (points in the same direction) with a given vector **v**. A typical problem will illustrate the concept in the manner in which it is frequently encountered.

PROBLEM If the angle between **u** and **v** is 35°, find the vector **w** that is coincident with **v** and forms the base of a right triangle, as shown in the following figure.

SOLUTION Using the Pythagorean theorem, $|\mathbf{u}| = \sqrt{(1.6)^2 + 4^2} \approx 4.3$ and $|\mathbf{v}| = \sqrt{36 + 16} \approx 7.2$. Also $|\mathbf{w}| = 4.3\cos(35°) \approx 3.5$. Thus the vector of magnitude $|\mathbf{w}| \approx 3.5$ in the direction of **v** is $\mathbf{w} = |\mathbf{w}|\frac{\mathbf{v}}{|\mathbf{v}|} \approx (3.5)\frac{\langle 6, 4 \rangle}{7.2} = \langle 2.9, 1.9 \rangle$.

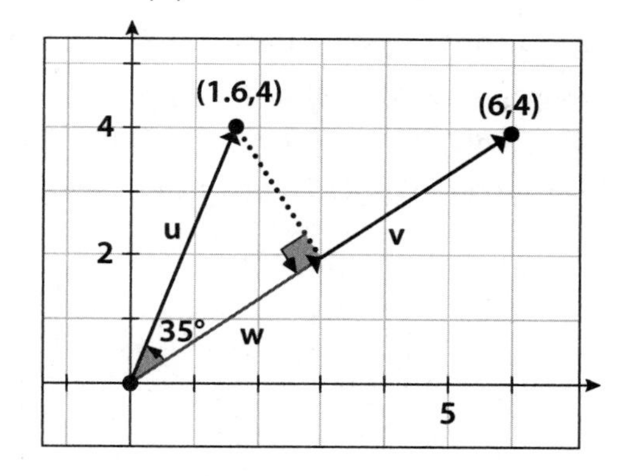

This process is referred to as *finding the projection of* **u** *onto* **v**. In general, if θ is the angle between **u** and **v**, then $u_v = |\mathbf{u}|\cos\theta$ is the component of **u** in the direction of **v**. Note that u_v is a scalar quantity (not a vector) that is the magnitude of the projection. In many instances, this quantity is all that is needed to answer a question. The entire vector in the direction of **v** may not be needed. An example involving an inclined plane illustrates the need for only u_v.

PROBLEM A crate weighing 500 pounds (lb) is on an inclined plane. What force is necessary to keep the crate from sliding down the plane?

SOLUTION The force of gravity, **g**, is pulling down on the plane and makes an angle of 90° with the base of the inclined plane. A force **u** is acting perpendicular to the plane, causing the crate to press against the plane. Since placement in the plane is irrelevant for vectors, **u** has been moved for clarity (dashed line). Angle β is 65° since it is complementary to 25°. Thus the component of **g** along the inclined plane is $g_p = 500\cos(65°) \approx 211$ pounds. A force of 211 pounds is needed to keep the crate from sliding down the plane.

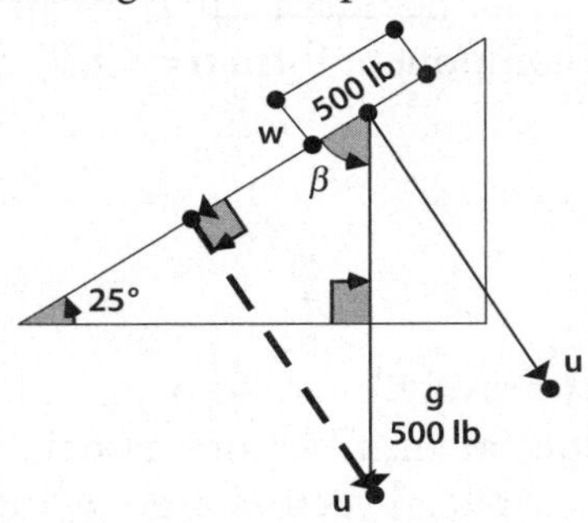

Notice that the vector **g** is the sum of two vectors **u** and **w** that are perpendicular to each other. This is an important observation and one that is developed further in the next few sections. The force vectors **u** and **w** are said to be acting on the single point representing the crate.

1. Sketch the vectors $\mathbf{u} = \langle 5, 2 \rangle$, $\mathbf{v} = \langle -3, 2 \rangle$, and $\mathbf{u} + \mathbf{v}$ on the coordinate plane.
2. If $\mathbf{u} = \langle 9, 0 \rangle$ and $\mathbf{v} = \langle 5, -4 \rangle$ then compute $\mathbf{u} - \mathbf{v}$, $4\mathbf{v}$, $\mathbf{v} - \mathbf{u}$, and $3\mathbf{u} - 4\mathbf{v}$.
3. Write the vector $\langle 5, 8 \rangle$ in $a\mathbf{i} + b\mathbf{j}$ form.
4. Find the magnitude of $\mathbf{v} = 5\mathbf{i} - 3\mathbf{j}$.
5. If $\mathbf{u} = 3\mathbf{i} + 6\mathbf{j}$ and $\mathbf{v} = -4\mathbf{i} + 2\mathbf{j}$, compute $\mathbf{u} + \mathbf{v}$.
6. Find the horizontal and vertical components of $\mathbf{w} = 2\mathbf{j} - 3\mathbf{i}$.
7. Draw a sketch of the force vectors $\mathbf{u} = -2\mathbf{i} + 3\mathbf{j}$, $\mathbf{v} = -\mathbf{i} - \mathbf{j}$, and $\mathbf{w} = 2\mathbf{i} + \mathbf{j}$ acting on a single point at the origin of the coordinate plane, and show the resultant vector **r**.

*For 8–10, find the component of **u** in the direction of **v** in the figures shown.*

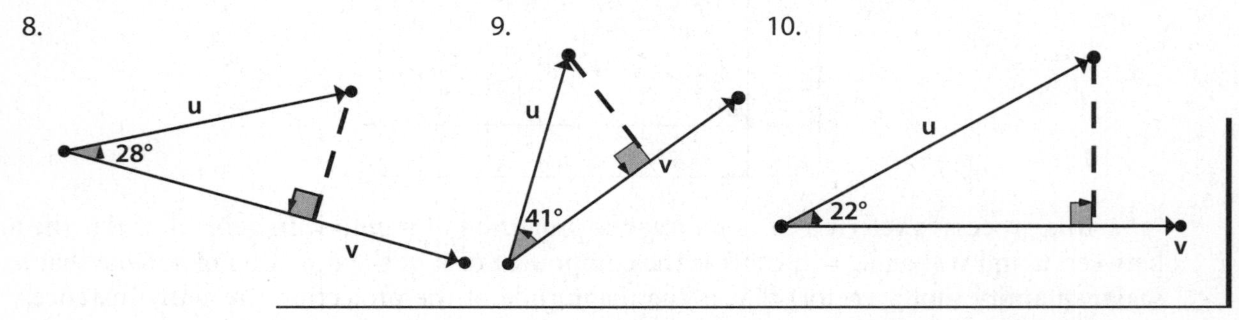

Dot product and equilibrium

In vector terms, equilibrium has special meaning. If several vector forces $\mathbf{u}_1, \mathbf{u}_2, \mathbf{u}_3, \ldots, \mathbf{u}_n$, are acting on single point then the resultant force is $\mathbf{u} = \mathbf{u}_1 + \mathbf{u}_2 + \mathbf{u}_3 + \ldots + \mathbf{u}_n$. The vector force needed to prevent the point from moving is $-\mathbf{u}$. This is the equilibrium force or the **equilibrant**. The intuitive notion is that $-\mathbf{u}$ "counterbalances" $\mathbf{u}$. The inclined plane example is an application of finding an equilibrant.

An extremely important concept used in working with vectors is that of a special type of vector product.

The *dot product of vectors*: If $\mathbf{u} = \langle a, b \rangle$ and $\mathbf{v} = \langle c, d \rangle$, then the dot product of $\mathbf{u}$ and $\mathbf{v}$ is $\mathbf{u} \cdot \mathbf{v} = ac + bd$. The dot product is a *scalar* and *not* a vector! The dot product is useful in many applications, of which one is to find the angle between two vectors.

The *angle θ between two vectors*: If $\mathbf{u}$ and $\mathbf{v}$ are nonzero vectors, then the angle θ between $\mathbf{u}$ and $\mathbf{v}$ is determined by the equation $\theta = \cos^{-1}\left(\dfrac{\mathbf{u} \cdot \mathbf{v}}{|\mathbf{u}||\mathbf{v}|}\right)$. It is also written as $\cos\theta = \dfrac{\mathbf{u} \cdot \mathbf{v}}{|\mathbf{u}||\mathbf{v}|}$ or $\mathbf{u} \cdot \mathbf{v} = |\mathbf{u}||\mathbf{v}|\cos\theta$, which is an alternate form of the dot product.

The following are arithmetic properties of the dot product:

1. $\mathbf{u} \cdot \mathbf{v} = \mathbf{u} \cdot \mathbf{v}$
2. $\mathbf{u} \cdot \mathbf{u} = |\mathbf{u}|^2$
3. $\mathbf{w} \cdot (\mathbf{u} + \mathbf{v}) = \mathbf{w} \cdot \mathbf{u} + \mathbf{w} \cdot \mathbf{v}$
4. $k(\mathbf{u} \cdot \mathbf{v}) = (k\mathbf{u}) \cdot \mathbf{v} = \mathbf{u} \cdot (k\mathbf{v})$ k **a scalar**
5. $\mathbf{0} \cdot \mathbf{u} = \mathbf{u} \cdot \mathbf{0} = \mathbf{0}$ where $\mathbf{0}$ is the zero vector
6. $\dfrac{\mathbf{u}}{|\mathbf{u}|} \cdot \dfrac{\mathbf{v}}{|\mathbf{v}|} = \dfrac{\mathbf{u} \cdot \mathbf{v}}{|\mathbf{u}||\mathbf{v}|}$

You can use the dot product to find the projection of a vector $\mathbf{u}$ onto a vector $\mathbf{v}$ by the following equation: The projection vector $\mathbf{w}$, of $\mathbf{u}$ onto $\mathbf{v}$, is $\mathbf{w} = \left(\dfrac{\mathbf{u} \cdot \mathbf{v}}{|\mathbf{v}|^2}\right)\mathbf{v}$. One advantage of the projection vector is that the vector $\mathbf{u}$ can be written as (resolved into) the sum of two vectors that are perpendicular (orthogonal). One of the vectors is $\mathbf{w}$, and the other is $\mathbf{u} - \mathbf{w}$, as depicted in Figure 19.6.

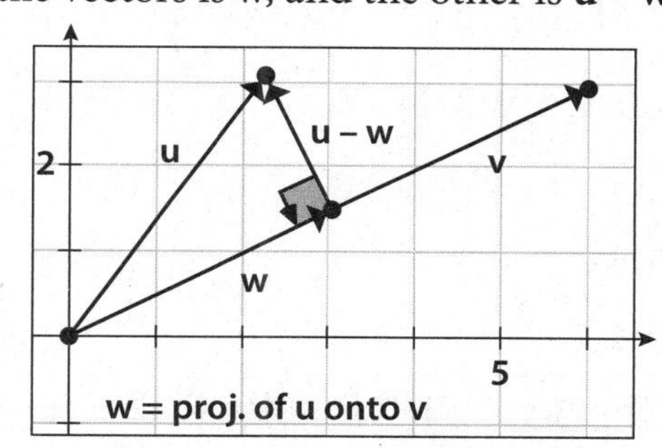

Figure 19.6 Vector **u** resolved into orthogonal components

One other consequence of the dot product is the following: If $\mathbf{u}$ and $\mathbf{v}$ are nonzero vectors, then $\mathbf{u} \cdot \mathbf{v} = 0$ if and only if $\mathbf{u}$ and $\mathbf{v}$ are perpendicular (orthogonal).

PROBLEM Find the projection vector, $\mathbf{w}$, of $\mathbf{u} = 3\mathbf{i} - 2\mathbf{j}$ onto $\mathbf{v} = 4\mathbf{i} + 3\mathbf{j}$ and resolve $\mathbf{u}$ into its orthogonal component vectors.

SOLUTION

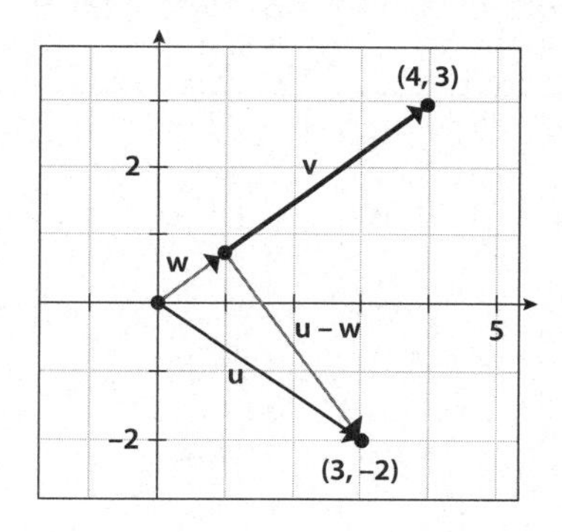

$$\mathbf{w} = \left(\frac{\mathbf{u}\cdot\mathbf{v}}{|\mathbf{v}|^2}\right)\mathbf{v} = \left(\frac{(3\mathbf{i}-2\mathbf{j})\cdot(4\mathbf{i}+3\mathbf{j})}{|(4\mathbf{i}+3\mathbf{j})|}\right)(4\mathbf{i}+3\mathbf{j}) = \frac{12-6}{16+9}(4\mathbf{i}+3\mathbf{j}) = \frac{6}{25}(4\mathbf{i}+3\mathbf{j})$$

$= \frac{24}{25}\mathbf{i} + \frac{18}{25}\mathbf{j}$. The component along $\mathbf{v}$ is $\mathbf{w} = \frac{24}{25}\mathbf{i} + \frac{18}{25}\mathbf{j}$, and the other component is $\mathbf{u} - \mathbf{w} = (3\mathbf{i} - 2\mathbf{j}) - \left(\frac{24}{25}\mathbf{i} + \frac{18}{25}\mathbf{j}\right) = \frac{51}{25}\mathbf{i} - \frac{68}{25}\mathbf{j}$. To verify that these are orthogonal, compute their dot product $\mathbf{w}\cdot(\mathbf{u}-\mathbf{w}) = \left(\frac{24}{25}\mathbf{i} + \frac{18}{25}\mathbf{j}\right)\cdot\left(\frac{51}{25}\mathbf{i} - \frac{68}{25}\mathbf{j}\right)$ $= \frac{1224}{625} - \frac{1224}{625} = 0$.

Vectors in three dimensions are encountered late in the study of calculus, but if you have a solid foundation in two-dimensional vectors, vectors in an added dimension can be assimilated fairly easily.

The exercises are designed to give you practice working with vector arithmetic, dot products, and projections. If you encounter these ideas in calculus or other scientific courses, you will have a good background with these concepts.

EXERCISE 19·2

For 1–3, find the angle between the two given vectors.

1. $\mathbf{u} = 5\mathbf{i} + 2\mathbf{j}$ and $\mathbf{v} = -3\mathbf{i} - 3\mathbf{j}$
2. $\mathbf{u} = \langle -3, 6\rangle$ and $\mathbf{v} = \langle 2, -5\rangle$
3. $\mathbf{u} = -4\mathbf{i} + 3\mathbf{j}$ and $\mathbf{v} = -6\mathbf{i} - 8\mathbf{j}$

*For 4–6, find the projection **w** of **u** onto **v**.*

4. $\mathbf{u} = 2\mathbf{i} + 6\mathbf{j}$ and $\mathbf{v} = 8\mathbf{i} + 3\mathbf{j}$
5. $\mathbf{u} = \langle -3, 8\rangle$ and $\mathbf{v} = \langle -12, 3\rangle$
6. $\mathbf{u} = -2\mathbf{i} - 8\mathbf{j}$ and $\mathbf{v} = -6\mathbf{i} + \mathbf{j}$
7. If the three vectors $\mathbf{u} = 5\mathbf{i} + 2\mathbf{j}, \mathbf{v} = -2\mathbf{i} - 6\mathbf{j}$, and $\mathbf{w} = 4\mathbf{i}$ are acting on a single point, find the equilibrant vector.
8. If the vectors $\mathbf{u} = \langle 4, 2\rangle, \mathbf{v} = \langle 0, -2\rangle$, and $\mathbf{w} = \langle 5, 3\rangle$ are acting on a single point, what is the resultant force vector on the point and what is its direction angle?
9. Prove that $\mathbf{u} = \mathbf{i}$ and $\mathbf{v} = \mathbf{j}$ are orthogonal.
10. Find all vectors $\mathbf{u}$ such that $|\mathbf{u}| = 3$ and $\mathbf{u}$ is perpendicular to $\mathbf{v} = 2\mathbf{i} + 5\mathbf{j}$.

ANALYTIC GEOMETRY

An entire semester course devoted to the study of analytic geometry is still taught at some colleges and universities. The study is rich in techniques of analyzing the conic sections. These conic sections are two-dimensional figures realized as the result of cutting a double-napped right circular cone with a plane. The four basic shapes of conic sections are the circle, ellipse, parabola, and hyperbola. They are formed by altering the angle of the cutting plane (see Figure 20.1).

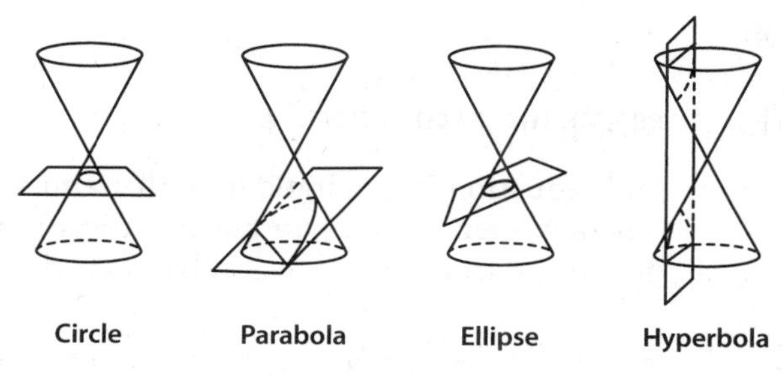

Figure 20.1 Conic sections

An in-depth analysis of these shapes is not the purpose of this section, but rather an attempt to familiarize you with them as graphs in the plane. When you encounter them in the study of calculus, you should recognize them on sight and be comfortable with their basic equations. As functions or relations, all four are quadratic in form and have unique properties. Moreover, each has a special, associated vocabulary. This special vocabulary is an important part of their recognition, and you should be able to use it with ease. The study here assumes the graphs are on the *xy* rectangular coordinate plane.

·20· Conics

Circle and ellipse

The standard equation of a circle with radius r and center at (h, k) is $(x-h)^2+(y-k)^2=r^2$.

Actually, this equation tells you everything you need to know about the circle. However, working with equations such as this one is where some of your algebra skills get a workout. The skill of completing the square is used to advantage. If you are rusty on that skill, you may need some review. First, notice that the circle is quadratic in both x and y. This observation is a starting point. An example will illustrate the point.

PROBLEM Graph the equation $x^2-4x+2y+y^2=4$.

SOLUTION This equation is quadratic in both x and y, so it may be a circle. The key then is to try to put this disguised form into standard form, and from there the graph is a cinch.

$x^2-4x+y^2+2y=4$	rewriting the equation
$x^2-4x+4+y^2+2y+1=4+4+1$	completing the square on both quadratics and balancing the equation
$(x-2)^2+(y+1)^2=9=3^2$	factoring and writing in standard form

The graph is recognizable now as a circle of radius 3 and center at (2, –1).

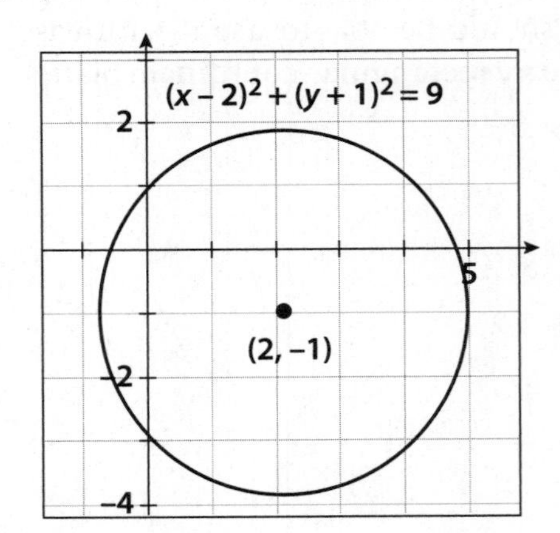

Since the circle is not the graph of a function, a graphing calculator cannot graph it directly. You must either use parametric equations or break the graph in two pieces and graph each piece. Hence equation recognition is invaluable. For example, given the circle $x^2+y^2=1$, solve for y to obtain $y=\pm\sqrt{1-x^2}$. Now, you can graph the upper and lower halves of the circle on the same graph to get the circle.

The standard form of the equation of an ellipse with center (h, k) in factored form is $\frac{(x-h)^2}{a^2}+\frac{(y-k)^2}{b^2}=1$. The constants a and b are the horizontal and vertical distances, respectively, from the center to the graph. If $a = b$, the ellipse is a circle of radius a. If $a>b$, then the *major axis* is horizontal with length $2a$ and the *minor axis* is vertical of length $2b$. If $b>a$, the major and minor axes are reversed. The vertices are the endpoints of the major axis (Figure 20.2).

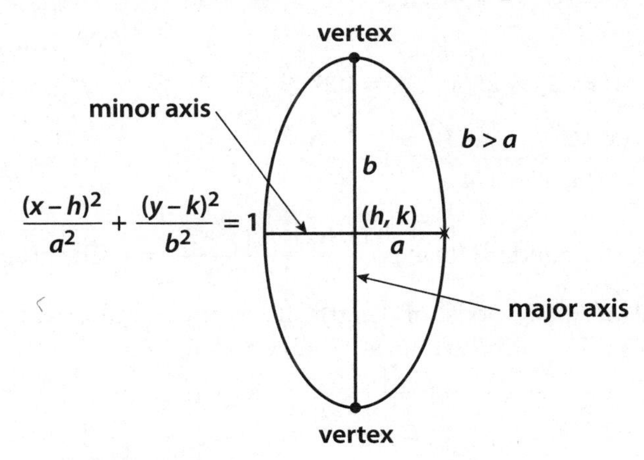

Figure 20.2 Ellipse

The ellipse equation is also quadratic in both x and y. The algebraic maneuver of completing the square may lead to an ellipse rather than a circle. In either case, once it is in standard form, the graph is easily drawn.

PROBLEM Graph the equation $16x^2+9y^2=144$.

SOLUTION Notice that the coefficients of x^2 and y^2 are different. This observation is the clue that the graph will be an ellipse; otherwise it will be a circle.

$16x^2+9y^2=144$ given equation

$x^2+\frac{9y^2}{16}=9$ divide each side by 16

$\frac{x^2}{9}+\frac{y^2}{16}=1$ divide each side by 9 to get the standard form

The graph is an ellipse with center (0, 0), vertical major axis of length 8, horizontal minor axis of length 6, and vertices at (0, 4) and (0, –4).

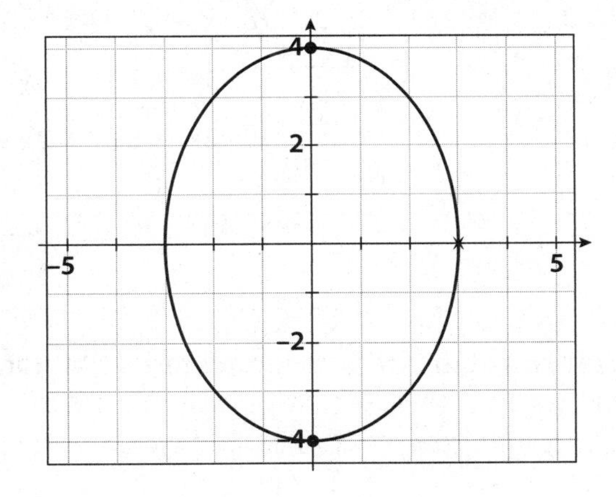

PROBLEM Draw the graph of $4x^2+25y^2-16x+150y+141=0$.

SOLUTION This is a good opportunity to practice your algebra skills.

$4x^2+25y^2-16x+150y+141=0$	given equation
$4x^2-16x+25y^2+150y=-141$	group like terms and subtract 141 from both sides
$4(x^2-4x)+25(y^2+6y)=-141$	factor out leading coefficients
$4(x^2-4x+4)+25(y^2+6y+9)=-141+16+225$	complete the square inside each parentheses and balance the equation
$4(x-2)^2+25(y+3)^2=100$	factor
$\dfrac{(x-2)^2}{25}+\dfrac{(y+3)^2}{4}=1$	divide each side by 100

The standard form is $\dfrac{(x-2)^2}{5^2}+\dfrac{(y+3)^2}{2^2}=1$. The graph is an ellipse with horizontal major axis of length 10, vertical minor axis of length 4, centered at (2, –3), with vertices at (7, –3) and (–3, –3).

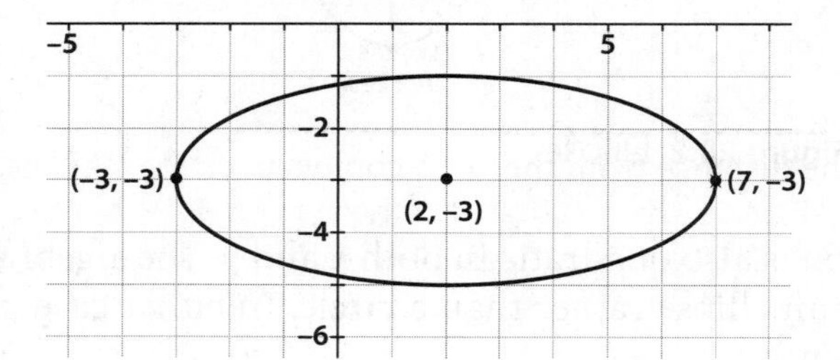

EXERCISE
20·1

In 1–4, write the equation of the graph satisfying the given conditions.

1. center at (0, 0), radius 7
2. center at (5, 0), radius $\sqrt{3}$
3. diameter has endpoints (4, 9) and (–2, 1)
4. center at (–1, 2), horizontal major axis of 8, and minor axis of 4

In 5–10, write each equation in standard form, then identify the center and vertices where applicable.

5. $x^2+y^2-12x-10y+52=0$
6. $2(x-2)^2+2(y+4)^2=18$
7. $x^2+4y^2-8y+4x-8=0$
8. $x^2+y^2-4x+10y=-4$
9. $5x^2+2y^2+20y-30x+75=0$
10. $2x^2+5y^2-12x+20y-12=0$

Hyperbola

The hyperbola is studied in two standard forms: the horizontal form and the vertical form. See Figure 20.3 and Table 20.1.

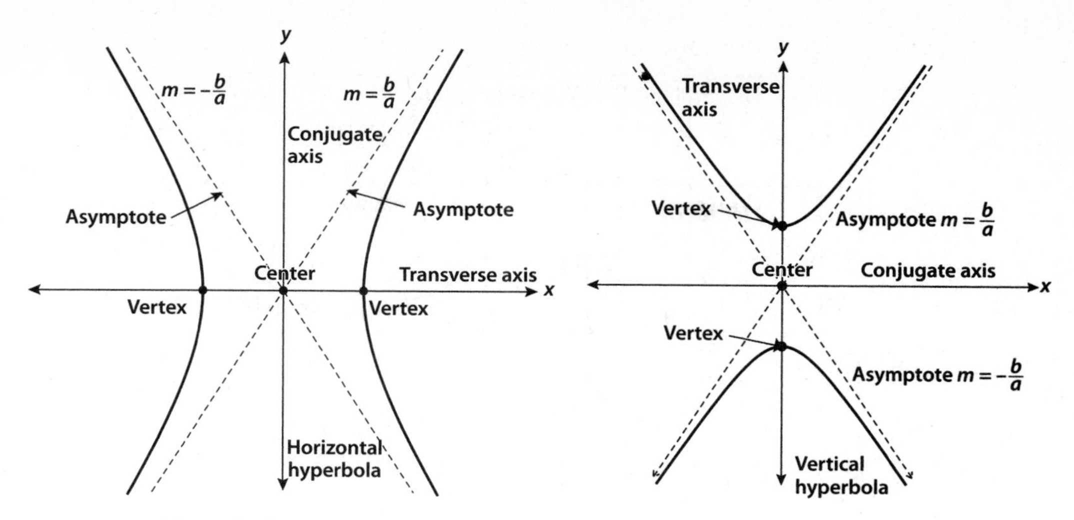

Figure 20.3 Hyperbolas

Table 20.1 Hyperbola standard equation forms

Horizontal form	Vertical form
$\frac{(x-h)^2}{a^2} - \frac{(y-k)^2}{b^2} = 1$	$\frac{(y-k)^2}{b^2} - \frac{(x-h)^2}{a^2} = 1$
The center is at (h, k) with transverse axis $y = k$ and conjugate axis $x = h$. The distance from the center to the vertices is a.	The center is at (h, k) with transverse axis $x = h$ and conjugate axis $y = k$. The distance from the center to the vertices is b.

If needed, asymptotes are lines through (h, k) with slopes $m = \pm\frac{b}{a}$.

As with circles and ellipses, the graphs can be drawn using parametric equations or in two pieces.

PROBLEM Graph the hyperbola $9y^2 - x^2 + 54y + 4x + 68 = 0$. Find the center, the vertices, and the equations of the asymptotes.

SOLUTION

$9y^2 - x^2 + 54y + 4x + 68 = 0$	given equation
$9(y^2 + 6y) - 1(x^2 - 4x) = -68$	combining terms, factoring, and subtracting 68
$9(y^2 + 6y + 9) - 1(x^2 - 4x + 4) = -68 + 81 - 4$	completing the square
$9(y+3)^2 - 1(x-2)^2 = 9$	factoring
$\frac{(y+3)^2}{1} - \frac{(x-2)^2}{3^2} = 1$	dividing by 9 and writing in standard form

The graph is a vertical hyperbola with center at (2, –3). The transverse axis is $x = 2$, and the conjugate axis is $y = -3$. Vertices are at (2, –3 + 1) and (2, –3 –1) → (2, –2) and (2, –4). The asymptote equations are found using the point slope form of the equation of a line $\frac{y+3}{x-2} = \pm\frac{1}{3}$. The equations are $y = \frac{1}{3}x - \frac{11}{3}$ and $y = -\frac{1}{3}x - \frac{7}{3}$. Solve $\frac{(y+3)^2}{1} - \frac{(x-2)^2}{3^2} = 1$ for y to get $y = \pm\frac{1}{3}\sqrt{9 + (x-2)^2} - 3$. Graph each piece.

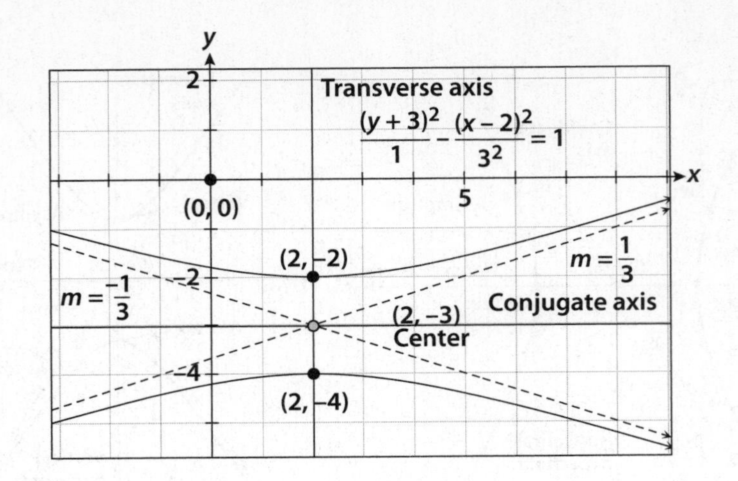

As you can see, the need for good algebra skills is critical for the study of precalculus.

One special hyperbola used often in calculus is neither horizontal nor vertical but is "slanted" at 45°. The equation is not in standard form and is very simple: $xy = 1$. This is seen extensively in the form $y = \frac{1}{x}$. See Figure 20.4.

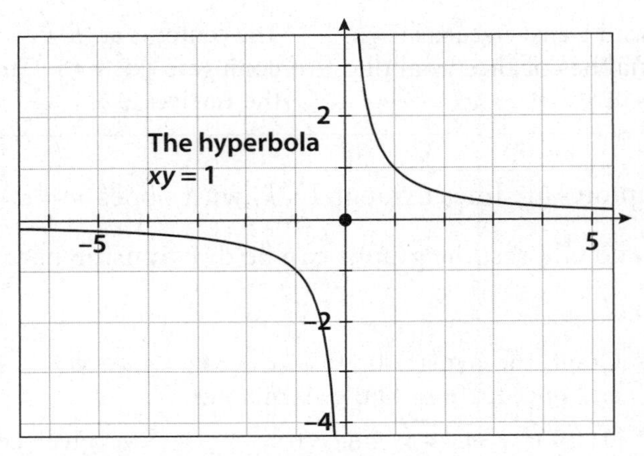

Figure 20.4 The hyperbola $xy = 1$

EXERCISE
20·2

For 1–4, identify each equation as that of a circle, ellipse, or hyperbola, and name the center of each.

1. $y^2 = 36 + 9x^2$
2. $16x^2 = 64 - 16y^2$
3. $4(x+5)^2 - 36 = 9(y-3)^2$
4. $3(x+2)^2 + 8(y-3)^2 = 24$

For 5–6, find the center, vertices, and transversal and conjugate axes, and sketch the graph including the asymptotes.

5. $4x^2 - 9y^2 - 24x + 72y - 144 = 0$
6. $16y^2 - 4x^2 + 24x = 100$

Parabola

The parabola is the only conic section that can be graphed directly as a function. As such, you see parabolas early in your study of mathematics. Similar to hyperbolas, there are vertical and horizontal parabolas. See Table 20.2.

Table 20.2 Standard equations for parabolas

Vertical parabola $y = f(x) = ax^2 + bx + c,\ a \neq 0$	**Horizontal parabola** $x = g(y) = ay^2 + by + c,\ a \neq 0$
If $a > 0$, the parabola opens upward; and if $a < 0$, the parabola opens downward. The axis of symmetry is $x = \frac{-b}{2a}$, and the vertex is at $\left(\frac{-b}{2a}, f\left(\frac{-b}{2a}\right)\right)$.	If $a > 0$, the parabola opens to the right; and if $a < 0$, the parabola opens to the left. The axis of symmetry is $y = \frac{-b}{2a}$, and the vertex is at $\left(g\left(\frac{-b}{2a}\right), \frac{-b}{2a}\right)$.

These equations differ from the other three conic sections in that there is only one second-degree term. Sample graphs of each type are illustrated in Figure 20.5.

The horizontal parabola as it is graphed is not a function of x. However, it is a function of y, and any functional study needed can be done on the corresponding vertical function by interchanging the x and y.

Considerable analysis of the parabolic function was done in Chapter 8, and no extra problems are needed here. This discussion was included for completeness of the conic sections.

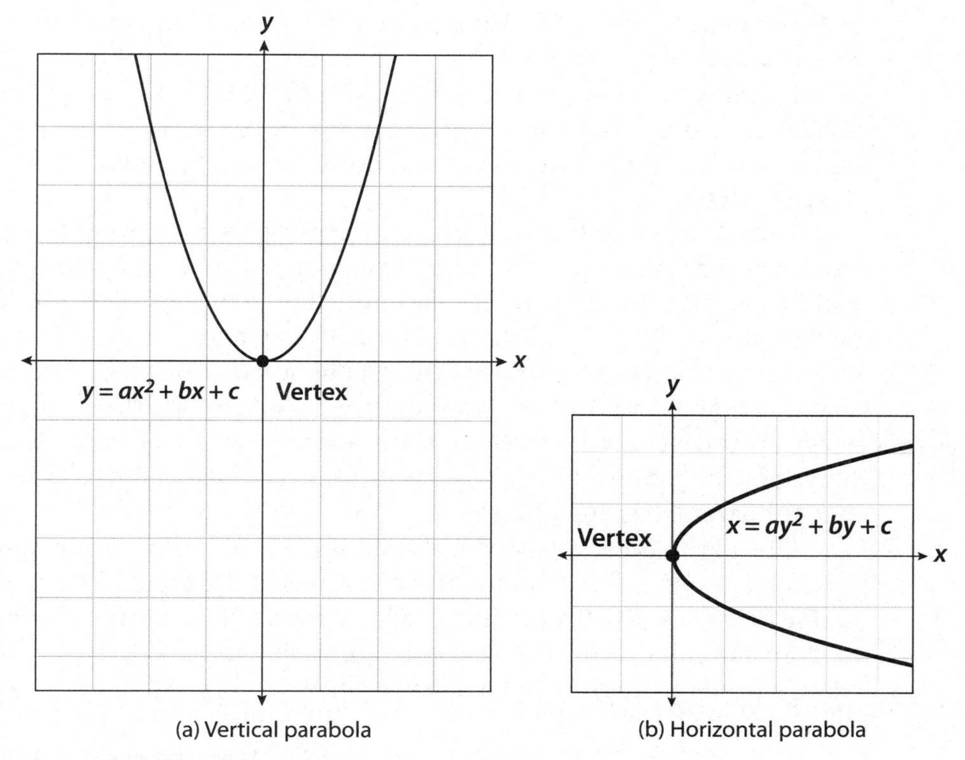

Figure 20.5 Parabolas

Parametric equations and polar coordinates

Parametric equations

For the most part, your study of mathematics to this point has been the study of applications that can be described by functions. There are, however, many applications that are described by nonfunctions. The study of the movement of planets about the Sun in elliptical orbits is one such application.

Description by **parametric equations** is a scheme of expressing *nonfunctions*, in a roundabout way, by using *functions*. Doing so enables you to use all the powerful tools developed for functions after all.

The driving force behind parametric equations is the question, "Is there another way of describing the nature of what I am studying?" A good example is the unit circle. The standard form of the equation of the unit circle is $x^2 + y^2 = 1$. One of the basic trig identities is $\sin^2\theta + \cos^2\theta = 1$. On the unit circle, $x = \cos\theta$ and $y = \sin\theta$. Thus, the points on the circle and hence the circle can be described by **functions** of the angle θ. The equations $x = \cos\theta$ and $y = \sin\theta$ are the **parametric equations** of the circle.

In a more general setting, if S is a set of points $P(x, y)$ and f and g are functions such that $x = f(t)$ and $y = g(t)$ for t in some set T, then the equations $x = f(t)$ and $y = g(t)$ are the parametric equations of the set of points in S. The variable t is the **parameter**.

With today's technology, you can easily graph parametric equations. The hard part is deducing or otherwise coming up with the parametric equations that will do the trick. In this vein, the quote attributed to Isaac Newton, "If I have seen farther, it is only by standing on shoulders of giants," rings so true. Parametric equations are known for most of the graphs studied today, and the chore is to be able to know where to find them and be able to relate them to your graphing needs. You have just seen where a trig identity was used to deduce parametric equations for a circle. Other trig identities have also been useful in the development of parametric equations.

One of the skills you need is the ability to change between rectangular equations of a graph and the parametric equations of the graph. The process of *eliminating the parameter* is a method used to convert parametric equations of a graph to its rectangular form. You also need to be able to go in the other direction and change from rectangular to parametric form.

PROBLEM Eliminate the parameter in the parametric equations $x = t^2 - 3$ and $y = 2t + 1$ to obtain the rectangular form.

SOLUTION Solve $y=2t+1$ for t: $t=\frac{y-1}{2}$. Substitute this value into the other equation $x=\left(\frac{y-1}{2}\right)^2-3=\frac{1}{4}(y-1)^2-3$. Simplify to $(y-1)^2=4(x+3)$. This is the equation of a horizontal parabola opening to the right with vertex at $(-3,1)$.

PROBLEM Eliminate the parameter in the parametric equations $x=2\cos\theta$ and $y=3\sin\theta$ to obtain the rectangular form.

SOLUTION You know that $\frac{x}{2}=\cos\theta$ and $\frac{y}{3}=\sin\theta$. From this result it follows that $\frac{x^2}{2^2}+\frac{y^2}{3^2}=\sin^2\theta+\cos^2\theta=1$. The graph of this equation is a vertical ellipse with center at (0, 0) with major axis of length 6. The vertices are at $(0,\pm 3)$.

PROBLEM Write parametric equations for $y=9(x-2)^2+1$.

SOLUTION Three different forms of a solution will show the richness and diversity of parametric equations:

1. Let $x=t$. Then $y=9(t-2)^2+1$. This is rather simple, and it somewhat begs the issue.
2. Let $x=t-2$. Then $y=9t^2+1$. This may be a little more advantageous.
3. The motivation for the next solution comes from always being on the lookout for things that might be useful in the future. In solution 2 above, if the 9 were missing, the y equation would have the *appearance* of *part* of the trig identity $\tan^2\theta+1=\sec^2\theta$.

   Let $x-2=\left(\frac{\tan\theta}{3}\right)$. That is, let $x=\left(\frac{\tan\theta}{3}\right)+2$. Then $y=9\left(\frac{\tan\theta}{3}\right)^2+1=\tan^2\theta+1$ or $y=\sec^2\theta$.

EXERCISE

21·1

For 1–5, eliminate the parameter to obtain the rectangular form, and identify the type of graph.

1. $x=2t-3$ and $y=4t-1$
2. $x=1-\cos\theta$ and $y=2+3\sin\theta$
3. $x=t+2$ and $y=t^2-1$
4. $x=2\sin\theta$ and $y=-3\cos\theta$
5. $x=4\cos\theta$ and $y=4\sin\theta$

For 6–7, find parametric equations for the graph with one being a trigonometric form.

6. $y=(x+3)^2+1$
7. $y=3x-2$

For 8–10, find parametric equations for the graph.

8. $\frac{(x-2)^2}{9}+\frac{(y+3)^2}{16}=1$
9. $(x+2)^2+(y-3)^2=16$
10. $\frac{x^2}{9}-\frac{(y-4)^2}{16}=1$

Polar coordinates

The use of rectangular coordinates is a convenient means of locating points in a plane based on directed distances from two fixed lines. In many cases a different way of locating points is more applicable. One method is the use of polar coordinates. Polar coordinates are based on a directed distance and an angle relative to a fixed point. Applications such as navigation that is based on compass headings are better served by this coordinate system. As illustrated in Figure 21.1, a fixed ray, called the **polar axis**, emanating from a fixed point **O**, called the origin, is the basis for the coordinate system. A point in the plane is then located by **polar coordinates** (r,θ).

One drawback of this coordinate system is that, unlike rectangular coordinates in which the coordinate is unique, a point has an unlimited number of representations in polar coordinates. For instance, $(r,\theta)=(r,\theta+2n\pi), n=1, 2, 3,\ldots$. Also, $(r, \theta)=(-r, \pi+\theta)=(-r, -\pi+\theta)$ (see Figure 21.2).

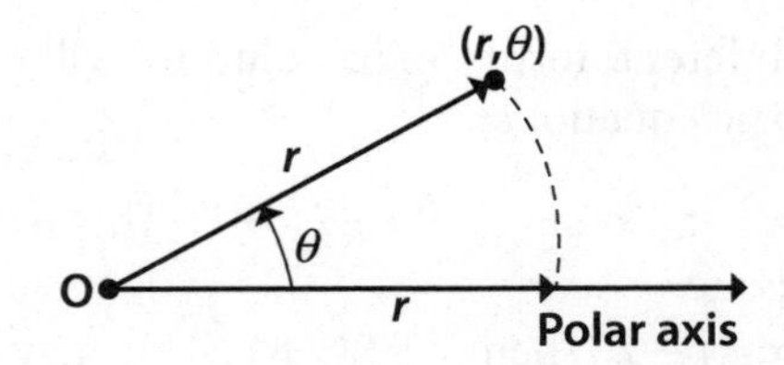

Figure 21.1 Polar coordinates

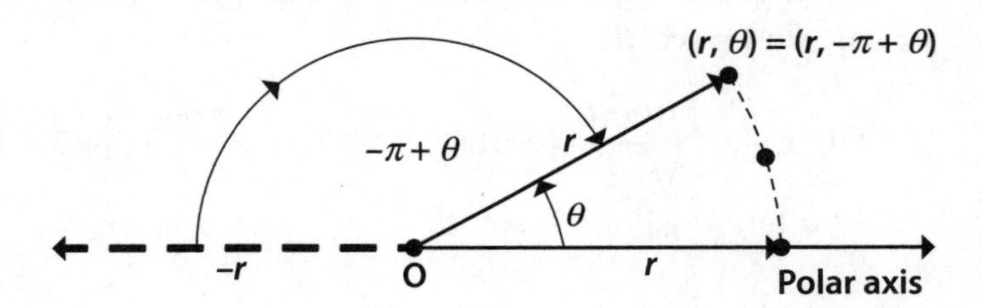

Figure 21.2 Non-unique representation of polar coordinates

PROBLEM Plot the points $\left(3, \dfrac{\pi}{4}\right)$, $(2, 30°)$, $(4, 75°)$, $(-3, \pi)$, and $(3, -60°)$ on the polar plane.

SOLUTION

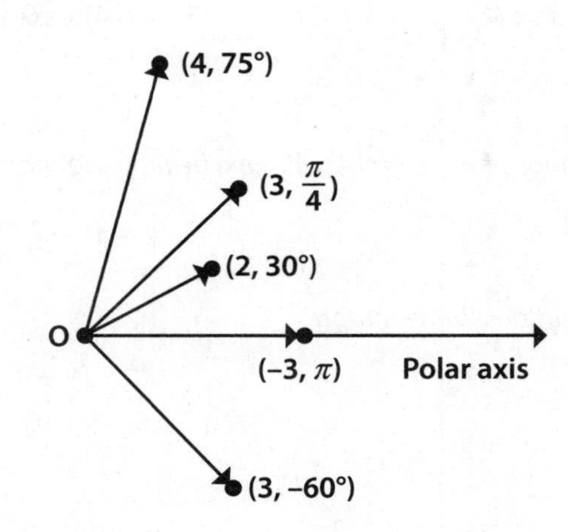

EXERCISE

21·2

Plot the following points on the polar plane.

$\left(2, \frac{\pi}{3}\right)$, $\left(3, \frac{\pi}{2}\right)$, $(-2, -90°)$, $(-4, 120°)$, and $\left(4, \frac{3\pi}{2}\right)$

Converting coordinate systems

The right triangle trigonometric relationships are in fact the conversion components for converting between coordinate systems (see Figure 21.3).

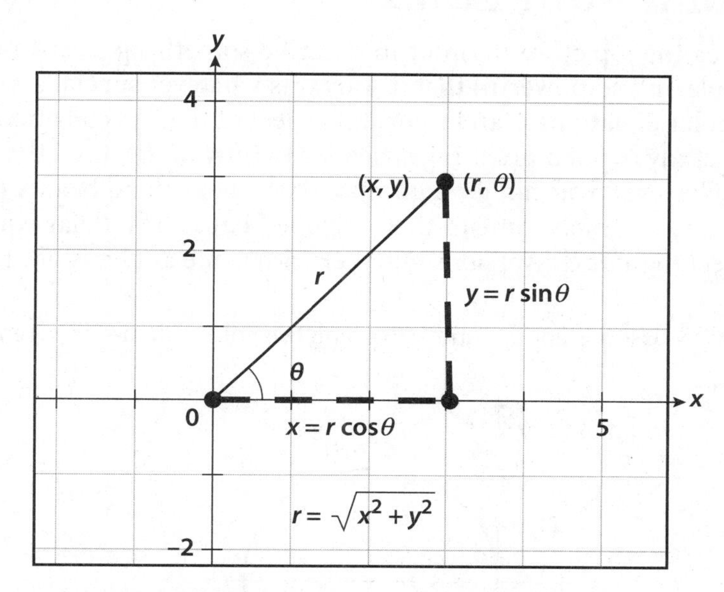

Figure 21.3 Polar-to-rectangular conversion

The conversion equations $x = r\cos\theta$ and $y = r\sin\theta$ enable you to convert from polar coordinates to rectangular coordinates.

On the other hand, referring again to Figure 21.3, the conversion equations from rectangular to polar coordinates are $r = \sqrt{x^2 + y^2}$ and $\theta = \tan^{-1}\frac{y}{x}, x \neq 0$.

PROBLEM Convert the polar coordinates $(6, 240°)$ to rectangular form.

SOLUTION $x = 6\cos(240°) = 6\left(-\frac{1}{2}\right) = -3$ and $y = 6\sin(240°) = 6\left(-\frac{\sqrt{3}}{2}\right) = -3\sqrt{3}$.

PROBLEM Convert the rectangular coordinates $(-3, 6)$ to polar form.

SOLUTION The point $(-3, 6)$ is in quadrant II. $r = \sqrt{(-3)^2 + 6^2} = \sqrt{45} \approx 6.7$.

$\theta_r = \tan^{-1}\left(\frac{6}{-3}\right) = \tan^{-1}(-2) \approx -63.4°$. Thus $\theta \approx 116.6°$.

EXERCISE

21·3

For 1–3, convert from rectangular to polar form.

1. $(4, 4)$
2. $(4\sqrt{3}, 4)$
3. $(-3.5, 12)$

For 4–6, convert from polar to rectangular form.

4. $(8, 45°)$
5. $\left(-2, \frac{7\pi}{6}\right)$
6. $(-4, -30°)$

Graphs of polar equations

The concept of creating a picture in order to describe something as abstract as a mathematical formula is so simple and so powerful that it warrants whatever special attention you can give it. The graphing of polar equations is an important aspect of that special attention.

An in-depth study of polar graphing is needed to fully understand the special equations and techniques used. However, you can get a glimpse of the power and beauty of polar graphs by using a graphing utility to graph some of the developed equations. Polar equations are developed using polar and rectangular conversions together with acute analyses of function behavior.

PROBLEM Use a graphing utility to graph the polar equation $r = 4\cos\theta$.

SOLUTION

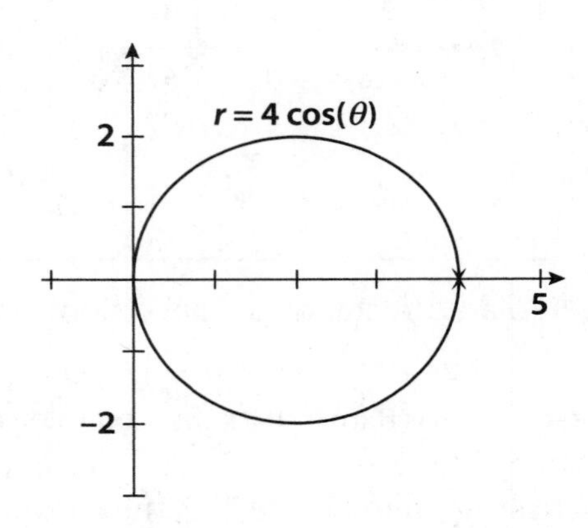

PROBLEM Use a graphing utility to graph the polar equation $r = 4\sin(2\theta)$.

SOLUTION

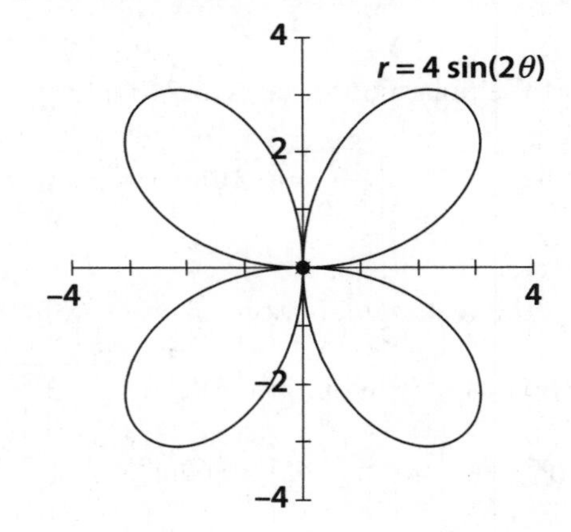

EXERCISE
21·4

Use a graphing utility to graph the polar equation.

1. $r = f(\theta) = 4\cos(2\theta)$
2. $r = f(\theta) = 4\sin(3\theta)$
3. $r = f(\theta) = 4\sin\theta$
4. $r = f(\theta) = 2(1 - \cos\theta)$
5. $r = f(\theta) = \left(\frac{1}{4}\right)\theta$
6. $r = f(\theta) = 2 - \cos\theta$
7. $r = f(\theta) = 4\sec\theta$
8. $r = f(\theta) = \frac{4}{3 - 2\sin\theta}$
9. $r = f(\theta) = 2$

APPENDIX A

Synthetic division

Synthetic division is a shortcut method for dividing a polynomial by a binomial, $x-r$. You simplify the process by working only with r and the coefficients of the polynomial—being careful to use 0 as a coefficient for missing powers of x. Here are the steps:

PROBLEM Divide $p(x)=2x^3+x^2-13x+6$ by $x-4$.

SOLUTION **Step 1.** Write the polynomial in descending powers of x, using a coefficient of 0 when a power of x is missing, if needed.

$$2x^3+x^2-13x+6$$

Step 2. Write only the coefficients as shown here.

$$2 \quad 1 \quad -13 \quad 6$$

Step 3. Write $r=4$ as shown here.

$$\begin{array}{r|rrrr} 4 & 2 & 1 & -13 & 6 \\ & & & & \\ \hline \end{array}$$

Step 4. Bring down the first coefficient.

$$\begin{array}{r|rrrr} 4 & 2 & 1 & -13 & 6 \\ & & & & \\ \hline & 2 & & & \end{array}$$

Step 5. Multiply the first coefficient by $r=4$, write the product under the second coefficient, and then add.

$$\begin{array}{r|rrrr} 4 & 2 & 1 & -13 & 6 \\ & & 8 & & \\ \hline & 2 & 9 & & \end{array}$$

Step 6. Multiply the sum by $r=4$, write the product under the third coefficient, and then add.

$$\begin{array}{r|rrrr} 4 & 2 & 1 & -13 & 6 \\ & & 8 & 36 & \\ \hline & 2 & 9 & 23 & \end{array}$$

Step 7. Repeat step 6 until you use up all the coefficients in the polynomial.

$$\begin{array}{r|rrrr} 4 & 2 & 1 & -13 & 6 \\ & & 8 & 36 & 92 \\ \hline & 2 & 9 & 23 & 98 \end{array}$$

Step 8. Separate the final sum, which is the remainder, as shown here.

$$\begin{array}{r|rrr|r} 4 & 2 & 1 & -13 & 6 \\ & & 8 & 36 & 92 \\ \hline & 2 & 9 & 23 & \underline{98} \end{array}$$

Step 9. Write the quotient and remainder, using the coefficients.

$2x^3 + x^2 - 13x + 6$ divided by $x = 4$ is $2x^2 + 9x + 23$ R 98

APPENDIX B
Trigonometric identities

Reciprocal	Ratio	Pythagorean	Symmetry
$\sec\theta = \dfrac{1}{\cos\theta}$	$\tan\theta = \dfrac{\sin\theta}{\cos\theta}$	$\sin^2\theta + \cos^2\theta = 1$	$\sin(-\theta) = -\sin\theta$
$\csc\theta = \dfrac{1}{\sin\theta}$	$\cot\theta = \dfrac{\cos\theta}{\sin\theta}$	$1+\tan^2\theta = \sec^2\theta$	$\cos(-\theta) = \cos\theta$
$\cot\theta = \dfrac{1}{\tan\theta}$		$1+\cot^2\theta = \csc^2\theta$	$\tan(-\theta) = -\tan\theta$

Cofunction		Sum and difference
$\sin\left(\dfrac{\pi}{2}-\theta\right) = \cos\theta$	$\tan\left(\dfrac{\pi}{2}-\theta\right) = \cot\theta$	$\sin(\alpha \pm \beta) = \sin\alpha\cos\beta \pm \cos\alpha\sin\beta$
$\cos\left(\dfrac{\pi}{2}-\theta\right) = \sin\theta$	$\cot\left(\dfrac{\pi}{2}-\theta\right) = \tan\theta$	$\cos(\alpha \pm \beta) = \cos\alpha\cos\beta \mp \sin\alpha\sin\beta$
$\sec\left(\dfrac{\pi}{2}-\theta\right) = \csc\theta$	$\csc\left(\dfrac{\pi}{2}-\theta\right) = \sec\theta$	$\tan(\alpha \pm \beta) = \dfrac{\tan\alpha \pm \tan\beta}{1 \mp \tan\alpha\tan\beta}$

Double-angle	Half-angle	Power reduction
$\sin(2\theta) = 2\sin\theta\cos\theta$	$\sin\left(\dfrac{x}{2}\right) = \pm\sqrt{\dfrac{1-\cos x}{2}}$	$\sin^2\theta = \dfrac{1-\cos(2\theta)}{2}$
$\cos(2\theta) = \cos^2\theta - \sin^2\theta$	$\cos\left(\dfrac{x}{2}\right) = \pm\sqrt{\dfrac{1+\cos x}{2}}$	$\cos^2\theta = \dfrac{1+\cos(2\theta)}{2}$
$\cos(2\theta) = 2\cos^2\theta - 1$	$\tan\left(\dfrac{x}{2}\right) = \pm\sqrt{\dfrac{1-\cos x}{1+\cos x}}$	$\tan^2\theta = \dfrac{1-\cos(2\theta)}{1+\cos(2\theta)}$
$\cos(2\theta) = 1 - 2\sin^2\theta$	$\tan\left(\dfrac{x}{2}\right) = \dfrac{1-\cos x}{\sin x}$	
$\tan(2\theta) = \dfrac{2\tan\theta}{1-\tan^2\theta}$	$\tan\left(\dfrac{x}{2}\right) = \dfrac{\sin x}{1+\cos x}$	

Law of sines	Law of cosines
$\dfrac{\sin A}{a} = \dfrac{\sin B}{b} = \dfrac{\sin C}{c}$	$a^2 = b^2 + c^2 - 2bc\cos A$
	$b^2 = a^2 + c^2 - 2ac\cos B$
	$c^2 = a^2 + b^2 - 2ab\cos C$

Answer key

REVIEW OF BASIC CONCEPTS

1 Real numbers

1·1 1. 10—natural numbers, whole numbers, integers, rationals, reals 2. $\sqrt{0.64}=0.8$—rationals, reals 3. $\sqrt[3]{\frac{8}{125}}=\frac{2}{5}$—rationals, reals 4. $-\pi$—irrationals, reals 5. $-1{,}000$—integers, rationals, reals 6. $\sqrt{2}$ — irrationals, reals 7. $-\sqrt{\frac{3}{4}}$—irrationals, reals 8. $-\sqrt{\frac{9}{4}}=-\frac{3}{2}$—rationals, reals 9. 1—natural numbers, whole numbers, integers, rationals, reals 10. $\sqrt[3]{0.001}=0.1$—rationals, reals 11. $\sqrt{25}=5$—rational 12. $\sqrt[3]{15}$—irrational 13. $\sqrt[3]{-\frac{64}{125}}=-\frac{4}{5}$ — rational 14. $\sqrt{41}$—irrational 15. $-\frac{\pi}{2}$—irrational

1·2 1. $(-\infty,\infty)$, unbounded, open 2. $(-\infty,0)$, unbounded, open 3. $(-\infty, 3.5)$, unbounded, open 4. $[-10, 30]$, bounded, closed 5. $[2.75, \infty)$, unbounded, half-open

1·3 1. closure property for multiplication 2. commutative property for addition 3. multiplicative inverse property 4. closure property for addition 5. associative property for addition 6. distributive property 7. additive inverse property 8. zero factor property 9. distributive property 10. associative property for multiplication 11. zero factor property 12. commutative property for addition 13. associative property for multiplication 14. multiplicative inverse property 15. distributive property

1·4 1. 45 2. 5.8 3. $5\frac{2}{3}$ 4. -60 5. 0 6. $\frac{\pi}{2}$ 7. $-a$ 8. $-a$ 9. a 10. a 11. $<$ 12. $>$ 13. $=$ 14. $<$ 15. $<$

1·5 1. -120 2. -0.7 3. $-\frac{1}{3}$ 4. -6 5. 800 6. $\frac{81}{5}=16\frac{1}{5}$ 7. 4 8. 385.12 9. $\frac{11}{8}=1\frac{3}{8}$ 10. -80

1·6 1. -32 2. 6 3. $\frac{27}{125}$ 4. $\frac{81}{36}=\frac{9}{4}$ 5. $\frac{9}{32}$ 6. -56 7. $-64^{\frac{2}{3}}=-\left(\sqrt[3]{64}\right)^2=-16$ 8. 1 9. $\frac{3}{2}$ 10. $\frac{6}{5}$ 11. $a^{\frac{1}{2}}a^{\frac{3}{4}}=a^{\frac{1}{2}+\frac{3}{4}}=a^{\frac{5}{4}}$ 12. b^4 13. $(y^3)^4=y^{3\cdot 4}=y^{12}$ 14. $\frac{y^3}{x^3}$

15. $(x+3)^3=(x+3)(x+3)(x+3)=(x+3)(x^2+6x+9)=x^3+9x^2+27x+27$

16. $\left(x^3y^6\right)^{\frac{1}{3}}=x^{3\cdot\frac{1}{3}}y^{6\cdot\frac{1}{3}}=xy^2$ 17. $\frac{\left(a^2\right)^3b^8}{a^5\left(b^3\right)^2}=\frac{a^6b^8}{a^5b^6}=ab^2$ 18. $\left(x^{-2}\right)^4y^7y^{-9}=x^{-8}y^{-2}=\frac{1}{x^8y^2}$

19. $\dfrac{\left(a^2b^{-5}\right)^{-4}}{\left(a^5b^{-2}\right)^{-3}}=\dfrac{\left(a^5b^{-2}\right)^3}{\left(a^2b^{-5}\right)^4}=\dfrac{a^{15}b^{-6}}{a^8b^{-20}}=a^7b^{14}$ 20. $\left(\left(x^{-2}y^3z\right)^3\right)^{-2}=\left(x^{-6}y^9z^3\right)^{-2}=x^{12}y^{-18}z^{-6}=\dfrac{x^{12}}{y^{18}z^6}$

1·7 1. 62 2. −98 3. 20 4. −4 5. −6 6. −109 7. 80 8. −3 9. 30 10. 0.25

2 Complex numbers

2·1 1. x, y 2. real, −1 3. $x = u, y = v$ 4. $0 + yi, y \neq 0$ 5. real 6. integers, rationals, reals, complex numbers 7. complex numbers 8. complex numbers 9. irrationals, reals, complex numbers 10. irrationals, reals, complex numbers 11. pure imaginary numbers, complex numbers 12. complex numbers 13. pure imaginary numbers, complex numbers 14. complex numbers 15. pure imaginary numbers, complex numbers 16. rational 17. pure imaginary 18. rational 19. pure imaginary 20. pure imaginary

2·2 1. $(-3-5i)+(4+5i)=1+0i=1$ 2. $(-1-i)-(-3-4i)=2+3i$ 3. $\left(1-i\sqrt{3}\right)+\left(3+2i\sqrt{3}\right)=4+i\sqrt{3}$

4. $(2.8-1.5i)-(-3.5+7i)=6.3-8.5i$ 5. $\left(-\frac{1}{2}+\frac{3}{4}i\right)+\left(-\frac{3}{2}-\frac{1}{4}i\right)=-2+\frac{1}{2}i$

6. $(9-5i)+(0+0i)=9-5i$ 7. $(5-i)-(-4i)=5+3i$ 8. $(4+2i)-(-4-2i)=8+4i$

9. $(4+2i)+(-4-2i)=0+0i=0$ 10. $\left(6+\frac{1}{6}i\right)+\left(-\frac{1}{2}-\frac{5}{6}i\right)=\frac{11}{2}-\frac{2}{3}i$

2·3 1. $(4+5i)(10-i)=4\cdot10-4i+50i-5i^2=45+46i$ 2. $(-2-3i)(4-8i)=-8+16i-12i+24i^2=-32+4i$

3. $\left(5-i\sqrt{3}\right)(2+i)=10+5i-2i\sqrt{3}-i^2\sqrt{3}=\left(10+\sqrt{3}\right)+\left(5-2\sqrt{3}\right)i$ 4. $(8+2i)(8-2i)=64+4=68$

5. $\left(\sqrt{2}-i\sqrt{5}\right)\left(\sqrt{2}+i\sqrt{5}\right)=2+5=7$ 6. $(-2+3i)(4-8i)=-8+16i+12i-24i^2=16+28i$

7. $\left(5-i\sqrt{3}\right)\left(1+i\sqrt{3}\right)=5+5i\sqrt{3}-i\sqrt{3}-3i^2=8+4i\sqrt{3}$

8. $(x+yi)\left(\dfrac{x}{x^2+y^2}+\dfrac{-y}{x^2+y^2}i\right)=\dfrac{x^2}{x^2+y^2}+\dfrac{-xy}{x^2+y^2}i+\dfrac{xy}{x^2+y^2}i+\dfrac{-y^2}{x^2+y^2}i^2=\dfrac{x^2+y^2}{x^2+y^2}=1$

9. $(2+i)\left[(3-4i)+(5+7i)\right]=(2+i)(3-4i)+(2+i)(5+7i)=13+14i$ 10. $i^{402}=i^{400}i^2=1\cdot-1=-1$

2·4 1. $\dfrac{1-i}{2+4i}=\dfrac{(1-i)}{(2+4i)}\cdot\dfrac{(2-4i)}{(2-4i)}=\dfrac{-2-6i}{20}=-\dfrac{1}{10}-\dfrac{3}{10}i$ 2. $\dfrac{4-2i}{2+3i}=\dfrac{(4-2i)}{(2+3i)}\cdot\dfrac{(2-3i)}{(2-3i)}=\dfrac{2-16i}{13}=\dfrac{2}{13}-\dfrac{16}{13}i$

3. $\dfrac{1}{(2+3i)}=\dfrac{1}{(2+3i)}\cdot\dfrac{(2-3i)}{(2-3i)}=\dfrac{2-3i}{13}=\dfrac{2}{13}-\dfrac{3}{13}i$ 4. $\dfrac{2i}{5+3i}=\dfrac{(2i)}{(5+3i)}\cdot\dfrac{(5-3i)}{(5-3i)}=\dfrac{6+10i}{34}=\dfrac{6}{34}+\dfrac{10}{34}i=\dfrac{3}{17}+\dfrac{5}{17}i$

5. $\dfrac{3-2i}{3+2i}=\dfrac{(3-2i)}{(3+2i)}\cdot\dfrac{(3-2i)}{(3-2i)}=\dfrac{5-12i}{13}=\dfrac{5}{13}-\dfrac{12}{13}i$

3 The Cartesian coordinate system

3·1 1. True 2. False 3. True 4. True 5. False 6. True 7. False 8. True 9. A: (5, 4) B: (2, −4) C(−3, 7) 10. A: (−4, −5) B: (1, 0) C: (−6, 3)

3·2 1. distance = 5 2. distance = $\sqrt{13}$ 3. distance = $\sqrt{74}$ 4. distance = 1 5. distance = $\sqrt{\dfrac{\pi^2+4}{4}}=\dfrac{1}{2}\sqrt{\pi^2+4}$

3·3 1. midpoint = (3, 5.5) 2. midpoint = $\left(0,-\dfrac{3}{2}\right)$ 3. midpoint = (−1, 1) 4. midpoint = $\left(\dfrac{7\sqrt{3}}{4},0\right)$

5. midpoint = $\left(\dfrac{3\pi}{4},\dfrac{1}{2}\right)$

3·4 1. rise 2. run 3. negative 4. positive 5. zero 6. $\frac{2}{3}$ 7. $-\frac{4}{3}$ 8. slope = −1 9. slope = $\frac{16}{15}$ 10. slope = $\frac{5}{8}$ 11. Slope is undefined. 12. slope = $\frac{4}{3}$ 13. slope = 0 14. slope = $-\frac{2}{3}$ 15. slope = $\frac{3}{2}$

PRECALCULUS ALGEBRA

4 Basic function concepts

4·1 1. False 2. True 3. True 4. False 5. True 6. False 7. independent, dependent 8. 3 9. $-\frac{1}{2}$ 10. 7 11. $y=f\left(\frac{3}{4}\right)=8\left(\frac{3}{4}\right)-10=6-10=-4$ 12. $y=f(3)=(3)^2+1=9+1=10$ 13. $y=f(-1)=4(-1)^5+2(-1)^4-3(-1)^3-5(-1)^2+(-1)+5=0$ 14. $y=f\left(\frac{\pi}{2}\right)=\left|\frac{\pi}{2}\right|=\frac{\pi}{2}$ 15. $y=f(2)$ when $y=f(2)=\frac{4(2)-5}{(2)^2+1}=\frac{3}{5}$ 16. $f(0)=\frac{2(0)-3}{(0)+1}=-3$ 17. $f\left(\frac{\pi}{2}\right)=\frac{2\left(\frac{\pi}{2}\right)-3}{\left(\frac{\pi}{2}\right)+1}=\frac{2\pi-6}{\pi+2}$ 18. $f(5a)=\frac{2(5a)-3}{(5a)+1}=\frac{10a-3}{5a+1}$ 19. $f(b-1)=\frac{2(b-1)-3}{(b-1)+1}=\frac{2b-5}{b}$ 20. $f(x^2-1)=\frac{2(x^2-1)-3}{(x^2-1)+1}=\frac{2x^2-5}{x^2}$

4·2 1. $D_f=\{3, 4, 10, 5.2\}$, $R_f=\left\{12,\sqrt{2},-\frac{3}{4},-1\right\}$ 2. $D_f=\left\{x\middle|x\geq-\frac{3}{2}\right\}$, $R_f=\{y|y\geq 0\}$ 3. $D_g=\{x|x\neq 1\}$, $R_g=\{y|y\neq 0\}$ 4. $D_h=(-\infty,\infty)$, $R_h=[0,\infty)$ 5. $D_v=(-\infty,\infty)$, $R_v=[1,\infty)$

4·3 1. No, *f* and *g* do not contain exactly the same set of ordered pairs. 2. No, *g* and *h* do not contain exactly the same set of ordered pairs. 3. No, $D_f\neq D_g$. 4. No, *f* and *g* do not contain exactly the same set of ordered pairs. 5. Yes, *f* and *g* contain exactly the same set of ordered pairs and $D_f=D_g=(-\infty,\infty)$.

4·4 1. $D_f=\{x|x\neq 0\}$, $D_g=(-\infty,\infty)$

2. domain of $f+g$ = domain of $f-g$ = domain of fg = domain of $\frac{f}{g}=\{x|x\neq 0\}$

3. $(f+g)(x)=f(x)+g(x)=\frac{1}{x}+x^3, x\neq 0$; $(f-g)(x)=f(x)-g(x)=\frac{1}{x}-x^3, x\neq 0$;

$(fg)(x)=f(x)g(x)=\left(\frac{1}{x}\right)(x^3)=x^2, x\neq 0$; $\left(\frac{f}{g}\right)(x)=\frac{f(x)}{g(x)}=\frac{1/x}{x^3}=\frac{1}{x^4}, x\neq 0$

4. $(f+g)(-5)=\frac{1}{(-5)}+(-5)^3=-125.2$, $(f-g)(2)=\frac{1}{2}-(2)^3=-7.5$, $(fg)(0)$ is undefined because 0 is not in the domain of *fg*, $\left(\frac{f}{g}\right)(-3)=\frac{1}{(-3)^4}=\frac{1}{81}$

5. $(f+g)\left(\frac{1}{2}\right)=\frac{1}{\left(\frac{1}{2}\right)}+\left(\frac{1}{2}\right)^3=2\frac{1}{8}$, $(f-g)(\pi)=\frac{1}{\pi}-(\pi)^3=\frac{1}{\pi}-\pi^3$, $(fg)(\sqrt[3]{5})=(\sqrt[3]{5})^2=5^{\frac{2}{3}}$,

$\left(\frac{f}{g}\right)(-1)=\frac{1}{(-1)^4}=1$

6. $D_f=[0,\infty)$, $D_g=[0,\infty)$

7. domain of $f+g$ = domain of $f-g$ = domain of $fg=[0,\infty)$; domain of $\dfrac{f}{g}=\{x|x\geq 0, x\neq 4\}$

8. $(f+g)(x)=f(x)+g(x)=(\sqrt{x}+2)+(\sqrt{x}-2)=2\sqrt{x}, x\geq 0;$

$(f-g)(x)=f(x)-g(x)=(\sqrt{x}+2)-(\sqrt{x}-2)=4, x\geq 0;$

$(fg)(x)=f(x)g(x)=(\sqrt{x}+2)(\sqrt{x}-2)=x-4, x\geq 0;$

$\left(\dfrac{f}{g}\right)(x)=\dfrac{f(x)}{g(x)}=\dfrac{\sqrt{x}+2}{\sqrt{x}-2}=\dfrac{\sqrt{x}+2}{\sqrt{x}-2}\cdot\dfrac{\sqrt{x}+2}{\sqrt{x}+2}=\dfrac{x+4\sqrt{x}+4}{x-4}, x\geq 0, x\neq 4$

9. $(f+g)(4)=2\sqrt{4}=4$, $(f-g)(-4)$ is undefined because -4 is not in the domain of $(f-g)$,

$(fg)\left(\dfrac{1}{4}\right)=\left(\dfrac{1}{4}\right)-4=-3\dfrac{3}{4}$, $\left(\dfrac{f}{g}\right)(4)$ is undefined because 4 is not in the domain of $\dfrac{f}{g}$

10. $(f+g)(\pi^2)=2\sqrt{\pi^2}=2\pi$, $(f-g)(0.25)=4$, $(fg)(16)=(16)-4=12$, $\left(\dfrac{f}{g}\right)(1)=\dfrac{(1)+4\sqrt{1}+4}{(1)-4}=-\dfrac{9}{3}=-3$

4·5

1. a. $(f\circ g)(x)=f(g(x))=\sqrt{g(x)+2}=\sqrt{(3x^2)+2}=\sqrt{3x^2+2}$ b. domain $=\{x|x\geq -2\}$

2. a. $(f\circ g)(x)=f(g(x))=|g(x)|=|4x-3|$ b. domain $=R$

3. a. $(f\circ g)(x)=f(g(x))=\sqrt{g(x)}=\sqrt{9x^2}=3|x|$ b. domain $=R$

4. a. $(f\circ g)(x)=f(g(x))=\dfrac{1}{g(x)+1}=\dfrac{1}{4x+1}$ b. domain $=\left\{x\middle|x\neq -\dfrac{1}{4}\right\}$

5. a. $(f\circ g)(x)=f(g(x))=\dfrac{1-g(x)}{3}=\dfrac{1-(1-3x)}{3}=\dfrac{3x}{3}=x$ b. domain $=R$

6. $\sqrt{10}-1$

7. 0

8. $\sqrt{2}-1$

9. $(g\circ f)(-4)=g(f(-4))=$ undefined because f is undefined at -4

10. 10

4·6

1. Yes 2. Yes 3. Yes 4. Yes 5. No

4·7

1. False 2. False 3. False 4. False 5. True

6. $f^{-1}(x)=-x$

7. $f^{-1}(x)=\dfrac{1}{x}$

8. $f^{-1}(x)=\sqrt[3]{x}$

9. $f^{-1}(x)=x+8$

10. $f^{-1}(x)=\dfrac{x}{5}$

11. $f^{-1}(x)=\dfrac{x+2}{x-1}$

12. $f^{-1}(x)=\sqrt{x-8}$

13. $f^{-1}(x)=\dfrac{9x}{5}+32$; thus, $f^{-1}(100)=\dfrac{9(100)}{5}+32=212$

14. $f^{-1}(x)=\dfrac{4x-20}{3}$; thus, $f^{-1}(11)=\dfrac{24}{3}=8$

15. $f^{-1}(x)=\sqrt[3]{2x+6}$; thus, $f^{-1}(-7)=\sqrt[3]{-8}=-2$

5 Graphs of functions

5·1

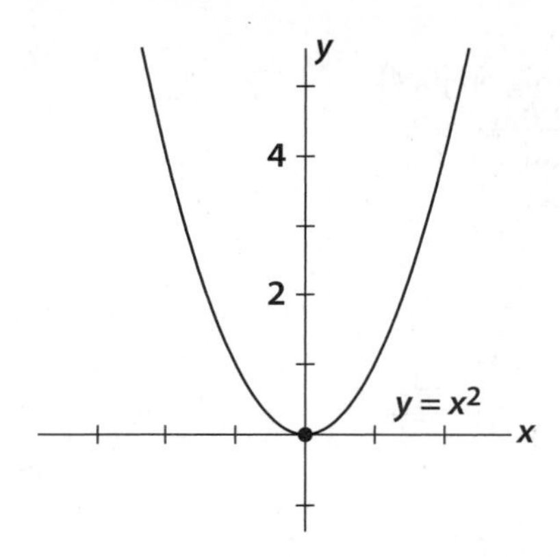

1. $y = x^2$, domain = $(-\infty,\infty)$, range = $\{y \mid y \ge 0\}$, it is a function

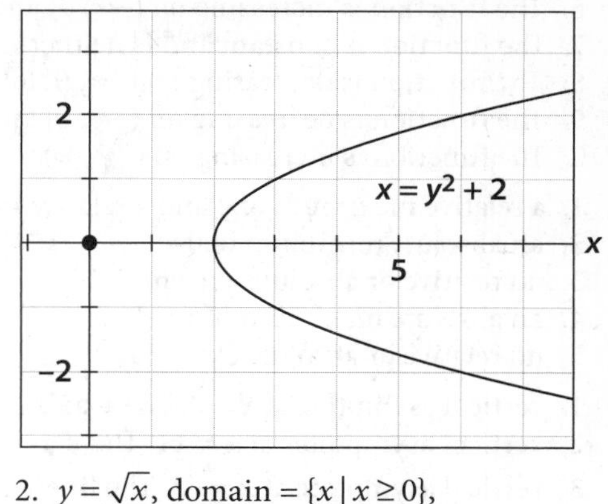

2. $y = \sqrt{x}$, domain = $\{x \mid x \ge 0\}$, range = $\{y \mid y \ge 0\}$, it is a function

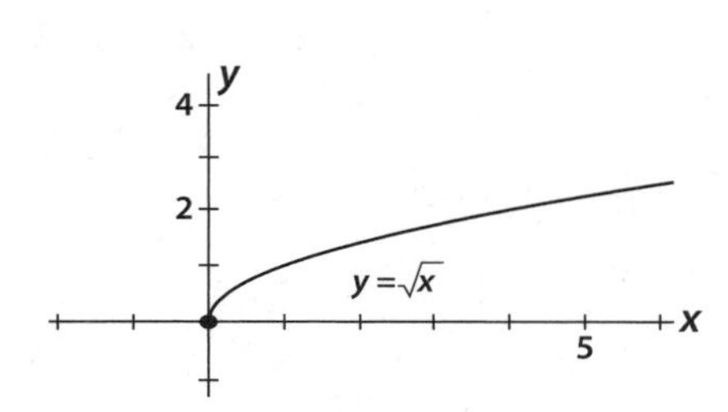

3. $x = y^2 + 2$, domain = $\{x \mid x \ge 2\}$, range = $(-\infty,\infty)$, it is not a function of x

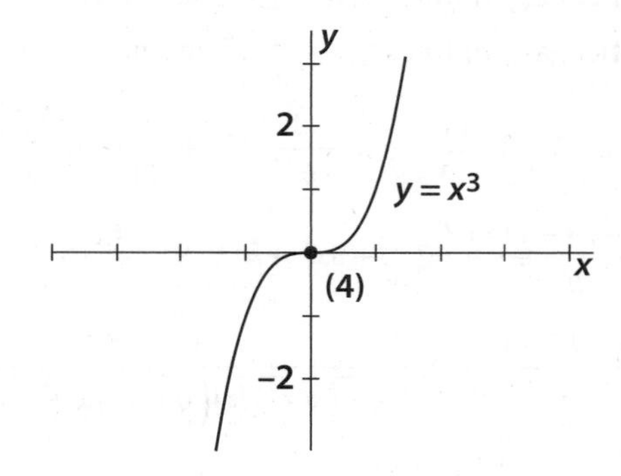

4. $y = x^3$, domain = $(-\infty,\infty)$, range = $(-\infty,\infty)$, it is a function

5. It is a function. It passes the vertical line test.

5·2

1. a. x-intercept: $\frac{1}{4}$; y-intercept: -1 b. $(0,-1),\left(\frac{1}{4},0\right)$
2. a. x-intercept: $\frac{5}{2}$; y-intercept: $-\frac{5}{3}$ b. $\left(0,-\frac{5}{3}\right),\left(\frac{5}{2},0\right)$
3. a. x-intercepts: 4, -3; y-intercept: -12 b. $(0,-12),(4,0),(-3,0)$
4. a. x-intercepts: 1, 4, -2, -5; y-intercept: 40 b. $(0,40),(1,0),(-5,0),(-2,0),(4,0)$
5. a. x-intercept: $\frac{3}{2}$; y-intercept: none b. $\left(\frac{3}{2},0\right)$
6. a. x-intercept: $\frac{9}{2}$; y-intercept: -3 b. $(0,-3),\left(\frac{9}{2},0\right)$
7. a. x-intercept: $-\frac{5}{3}$; y-intercept: 5 b. $(0,5),\left(-\frac{5}{3},0\right)$
8. a. x-intercept: none; y-intercept: 1 b. (0, 1)
9. a. x-intercepts: 4, 2; y-intercept: 8 b. (0, 8), (2, 0), (4, 0)
10. a. x-intercept: -1; y-intercept: 7 b. (0, 7), $(-1, 0)$

5·3

1. False. The values -1 and 2 are in $[-2, 2]$ and $-1 < 2$ and $(-1)^2 < (-2)^2$ but $f(x) = x^2$ is not decreasing on the interval.
2. True
3. True
4. True
5. The function is not defined on $[-3, \infty)$.
6. The function is increasing on $(-\infty, 0]$ and $[4, \infty)$ and decreasing on $[0, 4]$.
7. The function is constant and is neither increasing nor decreasing.
8. The function is decreasing on $(-\infty, 0]$ and increasing on $[0, \infty)$.
9. The function is decreasing on $(-\infty, 0]$ and increasing on $[0, \infty)$.
10. The function is increasing on $(-\infty, \infty)$.

5·4

1. a relative maximum of 4 and a relative minimum of -4
2. an absolute minimum of 0
3. no relative or absolute extrema
4. an absolute maximum of 1
5. no relative or absolute extrema

5·5

1. vertical asymptote at $x = 0$. The x-axis is a horizontal asymptote.
2. vertical asymptotes at $x = \pm 1$. The x-axis is a horizontal asymptote.
3. vertical asymptote at $x = -3$. The line $y = \frac{1}{2}$ is a horizontal asymptote.
4. vertical asymptotes at $x = \pm 3$. The x-axis is a horizontal asymptote.
5. vertical asymptotes at $x = \pm 1$. The line $y = 3$ is a horizontal asymptote.

5·6

1. -2 2. 2 3. $\frac{117}{3}$ 4. $-\frac{1}{50}$ 5. $-\frac{1}{120}$ 6. $\frac{f(x+h)-f(x)}{h} = -2$ 7. $\frac{f(x+h)-f(x)}{h} = 2x + h + 2$

8. $\frac{f(x+h)-f(x)}{h} = 3x^2 + 3xh + h^2$ 9. $\frac{f(x+h)-f(x)}{h} = -\frac{1}{x(x+h)}$

10. $\frac{f(x+h)-f(x)}{h} = -\frac{1}{\sqrt{x}\left(\sqrt{x+h}\right)\left(\sqrt{x}+\sqrt{x+h}\right)}$

6 Function transformations and symmetry

6·1

1. $g(x) = \sqrt{x} + 5$ 2. $g(x) = \sqrt{x-5}$ 3. $g(x) = |x| - \frac{3}{4}$ 4. $g(x) = \left|x + \frac{3}{4}\right|$ 5. $g(x) = \frac{1}{x-7}$ 6. $g(x) = \frac{1}{x} + 7$

7. a 8. c 9. b 10. d

6·2

1. $g(x) = -\frac{1}{x^3}$ 2. $g(x) = -|x+5|$ 3. $g(x) = |-x+5|$ 4. $g(x) = -3x^2 - 2x + 1$ 5. $g(x) = 3x^2 - 2x - 1$

6·3

1. b 2. a 3. a 4. c 5. d 6. $g(x) = \frac{1}{5x} + 3$ 7. $g(x) = 5\sqrt{x+1}$ 8. $g(x) = \left(\frac{1}{4}x\right)^2$ 9. $g(x) = \left|\frac{2}{3}x - 6\right|$

10. $g(x) = 100\sqrt{x} + 10{,}000$ 11. $\left(8, \frac{1}{16}\right)$ 12. $\left(8, \frac{1}{4}\right)$ 13. $(-2, 20)$ 14. $(-4, 1)$ 15. $\left(25, \sqrt{75}\right)$

6·4

1. $g(x) = 20\left[-\frac{1}{(x-10)^3}\right] + 5$ 2. $g(x) = -0.75|x + 4.5| + 9.25$ 3. $g(x) = 7[2(x-6)]^3 + 11$

4. $g(x) = 3(x+5)^2 + 2(x+5) + 9$ 5. $g(x) = 1000\sqrt{\frac{1}{3}(x+1)} - 5$

6·5

1. odd 2. neither 3. even 4. neither 5. even 6. odd 7. neither 8. even 9. even 10. neither

7 Linear functions

7·1 1. linear, slope, y-intercept 2. R, R 3. R, $\{b\}$ 4. nonvertical 5. slope-intercept 6. is not 7. zero 8. −3,−8, R, R 9. 0, −2, R, $\{-2\}$ 10. $x=-\frac{8}{3}, -\frac{8}{3}$

7·2 1. y-intercept = 2, x-intercept = $-\frac{2}{3}$, zero—$x=-\frac{2}{3}$

2. y-intercept = 9, x-intercept = 12, zero—x = 12

3. y-intercept = 0, x-intercept—all real numbers, zero—all real numbers

4. y-intercept = 100, x-intercept—none, zeros—none

5. y-intercept = −5, x-intercept = $\frac{5}{4}$, zero—$x=\frac{5}{4}$

7·3 1. $I(x) = x$, R, R 2. $\frac{1}{2}$ 3. 1,0,0,0 4. origin 5. I, f 6. constant 7. R, the single point, c 8. 0, b 9. no 10. horizontal, 15, below

7·4 1. constant of proportionality, R, R 2. k, origin,0, 0 3. 0 4. $\frac{1}{2\pi}$ 5. 4

7·5 1. is, slope 2. slope 3. $\frac{2}{3}$ 4. 1.25 5. $-\frac{12}{5}$

7·6 1. a. $y=-\frac{1}{2}x+3$, b. $m=-\frac{1}{2}$, y-intercept = 3 2. a. $y=\frac{3}{4}x+2$, b. $m=\frac{3}{4}$, y-intercept = 2

3. a. $y=-\frac{5}{6}x+\frac{8}{3}$, b. $m=-\frac{5}{6}$, y-intercept = $\frac{8}{3}$ 4. a. $y=x-\frac{8}{5}$, b. $m=1$, y-intercept = $-\frac{8}{5}$

5. a. $y=-3$, b. $m=0$, y-intercept = −3 6. $y=\frac{4}{5}x-\frac{11}{5}$ 7. $y=-x+10$ 8. $y=-\frac{1}{2}x+\frac{1}{2}$

9. $y=-3x-13$ 10. $y=-5$ 11. $y=x-2$ 12. $y=-\frac{5}{8}x+5$ 13. $y=-x$ 14. $y=-4$ 15. $y=\frac{4}{3}x$

8 Quadratic functions

8·1

1. $f(x)=ax^2+bx+c,\ a\neq 0$
2. $f(x)=a(x-h)^2+k,\ a\neq 0$
3. R, subset
4. parabola
5. y-axis
6. a. No real zeros b. $y = 28$ c. no x-intercepts
7. a. $x=5\pm\sqrt{2}$ b. $y=46$ c. x-intercepts at $x=5\pm\sqrt{2}$
8. a. $x = 5$ b. $y = 25$ c. x-intercept at $x = 5$
9. a. $1, \frac{1}{2}$ b. $y=-1$ c. $1, \frac{1}{2}$
10. a. no real zeros b. $y = 1$ c. no x-intercepts
11. $f(x)=\left(x-\frac{1}{2}\right)^2-\frac{25}{4}$
12. $g(x)=-(x-4)^2$
13. $h(x)=(x-5)^2$

14. $g(x)=-2\left(x-\frac{3}{4}\right)^2+\frac{1}{8}$

15. $h(x)=(x+1)^2+3$

8·2

1. upward, lowest, downward, highest
2. $\left(\frac{-b}{2a}, f\left(\frac{-b}{2a}\right)\right), (h,k)$
3. axis of symmetry
4. maximum
5. minimum
6. a. upward b. (5, 3) c. $x=5$ d. minimum = 3 e. Range is all $y \geq 3$. f. increasing on [5, ∞) and decreasing on (−∞, 5]
7. a. upward b. (5, −4) c. $x=5$ d. minimum = −4 e. Range is all $y \geq -4$. f. increasing [5, ∞) on and decreasing on (−∞, 5]
8. a. upward b. (5, 0) c. $x=5$ d. minimum = 0 e. Range is all $y \geq 0$. f. increasing on [5, ∞) and decreasing on (−∞, 5]
9. a. downward b. (−6, −1) c. $x=-6$ d. maximum = −1 e. Range is all $y \geq -1$. f. increasing on (∞,−6] and decreasing on [−6, ∞)
10. a. upward b. (0,1). c. $x=-0$ d. minimum = 1 e. Range is all $x \geq 1$. f. increasing on [0, ∞) and decreasing on (−∞, 0]

8·3

1. −4
2. 16
3. 2
4. −4
5. 12
6. $\frac{f(x+h)-f(x)}{h}=\frac{(x+h)^2+2(x+h)-x^2-2x}{h}=2x+h+2$
7. $\frac{f(x+h)-f(x)}{h}=\frac{(x+h)^2+2(x+h)+1-x^2-2x-1}{h}=2x+h+2$
8. $\frac{f(x+h)-f(x)}{h}=\frac{(x+h)^2+2(x+h)+100-x^2-2x-100}{h}=2x+h+2$
9. $\frac{f(x+h)-f(x)}{h}=\frac{\left[-2(x+h-3)^2+5\right]-\left[-2(x-3)^2+5\right]}{h}=-4(x-3)-2h$
10. $\frac{f(x+h)-f(x)}{h}=\frac{(x+h)^2+8(x+h)-2-x^2-8x+2}{h}=2x+h+8$

8·4

1. $5 \pm i\sqrt{3}$ 2. $5 \pm \sqrt{2}$ 3. 5 4. $-6 \pm i\frac{\sqrt{2}}{2}$ 5. $\pm i$

9 Polynomial functions

9·1

1. highest
2. power, zero, undefined
3. $r=(-\infty, \infty)$, odd, even

4. $p(r) = 0$, real

5. $p(0)$

6. a. 5 b. domain = R, range = R c. −3, −2, 1, 2, 4 d. −3, −2, 1, 2, 4 e. −96

7. a. 5 b. domain = R, range = R c. $-4, \pm 3, \pm\sqrt{5}$ d. $-4, \pm 3, \pm\sqrt{5}$ e. 180

8. a. 5 b. domain = R, range is $[-81, \infty)$ c. ±3 d. ±3 e. −81

9. a. 5 b. domain = R, range is $\left(-\infty, \frac{1}{8}\right]$ c. $\frac{1}{2}, 1$ d. $\frac{1}{2}, 1$ e. −1

10. a. 5 b. domain = R, range = R c. $-\frac{5}{3}$ d. $-\frac{5}{3}$ e. 5

9·2

1. continuous
2. turning point
3. none
4. none
5. one
6. a. (−2.39, 7.58), (3,0) b. relative maximum: 7.58, relative minimum: 0
7. a. no turning points b. no relative or absolute extrema
8. a. (3,5) b. absolute maximum: 5
9. a. (0,2) b. absolute minimum: 2
10. a. (−7.22, 40.31),(−2.42, −22.76),(1.97, 3.41),(5.27, −17.68) b. relative maxima: 40.31, 3.41; relative minima: −22.76, −17.68; no absolute extrema

9·3

1. $p(a)$ 2. $p(c) = 0$ 3. $(x-r)^k$, $(x-r)^{k+1}$ 4. n 5. 5 6. $p(2) = 0$ 7. $p(-2) = 20$ 8. −2,3,4

9. $c(x-5)(x+2)(x-\sqrt{3})(x+\sqrt{3})$, where c is any nonzero constant 10. $g(x) = 2(x-1)(x+1)(x-3)$

9·4

1. integral
2. positive
3. $p(-x)$
4. 4
5. 3
6. The number of variations in $p(x)$ is two. Thus there are two or zero positive real roots. The number of variations in $p(-x)$ is two. Thus there are two or zero negative real roots.
7. The number of variations in $p(x)$ is one. Thus there is at most one positive real root. The number of variations in $p(-x)$ is one. Thus there is at most one negative real root.
8. The number of variations in $p(x)$ is one. Thus there is at most one positive real root. The number of variations in $p(-x)$ is zero. Thus there are no negative real roots.
9. The number of variations in $p(x)$ is two. Thus there are two or zero positive real roots. The number of variations in $p(-x)$ is one. Thus there is at most one negative real root.
10. The number of variations in $p(x)$ is two. Thus there are two or zero positive real roots. The number of variations in $p(-x)$ is three. Thus there are three or one negative real roots.
11. $x = -\frac{3}{2}, -1, \frac{1}{3}, 1$
12. $x = \frac{1}{2}, -2$
13. $x = 1$
14. $x = -2, \frac{1}{2}, 3$
15. $x = \pm 1, \pm\sqrt{3}, -4$

9·5

1. complex, one
2. n
3. 2
4. 100
5. $5-3i$
6. a. $x=-4, \pm\sqrt{5}, \pm 6$ b. $p(x)=(x+4)(x+\sqrt{5})(x-\sqrt{5})(x+6)(x-6)$
7. a. $x=\pm 1, \frac{1\pm i\sqrt{3}}{2}, \frac{-1\pm i\sqrt{3}}{2}$ b. $p(x)=(x+1)(x-1)\left[x-\frac{1+i\sqrt{3}}{2}\right]\left[x-\frac{1-i\sqrt{3}}{2}\right]\left[x-\frac{-1+i\sqrt{3}}{2}\right]\left[x-\frac{-1-i\sqrt{3}}{2}\right]$
8. a. $x=1, \frac{1}{2}$ b. $p(x)=-2(x-1)\left(x-\frac{1}{2}\right)$
9. a. $x=-\frac{5}{3}, \pm 2, -1\pm i\sqrt{3}, 1\pm i\sqrt{3}$

   b. $p(x)=(3x+5)(x+2)(x-2)\left[x-\left(-1+i\sqrt{3}\right)\right]\left[x-\left(-1-i\sqrt{3}\right)\right]\left[x-\left(1+i\sqrt{3}\right)\right]\left[x-\left(1-i\sqrt{3}\right)\right]$
10. a. $x=\frac{1}{2}, 1+i, 1-i, 1+2i, 1-2i$ b. $p(x)=\left(x-\frac{1}{2}\right)(x-1-i)(x-1+i)(x-1-2i)(x-1+2i)$

9·6

1. a. positive b. positive
2. a. positive b. positive
3. a. positive b. negative
4. a. positive b. negative
5. a. positive b. negative

10 Rational functions

10·1

1. $\frac{p(x)}{q(x)}$, polynomials, $q(x)$
2. 0
3. numerator
4. $f(0)$
5. numerator
6. a. The domain is all real numbers except 0. b. There are no zeros. c. There are no intercepts.
7. a. The domain is all real numbers except $x=3$ and $x=-4$. b. The zeros are $x=-2$. c. The y-intercept is at $y=\frac{1}{3}$ and the x-intercepts are at $x=-2$.
8. a. The domain is all real numbers except $x=3$ and $x=-2$. b. The zero is $x=2$. c. The y-intercept is at $y=\frac{2}{3}$ and the x-intercept is at $x=2$.
9. a. The domain is all real numbers. b. The zero is $x=2$. c. The y-intercept is at $y=-3$ and the x-intercept is at $x=2$.
10. a. The domain is all real numbers except $x=3$ and $x=-4$. b. The zero is $x=-3$. c. The y-intercept is at $y=\frac{27}{4}$ and the x-intercept is at $x=-3$.

10·2

1. yes, no
2. yes, no
3. yes, yes
4. less than, $\frac{a_n}{b_n}$
5. an oblique
6. a. The domain is all real numbers except 0. b. no holes c. Vertical asymptote is at $x=0$ and the x-axis is a horizontal asymptote. d. no intercepts
7. a. The domain is all real numbers except $x=3$ and $x=-4$. b. no holes c. vertical asymptotes at $x=3$ and $x=-4$; horizontal asymptote at $y=1$ d. x-intercepts $=\pm 2$, y-intercept $=\frac{1}{3}$

8. a. The domain is all real numbers except $x = 3$ and $x = -2$. b. a hole at $x = -2$
c. vertical asymptotes at $x = 3$; horizontal asymptote at $y = 1$ d. x-intercept = 2, y-intercept = $\frac{2}{3}$
9. a. The domain is all real numbers. b. no holes c. no asymptotes d. x-intercept = 2, y-intercept = −3
10. a. The domain is all real numbers except $x = 3$ and $x = -4$. b. a hole at $x = 3$ c. vertical asymptote at $x = -4$. d. x-intercept = −3, y-intercept = $\frac{27}{4}$

10·3

1. a. As x approaches ∞, $f(x)$ approaches 0. b. As x approaches $-\infty$, $f(x)$ approaches 0.
2. a. As x approaches ∞, $f(x)$ approaches 1. b. As x approaches $-\infty$, $f(x)$ approaches 1.
3. a. As x approaches ∞, $f(x)$ approaches $-\infty$. b. As x approaches $-\infty$, $f(x)$ approaches ∞.
4. a. As x approaches ∞, $f(x)$ approaches $\frac{3}{2}$. b. As x approaches $-\infty$, $f(x)$ approaches $\frac{3}{2}$.
5. a. As x approaches ∞, $f(x)$ approaches 0. b. As x approaches $-\infty$, $f(x)$ approaches 0.

11 Exponential and logarithmic functions

11·1

1. $b \neq 0$ and $b > 1$ 2. $(-\infty,\infty)$ and $(0,\infty)$ 3. zeros 4. 1 5. $f(x) = e^x$ 6. 81 7. 256 8. 32 9. $\frac{8}{27}$ 10. 0.5

11·2

1. continuous
2. $(0, 1)$ and $(1, 10)$
3. $(-\infty, \infty)$ and $(0, \infty)$
4. 1
5. no
6. a. The domain is R and the range is $(0, \infty)$. b. There are no zeros. c. $x = 0$ is a horizontal asymptote. d. The y-intercept is $y = 1$ and there are no x-intercepts. e. $b = 5>1$, therefore f is increasing on R. f. As x approaches ∞, $f(x)$ approaches ∞ and as x approaches $-\infty$, $f(x)$ approaches 0.
7. a. The domain is R and the range is $(0, \infty)$. b. There are no zeros. c. $x = 0$ is a horizontal asymptote. d. The y-intercept is $y = 1$ and there are no x-intercepts. e. $b = 64>1$, therefore g is increasing on R. f. As x approaches ∞, $f(x)$ approaches ∞ and as x approaches $-\infty$, $f(x)$ approaches 0.
8. a. The domain is R and the range is $(0, \infty)$. b. There are no zeros. c. $x = 0$ is a horizontal asymptote. d. The y-intercept is 1 and there are no x-intercepts. e. $b = \frac{1}{2} < 1$, therefore f is decreasing on R. f. As x approaches ∞, $f(x)$ approaches 0 and as x approaches $-\infty$, $f(x)$ approaches ∞.
9. a. The domain is R and the range is $(0, \infty)$. b. There are no zeros. c. $x = 0$ is a horizontal asymptote. d. The y-intercept is 1 and there are no x-intercepts. e. $b = \frac{4}{9} < 1$, therefore h is decreasing on R. f. As x approaches ∞, $h(x)$ approaches 0 and as x approaches $-\infty$, $h(x)$ approaches ∞.
10. a. The domain is R and the range is $(0, \infty)$. b. There are no zeros. c. $x = 0$ is a horizontal asymptote. d. The y-intercept is 1 and there are no x-intercepts. e. $b = 0.25 < 1$, therefore g is decreasing on R. f. As x approaches ∞, $g(x)$ approaches 0 and as x approaches $-\infty$, $g(x)$ approaches ∞.

11·3

1. $\frac{e^7}{e^2} = e^{7-2} = e^5$ 2. $2^0 \cdot 2^1 = 2^{0+1} = 2^1 = 2$ 3. $10^{-x} = \frac{1}{10^x}$ 4. $\left(\frac{4}{9}\right)^{-1} = \frac{9}{4}$ 5. 4

11·4

1. $b>0$ and $b \neq 1$ 2. $(0,\infty)$ and $(-\infty,\infty)$ 3. y intercept 4. negative and 0 5. $f(x) = \log_e x = \ln x$ 6. 4 7. $\frac{4}{3}$ 8. −5 9. $\frac{3}{2}$ 10. $\frac{1}{2}$ 11. $g(x) = 8^x$ 12. $f(x) = \left(\frac{1}{4}\right)^x$ 13. $g(x) = \log_{10} x$ 14. $f(x) = \log_{1.05} x$ 15. $g(x) = e^x$

11·5

1. continuous
2. $(1, 0)$ and $(10, 1)$
3. $(0, \infty)$ and $(-\infty, \infty)$
4. $x = 1$
5. no

6. a. The domain is $(0,\infty)$ and the range is $(-\infty,\infty)$. b. $x=1$ is the only zero. c. The y-axis is a vertical asymptote. d. The x-intercept is (1,0) and there are no y-intercepts. e. $b=8>1$ so $f(x)$ is increasing on $(0,\infty)$. f. $b=8>1$ so as x approaches 0, $f(x)$ approaches $-\infty$ and as x approaches ∞, $f(x)$ approaches ∞.
7. a. The domain is $(0,\infty)$ and the range is $(-\infty,\infty)$. b. $x=1$ is the only zero. c. The y-axis is a vertical asymptote. d. The x-intercept is (1,0) and there are no y-intercepts. e. $b=\frac{1}{4}<1$ so $f(x)$ is decreasing on $(0,\infty)$. f. $b=\frac{1}{4}<1$ so as x approaches 0, $f(x)$ approaches ∞ and as x approaches ∞, $f(x)$ approaches $-\infty$.
8. a. The domain is $(-\infty,\infty)$ and the range is $(0,\infty)$. b. There are no zeros. c. The x-axis is a horizontal asymptote. d. The y-intercept is (0,1) and there are no x-intercepts. e. Since $b=10>1$, $h(x)$is increasing on. $(-\infty,\infty)$. f. When x approaches ∞, $h(x)$ approaches ∞ and when x approaches $-\infty$, $h(x)$ approaches 0. As x approaches 0, $h(x)$ approaches 1.
9. a. The domain is $(-\infty,\infty)$ and the range is $(0,\infty)$. b. There are no zeros. c. The x-axis is a horizontal asymptote. d. The y-intercept is (0,1)and there are no x-intercepts. e. Since $b=1.05>1$, $g(x)$ is increasing on $(-\infty,\infty)$. f. When x approaches ∞, $g(x)$ approaches ∞ and when x approaches $-\infty$, $g(x)$ approaches 0. As x approaches 0, $g(x)$ approaches 1.
10. a. The domain is $(0,\infty)$ and the range is $(-\infty,\infty)$. b. $x=1$ is the only zero. c. The y-axis is a vertical asymptote. d. The x-intercept is (1,0) and there are no y-intercepts. e. $b=e>1$ so $f(x)$ is increasing on $(0,\infty)$. f. $b=e>1$ so as x approaches 0, $f(x)$ approaches $-\infty$ and as x approaches ∞, $f(x)$ approaches ∞.

11·6

1. 100
2. $12\log_2 64=12(6)=72$
3. $\log_{10}(1000)-\log_{10}(0.00001)=\log_{10}(10^3)-\log_{10}(10^{-5})=3+5=8$
4. $\log_3 27+\log_3 81=3+4=7$
5. 200
6. $\frac{\ln 512}{\ln 8}=3$
7. $\frac{\ln(0.015625)}{\ln\left(\frac{1}{4}\right)}=3$
8. $\frac{\ln(15625)}{\ln 5}=6$
9. $\frac{\ln(2.5)}{\ln(1.05)}\approx 18.78$
10. $\frac{\ln 200}{\ln 2}\approx 7.64$

11·7

1. $\log_2(3x-1)=32;\ 2^{\log_2(3x-1)}=2^{32};\ 3x-1=2^{32};\ x=\frac{2^{32}+1}{3}$
2. $e^{\ln 8x}=e^{3.5}$; thus, $x=\frac{e^{3.5}}{8}$
3. $(1.005)^x=3;\ x=\frac{\ln 3}{\ln(1.005)}\approx 220.27$
4. $e^{.05x}=\frac{21}{10.5}=2;\ 0.05x=\ln 2;\ x=\frac{\ln 2}{0.05}\approx 13.86$
5. $5x+1=3x-4,\ \ x=-\frac{5}{2}$

12 Additional common functions

12·1 1. 1 2. 4 3. 3 4. 3 5. $\sqrt{11}$

12·2 1. 2 2. −3 3. 0 4. −12 5. −3

12·3 1. −2.4 2. 4 3. $\frac{25}{40}=\frac{5}{8}$ 4. $\frac{8}{16}=\frac{1}{2}$ 5. $|-10|=10$ 6. $\pm\frac{1}{2}$ 7. $-9.57<x<9.57$ 8. $x>40$ or $x<-40$

9. All real values for x are solutions since $|x|$ is always nonnegative. 10. There are no solutions since $|x|$ is always nonnegative.

12·4 1. 32 2. 125 3. 49 4. 4 5. ≈4,952.22 6. ≈18.93 7. no solution 8. no solution 9. 1 10. ≈0.1520

13 Function models and applications

13·1

1. a. $f(t)=70t$ b. 245
2. a. $f(t)=1000-120t$ b. 640
3. a. $f(t)=200-50t$ b. $62\frac{1}{2}$ miles
4. a. $F(x)=-6x$ b. −18 pounds
5. a. $f(t)=250+0.55x$ b. $544.80

13·2 1. $l=w=50$ feet 2. 150 feet 3. 60 thousand 4. $6,728,000 5. 2.5 seconds

13·3 1. 213.13 square centimeters 2. ≈3.37 grams 3. ≈2.8 4. ≈164.7 Earth-years 5. 0.495

14 Matrices and systems of linear equations

14·1 1. $\begin{bmatrix}1\\5\\7\end{bmatrix}$ 2. $\begin{bmatrix}1&0&0\\0&1&0\\0&0&1\end{bmatrix}$ 3. 3×2

14·2 1. $\begin{bmatrix}2&8\\-3&0\end{bmatrix}+\begin{bmatrix}5&4\\-2&3\end{bmatrix}=\begin{bmatrix}7&12\\-5&3\end{bmatrix}$ 2. $\begin{bmatrix}3&6&0\\4&-2&-5\\6&0&-1\end{bmatrix}-\begin{bmatrix}-1&6&4\\-4&3&-1\\5&4&2\end{bmatrix}=\begin{bmatrix}4&0&-4\\8&-5&-4\\1&-4&-3\end{bmatrix}$

3. $\begin{bmatrix}1&6&3\\0&5&4\end{bmatrix}\begin{bmatrix}2&4\\-3&1\\5&0\end{bmatrix}=\begin{bmatrix}-1&10\\5&5\end{bmatrix}$ 4. $\begin{bmatrix}1&0\\0&1\end{bmatrix}\begin{bmatrix}3&6\\-4&5\end{bmatrix}=\begin{bmatrix}3&6\\-4&5\end{bmatrix}$ 5. $\begin{bmatrix}0&0\\0&0\end{bmatrix}\begin{bmatrix}5&6\\4&9\end{bmatrix}=\begin{bmatrix}0&0\\0&0\end{bmatrix}$

14·3 1. −13 2. 0 3. $3\begin{vmatrix}7&-3\\1&5\end{vmatrix}-4\begin{vmatrix}0&-3\\-2&5\end{vmatrix}+2\begin{vmatrix}0&7\\-2&1\end{vmatrix}=3(38)-4(-6)+2(14)=166$

4. 1 5. 1 6. not applicable; not a square matrix

14·4 1. $A^{-1}=\frac{1}{3}\begin{bmatrix}1&-2\\1&1\end{bmatrix}$

2. $B^{-1}=\frac{1}{3}\begin{bmatrix}1&1\\-2&1\end{bmatrix}$

3. $(AB)^{-1}=\left(\begin{bmatrix}1&2\\-1&1\end{bmatrix}\begin{bmatrix}1&-1\\2&1\end{bmatrix}\right)^{-1}=\left(\begin{bmatrix}5&1\\1&2\end{bmatrix}\right)^{-1}=\frac{1}{9}\begin{bmatrix}2&-1\\-1&5\end{bmatrix}$

4. $A^{-1}B^{-1} = \frac{1}{9}\begin{bmatrix} 1 & -2 \\ 1 & 1 \end{bmatrix}\begin{bmatrix} 1 & 1 \\ -2 & 1 \end{bmatrix} = \frac{1}{9}\begin{bmatrix} 5 & -1 \\ -1 & 2 \end{bmatrix}$

5. $B^{-1}A^{-1} = \frac{1}{9}\begin{bmatrix} 1 & 1 \\ -2 & 1 \end{bmatrix}\begin{bmatrix} 1 & -2 \\ 1 & 1 \end{bmatrix} = \frac{1}{9}\begin{bmatrix} 2 & -1 \\ -1 & 5 \end{bmatrix}$

6. $A(-B^{-1}) = \begin{bmatrix} 1 & 2 \\ -1 & 1 \end{bmatrix}\left(-\frac{1}{3}\begin{bmatrix} 1 & 1 \\ -2 & 1 \end{bmatrix}\right) = \frac{1}{3}\begin{bmatrix} 1 & 2 \\ -1 & 1 \end{bmatrix}\begin{bmatrix} -1 & -1 \\ 2 & -1 \end{bmatrix} = \frac{1}{3}\begin{bmatrix} 3 & -3 \\ 3 & 0 \end{bmatrix} = \begin{bmatrix} 1 & -1 \\ 1 & 0 \end{bmatrix}$

7. $(-A)B^{-1} = \begin{bmatrix} 1 & -1 \\ 1 & 0 \end{bmatrix}$

8. They are equal.
9. They are equal.
10. They are not equal.

14·5

1. $D = 11,\ x = \frac{\begin{vmatrix} 16 & -3 \\ -4 & -2 \end{vmatrix}}{11} = \frac{-44}{11} = -4,\ y = \frac{\begin{vmatrix} 2 & 16 \\ 5 & -4 \end{vmatrix}}{11} = \frac{-88}{11} = -8$

2. $D = 8,\ x = \frac{24}{8} = 3,\ y = \frac{-16}{8} = -2$

3. $D = 32,\ x = \frac{20}{32} = \frac{5}{8},\ y = \frac{2}{32} = \frac{1}{16}$

4. $D = \begin{vmatrix} 1 & -2 & -3 \\ 2 & 4 & -5 \\ 3 & 7 & -4 \end{vmatrix} = (-16+35) + 2(-8+15) - 3(14-12) = 27$

$$x = \frac{\begin{vmatrix} -20 & -2 & -3 \\ 11 & 4 & -5 \\ 33 & 7 & -4 \end{vmatrix}}{27} = \frac{-20(19) + 2(-44+165) - 3(77-132)}{27} = \frac{27}{27} = 1$$

$$y = \frac{\begin{vmatrix} 1 & -20 & -3 \\ 2 & 11 & -5 \\ 3 & 33 & -4 \end{vmatrix}}{27} = \frac{121 + 20(7) - 3(33)}{27} = \frac{162}{27} = 6$$

$z = \frac{(1-12+20)}{3} = 3$ (solving the first equation for z)

5. $D = \begin{vmatrix} 1 & 2 & -1 \\ 4 & 3 & 2 \\ 9 & 8 & 3 \end{vmatrix} = -7 - 2(-6) - 5 = 0$

$X = \begin{vmatrix} 7 & 2 & -1 \\ 1 & 3 & 2 \\ 4 & 8 & 3 \end{vmatrix} = 7(-7) - 2(-5) - (1) = -40 \neq 0$. Thus, there is no solution.

6. $D = \begin{vmatrix} 2 & -4 & 7 \\ 3 & 2 & -1 \\ 1 & -10 & 15 \end{vmatrix} = 2(20) + 4(46) + 7(-32) = 0$

$X = \begin{vmatrix} 5 & -4 & 7 \\ 2 & 2 & -1 \\ 8 & -10 & 15 \end{vmatrix} = 5(20) + 4(38) + 7(-36) = 0$

$Y = \begin{vmatrix} 2 & 5 & 7 \\ 3 & 2 & -1 \\ 1 & 8 & 15 \end{vmatrix} = 2(38) - 5(46) + 7(22) = 0$

$Z = \begin{vmatrix} 2 & -4 & 5 \\ 3 & 2 & 2 \\ 1 & -10 & 8 \end{vmatrix} = 2(36) + 4(22) + 5(-32) = 0$

Solve $\begin{matrix} 2x - 4y = 5 - 7z \\ 3x + 2y = 2 + z \end{matrix}$.

$x = \dfrac{\begin{vmatrix} 5-7z & -4 \\ 2+z & 2 \end{vmatrix}}{16} = \dfrac{10 - 14z + 8 + 4z}{16} = \dfrac{18 - 10z}{16} = \dfrac{9 - 5z}{8}$

$y = \dfrac{\begin{vmatrix} 2 & 5-7z \\ 3 & 2+z \end{vmatrix}}{16} = \dfrac{4 + 2z - 15 + 21z}{16} = \dfrac{23z - 11}{16}$

The system solution is $\left(\dfrac{9-5z}{8}, \dfrac{23z-11}{16}, z\right)$.

14·6

1. $\begin{bmatrix} x \\ y \end{bmatrix} = X = A^{-1}C = \dfrac{1}{-8}\begin{bmatrix} -2 & -2 \\ -1 & 3 \end{bmatrix}\begin{bmatrix} 4 \\ 3 \end{bmatrix} = -\dfrac{1}{8}\begin{bmatrix} -14 \\ 5 \end{bmatrix}$. Thus, $x = \dfrac{7}{4}, y = -\dfrac{5}{8}$.

2. $\begin{bmatrix} x \\ y \end{bmatrix} = X = A^{-1}C = \dfrac{1}{12}\begin{bmatrix} 2 & 1 \\ -2 & 5 \end{bmatrix}\begin{bmatrix} 3 \\ 2 \end{bmatrix} = \dfrac{1}{12}\begin{bmatrix} 8 \\ 4 \end{bmatrix}$. Thus, $x = \dfrac{2}{3}, y = \dfrac{1}{3}$.

3. $\begin{bmatrix} 3 & 2 & 4 \\ 1 & -2 & 3 \end{bmatrix}$ Multiply the second row by (–3) and add to the first row to get

$\sim\begin{bmatrix} 0 & 8 & -5 \\ 1 & -2 & 3 \end{bmatrix}$. Interchange the two rows $\sim\begin{bmatrix} 1 & -2 & 3 \\ 0 & 8 & -5 \end{bmatrix}$. Multiply the second row by ¼ and add to the first row

$\sim\begin{bmatrix} 1 & 0 & \frac{7}{4} \\ 0 & 8 & -5 \end{bmatrix}$. Thus, $x = \dfrac{7}{4}, y = \dfrac{-5}{8}$.

4. $\begin{bmatrix} 1 & -3 & 1 & 2 \\ 2 & -1 & -2 & 1 \\ 3 & 2 & -1 & 5 \end{bmatrix}$ Multiply the first row by (–2) and add to the second row, and at the same time multiply the first row by (–3) and add to the third row to get

$\sim\begin{bmatrix} 1 & -3 & 1 & 2 \\ 0 & 5 & -4 & -3 \\ 0 & 11 & -4 & -1 \end{bmatrix}$. Multiply the second row by (–1) and add to the third row to get

$$\sim\begin{bmatrix}1 & -3 & 1 & 2\\0 & 5 & -4 & -3\\0 & 6 & 0 & 2\end{bmatrix}.$$ Multiply the third row by (−1) and add to second row to get

$$\sim\begin{bmatrix}1 & -3 & 1 & 2\\0 & -1 & -4 & -5\\0 & 6 & 0 & 2\end{bmatrix}\sim\begin{bmatrix}1 & 0 & 13 & 17\\0 & 1 & 4 & 5\\0 & 3 & 0 & 1\end{bmatrix}.$$ Solve this system. $y=\frac{1}{3}, \frac{1}{3}+4z=5$ or $z=\frac{7}{6}$. Finally, $x+13\left(\frac{7}{6}\right)=17$ or $x=\frac{11}{6}$.

5. $\begin{bmatrix}2 & -1 & 0 & 0\\0 & 2 & -1 & 0\\1 & 2 & -1 & 3\end{bmatrix}\sim\begin{bmatrix}0 & -5 & 2 & -6\\0 & 2 & -1 & 0\\1 & 2 & -1 & 3\end{bmatrix}\sim\begin{bmatrix}0 & -1 & 0 & -6\\0 & 2 & -1 & 0\\1 & 2 & -1 & 3\end{bmatrix}$. Solve this system. $y=6$, $x=3$, and $z=12$.

15 Sequences

15·1

1. $a_n=5+(n-1)1=n+4$ 2. $a_n=3+(n-1)3=3n$
3. $a_n=4(4)^{n-1}=4^n$ 4. $a_1=1,\ a_2=2,\ a_{n+1}=a_n+a_{n-1}$
5. $a_1=6,\ a_{n+1}=a_n+2$ 6. $a_1=5,\ a_{n+1}=a_n+5$

15·2

1. $s=2,\ d=5,\ a_n=2+(n-1)5=5n-3,\ a_{10}=47,\ a_{12}=57$
2. $s=\frac{3}{2},\ d=\frac{3}{4},\ a_n=\frac{3}{2}+(n-1)\frac{3}{4}=\frac{3}{4}(n+1),\ a_{10}=\frac{33}{4},\ a_{12}=\frac{39}{4}$
3. $s=3.10,\ d=0.15,\ a_n=3.10+(n-1)(0.15)=0.15n+2.95,\ a_{10}=4.45,\ a_{12}=4.75$
4. $s=8,\ d=-2,\ a_n=-2n+10,\ a_{10}=-10,\ a_{12}=-14$
5. $s=\frac{5}{7},\ d=-\frac{1}{2},\ a_n=-\frac{1}{2}n+\frac{17}{14},\ a_{10}=-\frac{53}{14},\ a_{12}=\frac{67}{14}$
6. $s=243,\ r=\frac{1}{3},\ a_n=243\left(\frac{1}{3}\right)^{n-1},\ a_{10}=243\left(\frac{1}{3}\right)^{9}$
7. $s=\frac{1}{3}, r=-3, a_n=\frac{1}{3}(-3)^{n-1}, a_{10}=-\frac{1}{3}(3)^9=-(3)^8$
8. $s=-\frac{7}{8},\ r=-2,\ a_n=-\frac{7}{8}(-2)^{n-1}=(-1)^n7\left(2^{n-4}\right),\ a_{10}=7\left(2^6\right)$
9. $s=2,\ r=2,\ a_n=2^n,\ a_{10}=2^{10}$
10. $s=\frac{1}{2}, r=\frac{2}{3}, a_n=\frac{1}{2}\left(\frac{2}{3}\right)^{n-1}, a_{10}=\frac{1}{2}\left(\frac{2}{3}\right)^9$

15·3

1. $\sum_{k=1}^{n}(5k-1)$
2. $\sum_{k=1}^{n}4\left(\frac{1}{3}\right)^{k-1}$
3. $\sum_{k=1}^{n}(6k-2)$
4. $\sum_{k=1}^{n}(-2k+7)$
5. $\frac{1}{6}\sum_{k=1}^{n}\left(\frac{2}{5}\right)^{k-1}$
6. $\sum_{k=0}^{n-1}(3k+9)$
7. $\sum_{k=1}^{n-2}(3k+6)+(3n+3)+(3n+6)=\sum_{k=1}^{n-2}(3k+6)+6n+9$
8. $\sum_{k=1}^{n}(3k+6)=3\sum_{k=1}^{n}k+6n$
9. $\sum_{k=1}^{18}4\left(\frac{1}{5}\right)^k+\sum_{k=19}^{n}4\left(\frac{1}{5}\right)^k$
10. $\sum_{j=3}^{n+2}4\left(\frac{1}{5}\right)^{j-2}$

15·4

1. $\sum_{k=1}^{30}(3k-4)=3\frac{30(31)}{2}-4(30)=1{,}275$

2. $\sum_{j=1}^{32}\left(\frac{1}{2}j-5\right)=\frac{1}{2}\left(\frac{32(33)}{2}\right)-5(32)=104$

3. $\sum_{n=2}^{30}(8-3n)=\sum_{k=1}^{29}(5-3k)=5(29)-3\left(\frac{29(30)}{2}\right)=-1{,}160$

4. $\sum_{k=2}^{10}3\left(\frac{1}{3}\right)^k=\sum_{j=1}^{9}\left(\frac{1}{3}\right)^j=\frac{1}{3}\left(\frac{1-\left(\frac{1}{3}\right)^9}{\frac{2}{3}}\right)=\frac{(3^9-1)}{2(3^9)}$

5. $\sum_{k=0}^{10}2^k=2+\left(\frac{2-2^{11}}{-1}\right)=2^{11}=2{,}048$

6. $\frac{1}{2}\sum_{k=1}^{15}(5k-29)=\frac{1}{2}\left[5\left(\frac{15(16)}{2}\right)-29(15)\right]=82.5$

7. $\sqrt{2}\sum_{k=1}^{20}k=\sqrt{2}(210)=210\sqrt{2}$

8. $s_{20}=\frac{1}{2}\sum_{k=1}^{20}(-2k+11)=\frac{1}{2}(-2(210)+20(11))=-100$

9. $s_8=\sum_{k=1}^{8}\left(\frac{1}{5}\right)^k=\left[\frac{\left(\frac{1}{5}\right)-\left(\frac{1}{5}\right)^9}{\frac{4}{5}}\right]=\frac{1}{4}\left(1-\left(\frac{1}{5}\right)^8\right)$

10. 5, 9, 13, 17. Solve $s_{10}=\frac{10(5+a_{10})}{2}=230$ for a_{10}; then solve $a_{10}=5+(10-1)d$ for $d=4$.

15·5 *Note*: For these proofs, the step for $n=1$ is not shown. The reader can easily supply that step.

1. If $2+4+\ldots+2k=k(k+1)$, then

$$\begin{aligned}2+4+\ldots+2k+2(k+1)&=k(k+1)+2(k+1)\ \text{IH}\\&=(k+1)(k+2)\end{aligned}$$

2. If $5+9+\ldots+(4k+1)=k(2k+3)$, then

$$\begin{aligned}5+9+\ldots+(4k+1)+[4(k+1)+1]&=k(2k+3)+4(k+1)+1\ \text{IH}\\&=2k^2+7k+5 && \text{collect terms}\\&=(k+1)(2k+5) && \text{factor}\\&=(k+1)[2(k+1)+3] && \text{rewrite}\end{aligned}$$

3. If $2+4+\ldots+2^k=2^{k+1}-2$, then

$$\begin{aligned}2+4+\ldots+2^k+2^{k+1}&=2^{k+1}-2+2^{k+1} && \text{IH}\\&=2^{k+1}(2)-2 && \text{factor}\\&=2^{k+2}-2 && \text{simplify}\end{aligned}$$

4. If $\frac{1}{3}+\frac{1}{15}+\frac{1}{35}+\ldots+\frac{1}{(2n-1)(2n+1)}=\frac{n}{2n+1}$, then

$$\frac{1}{3}+\frac{1}{15}+\frac{1}{35}+\ldots+\frac{1}{(2n-1)(2n+1)}+\frac{1}{(2n+1)(2n+3)} = \frac{n}{2n+1}+\frac{1}{(2n+1)(2n+3)} \quad \text{IH}$$

$$= \frac{1}{2n+1}\left[\frac{2n^2+3n+1}{2n+3}\right] \quad \text{factor and then add}$$

$$= \frac{(2n+1)(n+1)}{(2n+1)(2n+3)} \quad \text{factor}$$

$$= \frac{n+1}{2n+3} \quad \text{cancel}$$

5. If $3+9+27+\ldots+3^n=\frac{3(3^n-1)}{2}$, then

$$3+9+27+\ldots+3^n+3^{n+1} = \frac{3(3^n-1)}{2}+3^{n+1} \quad \text{IH}$$

$$= \frac{3^{n+1}-3+2(3^{n+1})}{2} \quad \text{add fractions}$$

$$= \frac{3^{n+1}(3)-3}{2} \quad \text{collect terms}$$

$$= \frac{3(3^{n+1}-1)}{2} \quad \text{factor}$$

6. If $2^k \geq 2k+1$, then

$$2^{k+1} \geq 2(2k+1) = 4k+2 \quad \text{multiply by 2; IH}$$

$$\geq 2k+2k+2 \quad \text{rewrite}$$

$$\geq 2(k+1)+2k \quad \text{rewrite}$$

$$\geq 2(k+1)+1 \quad \text{since } 2k>1$$

7. If $\frac{1}{2}+\frac{1}{6}+\frac{1}{12}+\ldots+\frac{1}{n(n+1)}=\frac{n}{n+1}$, then

$$\frac{1}{2}+\frac{1}{6}+\frac{1}{12}+\ldots+\frac{1}{n(n+1)}+\frac{1}{(n+1)(n+2)} = \frac{n}{n+1}+\frac{1}{(n+1)(n+2)} \quad \text{IH}$$

$$= \frac{1}{n+1}\left(\frac{n^2+2n+1}{n+2}\right) \quad \text{factor and then add}$$

$$= \frac{(n+1)^2}{(n+1)(n+2)} \quad \text{factor}$$

$$= \frac{n+1}{n+2} \quad \text{cancel}$$

8. If $s+sr+sr^2+sr^3+\ldots+sr^{n-1}=\frac{s(1-r^n)}{1-r}$, then

$$s+sr+sr^2+sr^3+\ldots+sr^{n-1}+sr^n = \frac{s(1-r^n)}{1-r}+sr^n \quad \text{IH}$$

$$= \frac{s-sr^n+sr^n-sr^{n+1}}{1-r} \quad \text{add fractions}$$

$$= \frac{s(1-r^{n+1})}{1-r} \quad \text{simplify}$$

9. If $1+8+27+64+\ldots+n^3=\frac{n^2(n+1)^2}{4}$, then

$$1+8+27+64+\ldots+n^3+(n+1)^3 = \frac{n^2(n+1)^2}{4}+(n+1)^3 \quad \text{IH}$$

$$=(n+1)^2\left(\frac{n^2+4n+1}{4}\right) \quad \text{factor and add}$$

$$=\frac{(n+1)^2(n+2)^2}{4} \quad \text{factor}$$

10. If $3(4^{n-1})\le 4^n-1$, then

$4^{n+1}-4\ge 3(4^n)$ multiply by 4; IH

$4^{n+1}-1\ge 3(4^n)+3$ add 3 to each side

$4^{n+1}-1\ge 3(4^n)$ since $3>0$

PRECALCULUS TRIGONOMETRY

16 Trigonometric functions

16·1

1. $1°=\frac{\pi}{180}\approx 0.0175$ radians
2. 1 radian $\approx 57.296°$
3. $30°=\frac{30\pi}{180}\approx 0.524$ radians
4. $\frac{\pi}{6}$ radians $=\frac{180}{\pi}\left(\frac{\pi}{6}\right)=30°$
5. $147°=147\left(\frac{\pi}{180}\right)\approx 2.566$ radians
6. 22 radians $=22\left(\frac{180}{\pi}\right)=1260.51°$
7. $27°53'25''=27+53\left(\frac{1}{60}\right)+25\left(\frac{1}{3600}\right)=27.89°$
8. $57.5692°=57°+(0.5692)60'=57°+34.152'=57°+34'+(0.152)60''=57°34'9''$
9. False
10. False

16·2

1. $\angle B=30°,\ b=35(\tan 30°)\approx 20.2,\ \sqrt{35^2+(20.2)^2}\approx 40.4$
2. $\angle A=55°,\ a=25(\sin 55°)\approx 20.5,\ b=25(\sin 35°)\approx 14.3$
3. $c=\sqrt{(126)^2+(200)^2}\approx 236.4, \sin(A)=\frac{200}{236.4}\approx 0.8460,\ \angle A=\sin^{-1}(0.8460)\approx 57.8°,\ \angle B\approx 32.2°$
4. $\angle B=60°,\ a=25(\sin 30°)=25\left(\frac{1}{2}\right)=\frac{25}{2},\ b=\frac{25\sqrt{3}}{2}$
5. $\angle A=45°,\ b=30,\ c=30\sqrt{2}$
6. $\sin A=\frac{3}{10}=0.3,\ A=\sin^{-1}(0.3)\approx 17.5°,\ \angle B\approx 72.5°,\ b=\sqrt{100-9}=\sqrt{91}$
7. $b=\sqrt{3},\ \angle A=30°,\ \angle B=60°$
8. $\angle A=\angle B=45°,\ c=\sqrt{2}$

9. $\angle A = 30°$, $\angle B = 60°$, $b = 60\sqrt{3}$

10. $\angle A = \angle B = 45°$, $c = 38\sqrt{2}$

16·3

1. 42°
2. 40°
3. $\sin A = \frac{1}{2}$, $\cos A = \frac{\sqrt{3}}{2}$, $\tan A = \frac{\sqrt{3}}{3}$, $\csc A = 2$, $\sec A = \frac{2\sqrt{3}}{3}$, $\cot A = \sqrt{3}$

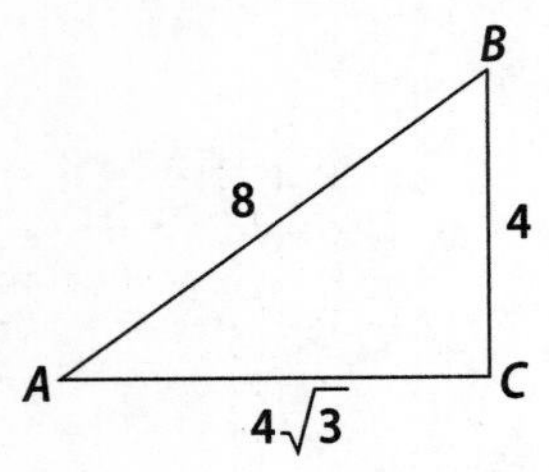

4. $\sin\left(\frac{4\pi}{3}\right) = -\frac{\sqrt{3}}{2}$, $\cos\left(\frac{4\pi}{3}\right) = -\frac{1}{2}$, $\tan\left(\frac{4\pi}{3}\right) = \frac{\sqrt{3}}{3}$
5. $\sin(390°) = \sin 30° = \frac{1}{2}$, $\cos 30° = \frac{\sqrt{3}}{2}$, $\tan 30° = \frac{\sqrt{3}}{3}$
6. The angle is in quadrant III. $r = 5$, $x = 4$, $y = -3$
7. Quadrant I, positive, $\sin(723°) = 0.052$
8. Quadrant IV, negative, $\tan(995°) = -11.43$
9. Quadrant II, negative, $\cos(528°) = -0.978$
10. Quadrant IV, negative, $\csc(687°) = -1.836$

16·4

1. $0, \frac{\pi}{6}, \frac{\pi}{4}, \frac{\pi}{3}, \frac{\pi}{2}, \frac{2\pi}{3}, \frac{3\pi}{4}, \frac{5\pi}{6}, \pi, \frac{3\pi}{2}, 2\pi$
2. $45° + n \cdot 180°$
3. $135° + n \cdot 360°$, $315° + n \cdot 360°$
4. 45°
5. $\frac{\pi}{4}$
6. negative
7. the Pythagorean theorem
8. yes
9. −1
10. 0

16·5

1. Amplitude = 3, period = π, zeros occur when $2x - \pi = (2n-1)\frac{\pi}{2}$ or when $x = (2n+1)\frac{\pi}{4}$, the graph is shifted to the right $\frac{\pi}{2}$ units. When $2x - \pi = 2n\pi$ or when $x = (2n+1)\frac{\pi}{2}$, the maxima occur. The minima occur when $2x - \pi = (2n-1)\pi$ or when $x = n\pi$.
2. Amplitude = 1, period = 12, zeros occur when $\frac{\pi}{6}x - \frac{\pi}{3} = n\pi$ or when $x = 6n + 2$, the graph is shifted to the right 2 units. When $\frac{\pi}{6}x - \frac{\pi}{3} = (4n-3)\frac{\pi}{2}$ or when $x = 12n - 7$, the maxima occur; and when $\frac{\pi}{6}x - \frac{\pi}{3} = (4n-1)\frac{\pi}{2}$ or when $x = 12n - 1$, the minima occur.
3. Amplitude is not applicable, period $= \frac{\pi}{4}$, zeros occur when $4x + 2\pi = n\pi$ or when $x = (n-2)\frac{\pi}{4}$, the graph is shifted $\frac{\pi}{2}$ units to the left. Maxima and minima are not applicable.

For 4–5, there are several correct answers. Two possible equations are given for each graph.

4. $y = 4\sin\theta$ or $y = 4\cos\left(\theta - \frac{\pi}{2}\right)$
5. $y = 2\sin(2\theta)$ or $y = 2\cos\left(2\theta - \frac{\pi}{2}\right)$

17 Analytic trigonometry

17·1

1. $\sin x \cot x = \sin x \dfrac{\cos x}{\sin x} = \cos x$
2. $\sin x(\csc x - \sin x) = \sin x \csc x - \sin^2 x = 1 - \sin^2 x = \cos^2 x$
3. $\tan x(\csc x + \cot x) = \tan x \csc x + \tan x \cot x = \dfrac{\sin x}{\cos x}\dfrac{1}{\sin x} + 1 = \dfrac{1}{\cos x} + 1 = \sec x + 1$
4. $\tan^2 x \csc^2 x - \tan^2 x = \tan^2 x(\csc^2 x - 1) = \tan^2 x \cot^2 x = 1$
5. $\dfrac{\sin x \cos x + \cos x}{\sin x + \sin^2 x} = \dfrac{\cos x(\sin x + 1)}{\sin x(1 + \sin x)} = \dfrac{\cos x}{\sin x} = \cot x$
6. $\dfrac{(\sin x + \cos x)^2}{\cos x} = \dfrac{\sin^2 x + 2\sin x \cos x + \cos^2 x}{\cos x} = \dfrac{1 + 2\sin x \cos x}{\cos x} = \dfrac{1}{\cos x} + \dfrac{2\sin x \cos x}{\cos x} = \sec x + 2\sin x$
7. $(1 + \sin x)(1 + \sin(-x)) = (1 + \sin x)(1 - \sin x) = 1 - \sin^2 x = \cos^2 x$
8. $\dfrac{\cos^2 x}{\sin x} + \dfrac{\sin x}{1} = \dfrac{\cos^2 x + \sin^2 x}{\sin x} = \dfrac{1}{\sin x} = \csc x$
9. $\dfrac{\tan x}{\csc x} - \dfrac{\sin x}{\cos x} = \dfrac{\dfrac{\sin x}{\cos x}}{\dfrac{1}{\sin x}} - \dfrac{\sin x}{\cos x} = \dfrac{\sin^2 x}{\cos x} - \dfrac{\sin x}{\cos x} = \dfrac{\sin x(\sin x - 1)}{\cos x} = \dfrac{\sin x - 1}{\dfrac{\cos x}{\sin x}} = \dfrac{\sin x - 1}{\cot x}$
10. Let $\theta = 0°$. Then $\cos(2\theta) = \cos 0° = 1$ and $2\cos 0° = 2$. Thus, it is not an identity.

17·2

1. $\sin(x + 2\pi) = \sin x \cos(2\pi) + \cos x \sin(2\pi) = (\sin x)1 + 0 = \sin x$
2. $\cos(x + 2\pi) = \cos x \cos(2\pi) - \sin x \sin(2\pi) = \cos x$
3. $\tan(x + \pi) = \dfrac{\tan x + \tan \pi}{1 - \tan x \tan \pi} = \dfrac{\tan x + 0}{1 - 0} = \tan x$
4. $\sin(\alpha + \beta)\sin(\alpha - \beta) = [\sin\alpha \cos\beta + \cos\alpha \sin\beta][\sin\alpha \cos\beta - \cos\alpha \sin\beta]$
   $= \sin^2\alpha \cos^2\beta - \cos^2\alpha \sin^2\beta = \sin^2\alpha \cos^2\beta + \sin^2\alpha \sin^2\beta - \sin^2\alpha \sin^2\beta - \cos^2\alpha \sin^2\beta$
   $= \sin^2\alpha(\cos^2\beta + \sin^2\beta) - \sin^2\beta(\cos^2\alpha + \sin^2\alpha) = \sin^2\alpha - \sin^2\beta$
5. $\cos\left(x + \dfrac{\pi}{4}\right) = \cos x \cos\left(\dfrac{\pi}{4}\right) - \sin x \sin\dfrac{\pi}{4} = \dfrac{\sqrt{2}}{2}\cos x - \dfrac{\sqrt{2}}{2}\sin x = \dfrac{\sqrt{2}}{2}(\cos x - \sin x)$
6. $\sin(2t) = \sin(t + t) = \sin t \cos t + \cos t \sin t = 2\sin t \cos t$
7. $\cos(\alpha + \beta) + \cos(\alpha - \beta) = \cos\alpha \cos\beta - \sin\alpha \sin\beta + \cos\alpha \cos\beta + \sin\alpha \sin\beta = 2\cos\alpha \cos\beta$
8. $\tan\left(x + \dfrac{\pi}{4}\right) = \dfrac{\tan x + \tan\dfrac{\pi}{4}}{1 - \tan x \tan\dfrac{\pi}{4}} = \dfrac{\tan x + 1}{1 - \tan x} = \dfrac{1 + \tan x}{1 - \tan x}$
9. $\tan\left(\dfrac{2\pi}{3} + \dfrac{\pi}{4}\right) = \dfrac{\tan\left(\dfrac{2\pi}{3}\right) + \tan\left(\dfrac{\pi}{4}\right)}{1 - \tan\left(\dfrac{2\pi}{3}\right)\tan\left(\dfrac{\pi}{4}\right)} = \dfrac{-\sqrt{3} + 1}{1 + \sqrt{3}} = \dfrac{1 - \sqrt{3}}{1 + \sqrt{3}}$

17·3

1. $\tan(2x) = \dfrac{\tan x + \tan x}{1 - \tan x \tan x} = \dfrac{2\tan x}{1 - \tan^2 x}$
2. $\cos(2x) = \cos(x + x) = \cos x \cos x - \sin x \sin x = \cos^2 x - \sin^2 x$
3. $\cos(8x) = \cos[2(4x)] = \cos^2(4x) - \sin^2(4x)$
4. $\dfrac{\cos(2x)}{\sin^2 x} = \dfrac{\cos^2 x - \sin^2 x}{\sin^2 x} = \dfrac{\cos^2 x}{\sin^2 x} - \dfrac{\sin^2 x}{\sin^2 x} = \cot^2 x - 1$

5. $\sin(4x) = \sin[2(2x)] = 2\sin(2x)\cos(2x) = 2(2\sin x\cos x)(1-2\sin^2 x)$
$= 4\sin x\cos x(1-2\sin^2 x)$

6. From problem 1, $\tan(2x) = \dfrac{2\tan x}{1-\tan^2 x} = \dfrac{2}{\frac{1-\tan^2 x}{\tan x}} = \dfrac{2}{\frac{1}{\tan x} - \frac{\tan^2 x}{\tan x}} = \dfrac{2}{\cot x - \tan x}$

7. $\tan(2x) = \dfrac{\sin(2x)}{\cos(2x)} = \dfrac{2\sin x\cos x}{\cos^2 x - \sin^2 x}$

8. $(\sin^2 x - 1)^2 = \sin^4 x - 2\sin^2 x + 1 = \sin^4 x + (1-2\sin^2 x) = \sin^4 x + \cos(2x)$

9. $\cos^2\left(\dfrac{x}{2}\right) - \sin^2\left(\dfrac{x}{2}\right) = \cos\left(2\left(\dfrac{x}{2}\right)\right) = \cos x$

10. $\dfrac{\cos(2x)}{\sin(2x)} = \dfrac{\cos^2 x - \sin^2 x}{2\sin x\cos x} = \dfrac{\frac{\cos^2 x - \sin^2 x}{\sin x\cos x}}{2} = \dfrac{\frac{\cos x}{\sin x} - \frac{\sin x}{\cos x}}{2} = \dfrac{\cot x - \tan x}{2}$

17·4

1. $\dfrac{\pi}{4}$
2. $-\dfrac{\sqrt{2}}{2}$
3. $\dfrac{3}{5}$
4. $\dfrac{\sqrt{5}}{3}$. Sketch the right triangle with sides $\sqrt{5}$ and 2.
5. $\dfrac{\sqrt{4x^2 - 25}}{2x}$. Sketch the right triangle with hypotenuse $2x$ and side adjacent to the angle equal to 5.
6. $\cos\theta = \pm\dfrac{\sqrt{3}}{2}$. The solutions in $[0,2\pi]$ are $\dfrac{\pi}{6}, \dfrac{5\pi}{6}, \dfrac{7\pi}{6}, \dfrac{11\pi}{6}$.
7. $\cos\theta = 0$ or $\cos\theta = -\dfrac{1}{2}$. The solutions in $[-\pi,\pi]$ are $\pm\dfrac{\pi}{2}, \pm\dfrac{2\pi}{3}$.
8. $\cos\theta = \dfrac{1}{3}$. The solution in $[0,\pi]$ is $\cos^{-1}\left(\dfrac{1}{3}\right) = 1.23$ radians.
9. $\sin(2\theta) = \dfrac{2}{3}$. $2\theta = \sin^{-1}\left(\dfrac{2}{3}\right) \approx 42°$. Thus the only solution in $[-90°, 90°]$ is $\theta \approx 21°$.
10. $\sec\theta = \dfrac{4}{\sqrt{2}}$. So $\cos\theta = \dfrac{\sqrt{2}}{4}$. $\theta = \cos^{-1}\left(\dfrac{\sqrt{2}}{4}\right) \approx 69.3°$ is the only solution in $[0, 180°]$.

17·5

1. $z_1 = 2(\cos 30° + i\sin 30°)$ and $z_2 = 2[\cos 60° + i\sin 60°]$, $z_1 z_2 = 4\,\text{cis}\,90° = 4i$, $\dfrac{z_1}{z_2} = \text{cis}(-30°) = \dfrac{\sqrt{3}}{2} - \dfrac{1}{2}i$
2. $z_1 z_2 = 40i$, $\dfrac{z_1}{z_2} = \dfrac{5\sqrt{3}}{4} + \dfrac{5}{4}i$
3. $z_1 z_2 = -10 - 10\sqrt{3}i$, $\dfrac{z_1}{z_2} = -\dfrac{5}{2}$
4. $z_1 z_2 = -10.9 + 12i$, $\dfrac{z_1}{z_2} = -1.6 - 4.8i$
5. $z_1 z_2 = -2.9 + 8.5i$, $\dfrac{z_1}{z_2} = 2.3 + 3.3i$
6. $(3+3i)^4 = (3\sqrt{2}\,cis\,45°)^4 = 324\,cis\,180° = -324$
7. $\dfrac{1}{2} + \dfrac{\sqrt{3}}{2}i$
8. -64
9. $\dfrac{1}{16}$
10. $-8i$

18 Solving triangles

18·1

1. $\angle C = 78°, b = \dfrac{a\sin B}{\sin A} = \dfrac{75\sin 64°}{\sin 38°} \approx 109.5, c = \dfrac{75\sin 102°}{\sin 38°} \approx 119.2$
2. $\angle C = 90°$, $a = 10$, $c = 20$. This is a quick solution since the triangle is a $30°$–$60°$–$90°$ triangle.
3. $\angle C = 25°$, $a = \dfrac{b\sin A}{\sin B} = \dfrac{385\sin 108°}{\sin 47°} \approx 500.7$, $c = \dfrac{385\sin 25°}{\sin 47°} \approx 222.5$
4. $\angle C = 55°$, $b = \dfrac{7.2\sin 98°}{\sin 27°} \approx 15.7$, $c = \dfrac{7.2\sin 55°}{\sin 27°} \approx 13.0$
5. $\angle C = 96.2°$, $a = \dfrac{12.9\sin 20.4°}{\sin 96.2°} \approx 4.5$, $b = \dfrac{4.5\sin 63.4°}{\sin 20.4°} \approx 11.5$
6. $\angle C = 145°$, $a = \dfrac{126\sin 13°}{\sin 145°} \approx 49.4$, $b = \dfrac{126\sin 22°}{\sin 145°} \approx 82.3$
7. $\angle C = 12°$, $a = \dfrac{0.8\sin 56°}{\sin 12°} \approx 3.2$, $b = \dfrac{0.8\sin 112°}{\sin 12°} \approx 3.6$
8. $\angle C = 90°$, $a = b = 15$. This is a quick solution since the triangle is a $45°$–$45°$–$90°$ triangle.
9. $\angle A = 57°$, $b = \dfrac{42.7\sin 103.4°}{\sin 57°} \approx 49.5$, $c = \dfrac{42.7\sin 19.6°}{\sin 57°} \approx 17.1$
10. $\sin B = \dfrac{b\sin C}{c} = \dfrac{67\sin 59°}{58} \approx 0.9902$, $\angle B_1 \approx 82°$ or $\angle B_2 \approx 98°$ using the inverse sine on the calculator. Then $\angle A \approx 39°$, $a = \dfrac{58\sin 39°}{\sin 59°} \approx 42.6$ or $\angle A \approx 23°$, $a = \dfrac{58\sin 23°}{\sin 59°} \approx 26.4$.

18·2

1. $a^2 = (3.2)^2 + (1.5)^2 - 2(3.2)(1.5)\cos 95.7° = 13.44$. Thus, $a \approx 3.7$.
   $\sin B = \dfrac{b\sin A}{a} = \dfrac{3.2\sin 95.7°}{3.7} \approx 0.8606$. $\angle B \approx 59.4°$, $\angle C \approx 24.9°$
2. $c^2 = 75^2 + 32^2 - 2(75)(32)\cos 38° \approx 2866.55$. Thus, $c \approx 53.5$.
   $\sin B = \dfrac{32\sin 38°}{53.5} \approx 0.3682$. $\angle B \approx 21.6°$, $\angle A \approx 120.4°$
3. $b^2 = (12.9)^2 + (25.8)^2 - 2(12.9)(25.8)\cos 30° = 255.6$. Thus, $b \approx 16$.
4. $b^2 = (6.7)^2 + (10.9)^2 - 2(6.7)(10.9)\cos 98° \approx 184$. Thus, $b \approx 13.6$.
5. $\cos A = \dfrac{b^2 + c^2 - a^2}{2bc} = \dfrac{(6\sqrt{3})^2 + (10\sqrt{3})^2 - (15\sqrt{3})^2}{2(6\sqrt{3})(10\sqrt{3})} \approx -0.7417$. Thus, $\angle A \approx 137.9°$.
6. $\cos A = \dfrac{(129)^2 + (300)^2 - (282)^2}{2(129)(300)} \approx 0.3503$. Thus, $\angle A \approx 69.5°$.

19 Vectors in two dimensions

19·1

1. 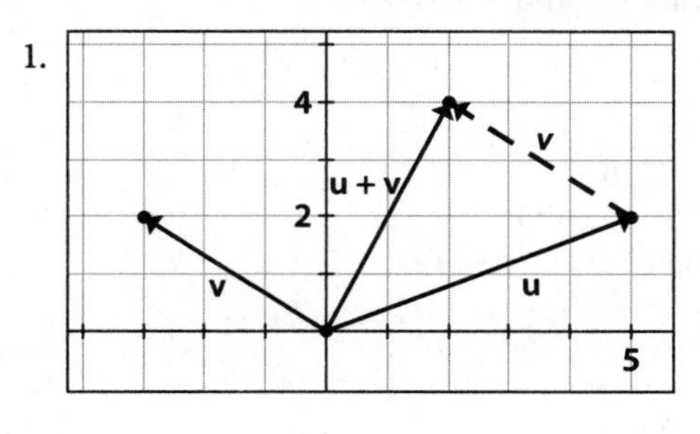

2. $\mathbf{u} - \mathbf{v} = \langle 4, 4\rangle, 4\mathbf{v} = \langle 20, -16\rangle$,
   $\mathbf{v} - \mathbf{u} = \langle -4, -4\rangle, 3\mathbf{u} - 4\mathbf{v} = \langle 7, 16\rangle$
3. $5\mathbf{i} + 8\mathbf{j}$
4. $|\mathbf{v}| = \sqrt{34}$
5. $\mathbf{u} + \mathbf{v} = -\mathbf{i} + 8\mathbf{j}$
6. The horizontal component is 2 and the vertical component is –3.

7. 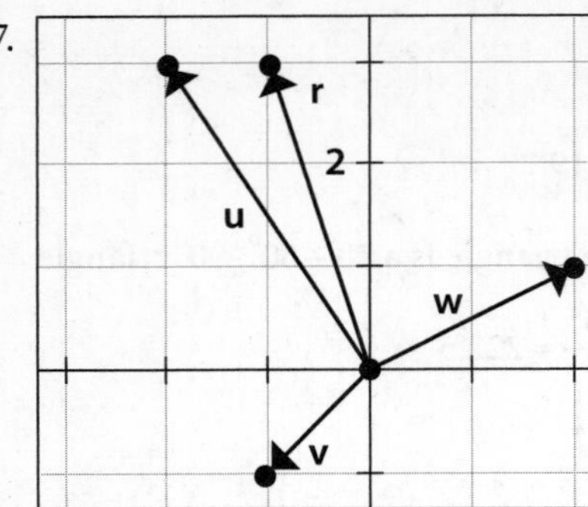

8. $\mathbf{u}_v = |\mathbf{u}|\cos 28° \approx 0.883|\mathbf{u}|$

9. $\mathbf{u}_v = |\mathbf{u}|\cos 41° \approx 0.755|\mathbf{u}|$

10. $\mathbf{u}_v = |\mathbf{u}|\cos 22° \approx 0.927|\mathbf{u}|$

19·2

1. $\cos\theta = \frac{\mathbf{u}\cdot\mathbf{v}}{|\mathbf{u}||\mathbf{v}|} = \frac{-15-6}{\left(\sqrt{5^2+2^2}\right)\left(\sqrt{(-3)^2+(-3)^2}\right)} = \frac{-21}{(\sqrt{29})(\sqrt{18})} \approx -0.9191$. Thus, $\theta \approx 203°$ since the angle is in quadrant III.
2. $\cos\theta = \frac{\mathbf{u}\cdot\mathbf{v}}{|\mathbf{u}||\mathbf{v}|} = \frac{-6-30}{(\sqrt{45})(\sqrt{29})} = \frac{-36}{(\sqrt{45})(\sqrt{29})} \approx -0.9966$. Thus, $\theta \approx 175.2°$ since the angle is in quadrant II.
3. $\mathbf{u}\cdot\mathbf{v} = 0$ so $\theta = 90°$.
4. $\mathbf{w} = \left(\frac{\mathbf{u}\cdot\mathbf{v}}{|\mathbf{v}|^2}\right)\mathbf{v} = 3.73\mathbf{i} + 1.2\mathbf{j}$
5. $\mathbf{w} = \left(\frac{\mathbf{u}\cdot\mathbf{v}}{|\mathbf{v}|^2}\right)\mathbf{v} = -4.71\mathbf{i} + 1.18\mathbf{j}$
6. $\mathbf{w} = \left(\frac{\mathbf{u}\cdot\mathbf{v}}{|\mathbf{v}|^2}\right)\mathbf{v} = -0.65\mathbf{i} + 0.11\mathbf{j}$
7. $-7\mathbf{i} + 4\mathbf{j}$
8. $\mathbf{r} = \mathbf{u}+\mathbf{v}+\mathbf{w} = \langle 9,3\rangle$. $\theta = \tan^{-1}\left(\frac{3}{9}\right) \approx 18.4°$
9. $\mathbf{u}\cdot\mathbf{v} = \mathbf{i}\cdot\mathbf{j} = \langle 1,0\rangle\cdot\langle 0,1\rangle = 1\cdot 0 + 0\cdot 1 = 0$
10. If $\mathbf{u} = (a\mathbf{i} + b\mathbf{j})$, then $(a\mathbf{i}+b\mathbf{j})\bullet(2\mathbf{i}+5\mathbf{j}) = 0$. Also, $|\mathbf{u}|^2 = a^2 + b^2 = 9$. Thus, $2a + 5b = 0$ or $a = -\frac{5}{2}b$. Substituting and solving the quadratic, $\left(-\frac{5b}{2}\right)^2 + b^2 - 9 = 0$, gives $b = \pm 1.1, a = \mp 2.75$. The vectors are $\pm(2.75\mathbf{i} - 1.1\mathbf{j})$.

IV ANALYTIC GEOMETRY

20 Conics

20·1

1. $x^2 + y^2 = 49$
2. $(x-5)^2 + y^2 = 3$
3. Using the distance formula, $d = 10$. The center is the midpoint of the diameter so the equation of the circle is $(x-1)^2 + (y-5)^2 = 25$.
4. $\frac{(x+1)^2}{4^2} + \frac{(y-2)^2}{2^2} = 1$, center at (−1,2) and vertices at (3,2) and (−5,2)
5. $(x-6)^2 + (y-5)^2 = 9$, center at (6,5)
6. $(x-2)^2 + (y+4)^2 = 9$, center at (2,−4)
7. $\frac{(x+2)^2}{16} + \frac{(y-1)^2}{4} = 1$, center at (−2,1) and vertices at (2,1) and (−6,1)
8. $(x-2)^2 + (y+5)^2 = 25$, center at (2,−5)

9. $\frac{(x-3)^2}{3}+\frac{(y+5)^2}{10}=1$, center at (3,–5) and vertices at $(3,-5+\sqrt{10})$ and $(3,-5-\sqrt{10})$

10. $\frac{(x-3)^2}{25}+\frac{(y+2)^2}{10}=1$, center at (3,–2) and vertices at (8,–2) and (–2,–2)

20·2

1. hyperbola, center at (0,0)
2. circle, center at (0,0)
3. hyperbola, center at (–5,3)
4. ellipse, center at (–2,3)
5. $\frac{(x-3)^2}{9}-\frac{(y-4)^2}{4}=1$. Vertices at (0,4) and (6,4), transverse-axis is $y = 4$ and conjugate-axis is $x = 3$.

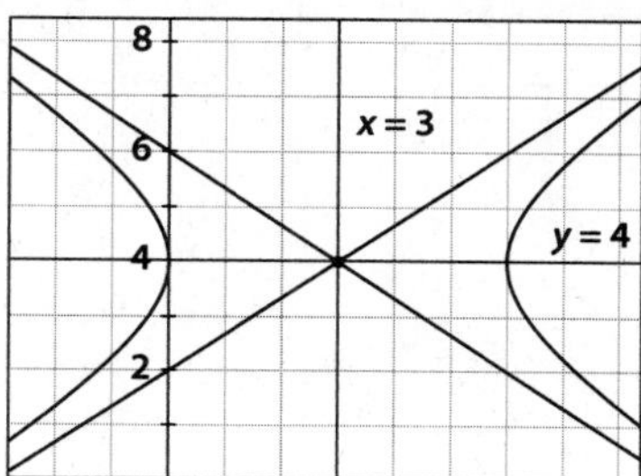

6. $\frac{y^2}{4}-\frac{(x-3)^2}{16}=1$. Vertices at (3,2) and (3–2), transverse-axis is $x = 3$ and conjugate-axis is $y = 0$.

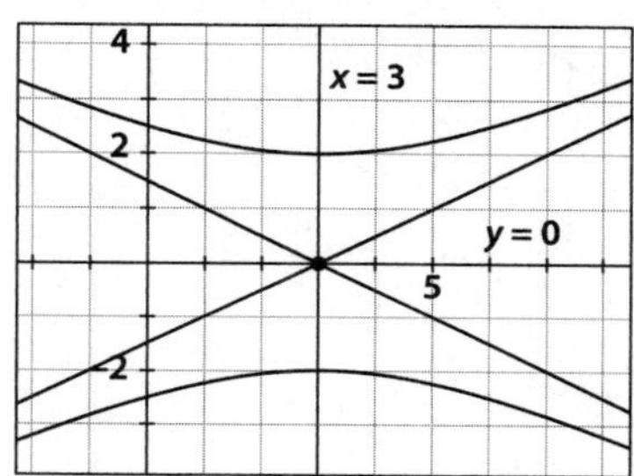

21 Parametric equations and polar coordinates

21·1

1. $\frac{x+3}{2}=\frac{y+1}{4}$ or $y=2x+5$, straight line
2. $(x-1)^2+\frac{(y-2)^2}{9}=1$, ellipse
3. $y=(x-2)^2+1$, parabola
4. $\frac{x^2}{4}+\frac{y^2}{9}=1$, ellipse
5. $x^2+y^2=16$, circle
6. $x=t-3, y=t^2+1$ and $x=\tan\theta-3, y=\sec^2\theta$
7. $x=t, y=3t-2$ and $x=\cos t, y=3\cos t-2$
8. $x=3\cos\theta+2, y=4\sin\theta-3$
9. $x=4\cos\theta-2, y=4\sin\theta+3$
10. $x=3\sec\theta, y=4\tan\theta+4$

21·2

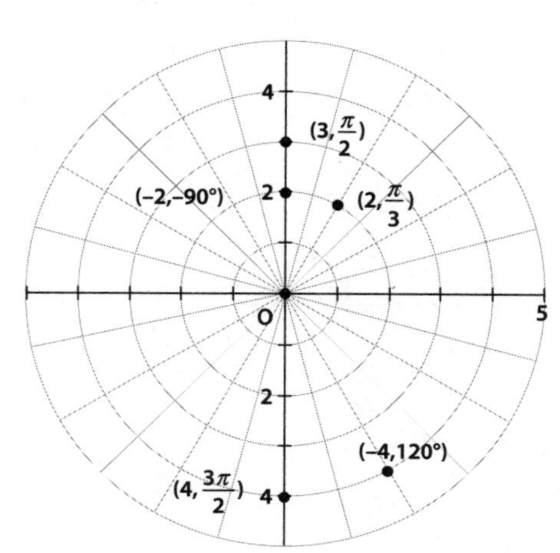

21·3

1. $(4, 45^\circ)$
2. $(8, 30^\circ)$
3. $(12.5, 106.3^\circ)$
4. $(4\sqrt{2}, 4\sqrt{2})$
5. $(\sqrt{3}, 1)$
6. $(-2\sqrt{3}, 2)$

21·4

1. 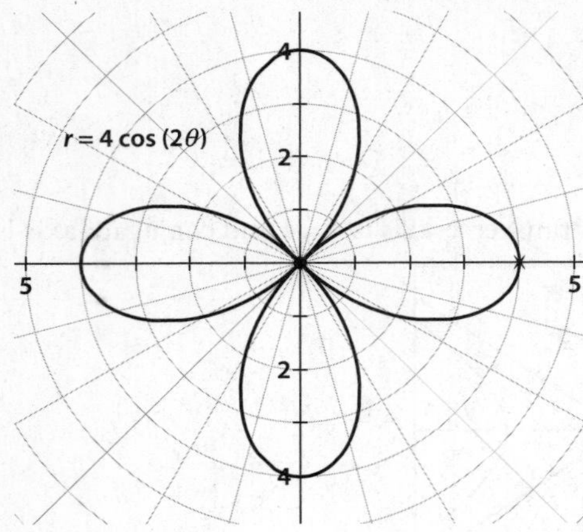

2. 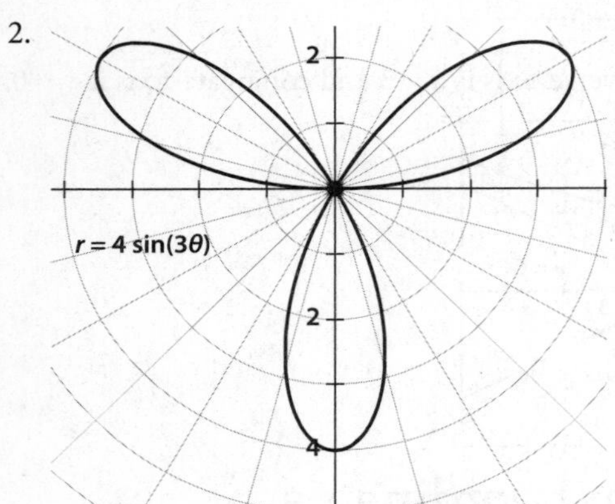

3. 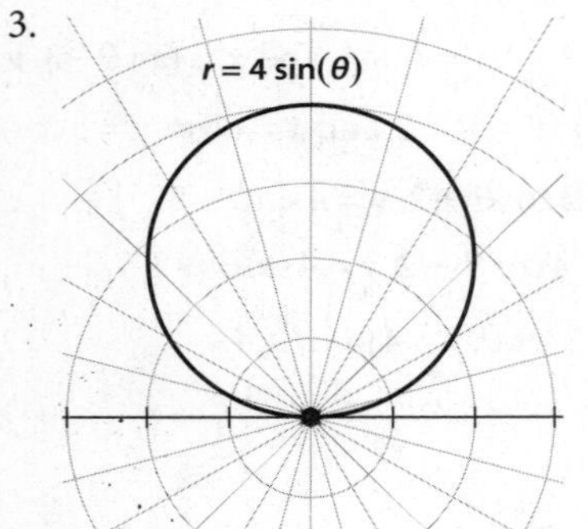

4. 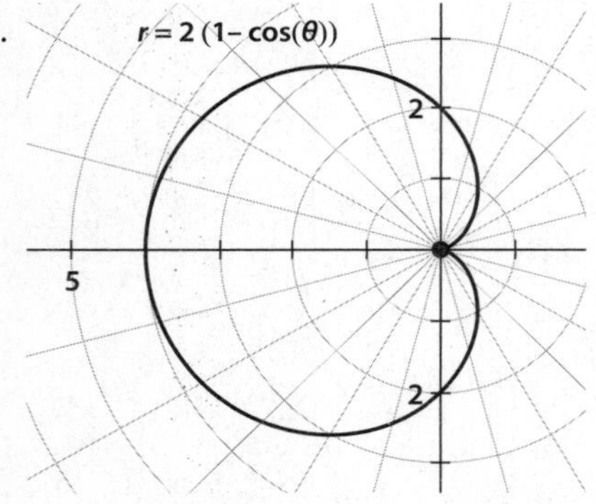

5. 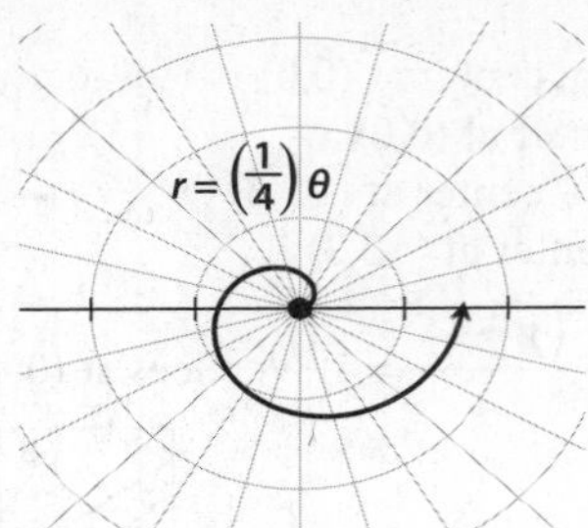

6. 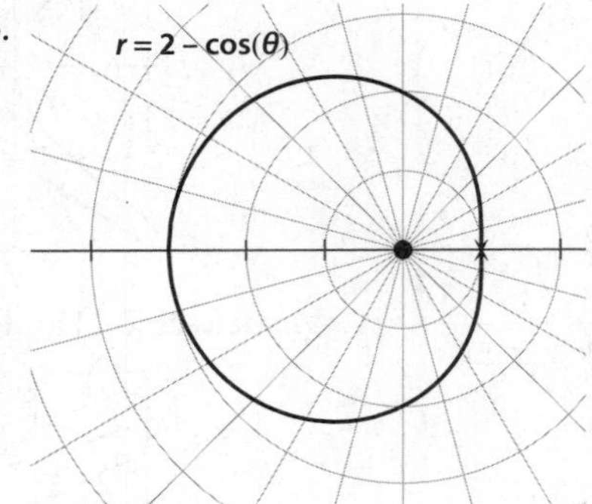

7. 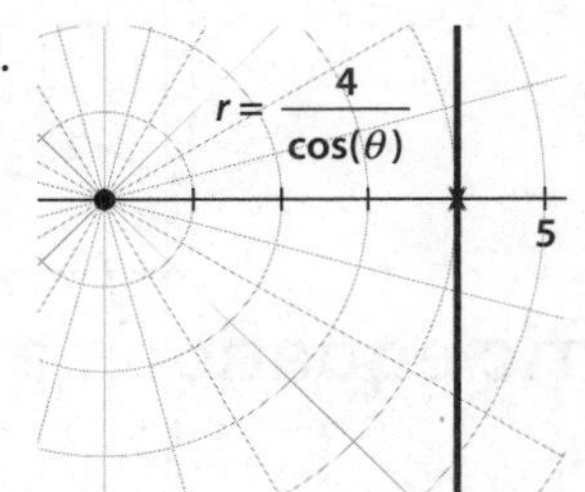

8. 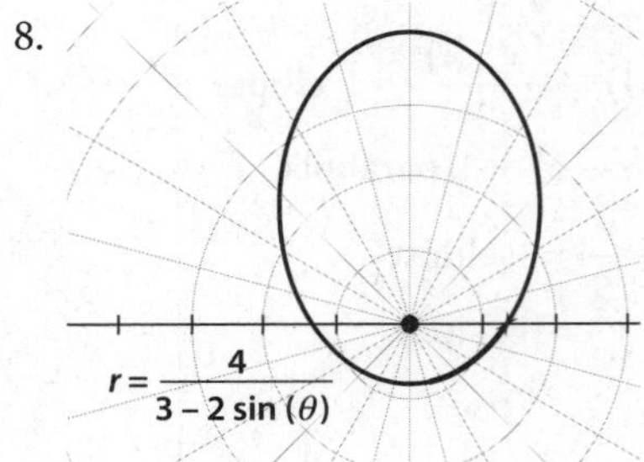

9. 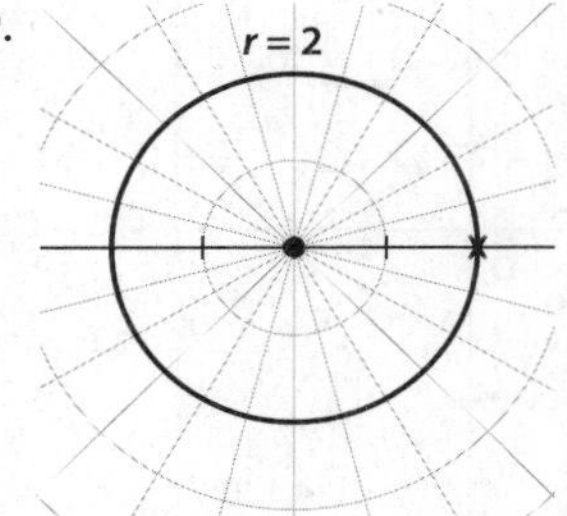